Current Topics in Behavioral Neurosciences

Volume 53

Current Topics in Behavioral Neurosciences provides critical and comprehensive discussions of the most significant areas of behavioral neuroscience research, written by leading international authorities. Each volume in the series represents the most informative and contemporary account of its subject available, making it an unrivalled reference source. Each volume will be made available in both print and electronic form.

With the development of new methodologies for brain imaging, genetic and genomic analyses, molecular engineering of mutant animals, novel routes for drug delivery, and sophisticated cross-species behavioral assessments, it is now possible to study behavior relevant to psychiatric and neurological diseases and disorders on the physiological level. The Behavioral Neurosciences series focuses on translational medicine and cutting-edge technologies. Preclinical and clinical trials for the development of new diagnostics and therapeutics as well as prevention efforts are covered whenever possible. Special attention is also drawn on epigenetical aspects, especially in psychiatric disorders.

CTBN series is indexed in PubMed and Scopus.

Susan L. Andersen
Editor

Sensitive Periods of Brain Development and Preventive Interventions

Springer

Editor
Susan L. Andersen
Harvard Medical School
McLean Hospital
Belmont, MA, USA

ISSN 1866-3370 ISSN 1866-3389 (electronic)
Current Topics in Behavioral Neurosciences
ISBN 978-3-031-04472-4 ISBN 978-3-031-04473-1 (eBook)
https://doi.org/10.1007/978-3-031-04473-1

This Springer imprint is published by the registered company Springer Nature Switzerland AG
The registered company address is: Gewerbestrasse 11, 6330 Cham, Switzerland

Preface

Development occurs in a series of overlapping stages and is influenced by experiences. Traditionally, these stages are described as critical periods and sensitive periods. During a critical period, environmental information is necessary for the development of a given structure/function. Similarly, yet different, the brain is responsive to environmental influences at a distinct stage in life during a sensitive period. The terms sensitive and critical are not independent and likely reflect a continuum of ways the immature brain is sculpted to match the demands of the environment. Dr. Bill Greenough, a pioneer in this field, framed the issue of differing influences of experience on maturation as a matter of dependency. Development of certain aspects of the nervous system is either dependent on information to proceed (experience-dependent) or expect the information to be present but recognizes that unique experiences influence individual development (experience-expectant). Within a sensitive period framework of how experience influences maturation, critical periods are on one end of the continuum and sensitive periods are on the other. Classically, critical period research was founded on the formation of ocular dominance columns and sensory systems. Here, the brain requires (or depends on) certain information (light) to develop within a defined period. Neurons make relevant synaptic connections, and then prune based on functional validation. Information that is idiosyncratic to the individual (e.g., diet, exercise, drug exposure, trauma) leads to an accommodation of these events into the fabric of our being. It is important to distinguish that the incorporation of experiences that occur during development is permanent by altering brain structure and function; in adulthood, experiences typically lead to a transient, compensatory reaction that often return to a pre-experience baseline of function with the passing of time.

This volume on sensitive periods is composed of chapters by authors interested in defining, understanding, and manipulating what happens during sensitive periods. The chapters will discuss how various factors influence brain development and how unique periods of plasticity can be exploited to improve the quality of life.

Traditional views are nuanced, or even challenged, and new viewpoints are introduced as science enables more sophisticated analyses.

In the first part of the volume, basic studies are presented to suggest that sensitive periods are not the result of the presence or absence of a single environmental event. Rather, the brain is refined by convergent information that is often from different cell types or brain areas to sculpt the adult topography. In many ways, this more complex synthesis makes sense as brain development should yield an organ that is plastic and adaptable to multiple and simultaneously occurring challenges. We find such examples in the chapters by Stacy and Van Hooser, Yang and Tseng, Dziabis and Bilbo, and Luciana and Collins. The occurrences effect a wide range of brain function that includes sensory, emotional, motivational, and cognitive systems. Maybe a finishing school for brain development, experiences mold the adolescent to prepare for adulthood. Conversely, adverse experiences can alter the trajectory of the brain to permanently change the individual.

The second part of the volume explores how external factors influence the brain. The age-dependent effects of exercise, diet, musical training, and direct brain injury on maturation are reviewed. The goal of the third section of the volume is to provide a blueprint for how sensitive periods can be applied to effect community change. These final two chapters are specifically aimed at identifying children and adolescents at-risk for negative consequences. Importantly, they both provide new and/or scientifically validated approaches to intervene by capitalizing on sensitive periods.

Part I

Classic critical periods include age-related changes in the visual system. Building upon the classic work of Hubel and Wiesel, Stacy and Van Hooser (Brandeis University) provide an updated review of how the visual system responds to environmental challenges. Their recent work shows that the visual system remains highly plastic beyond the original construction of the ocular dominance columns in response to light. By carefully dissecting the relationship between neurons in the lateral geniculate neurons in the thalamus and retinogeniculate cells, the authors reveal three phases of development that require: (1) no visual experience; (2) visually evoked activity for fine-scale refinement; and (3) a phase of retinogeniculate refinement that depends on normal visual experience. The interplay of these neurons shows us how early phases of development involve simple wiring that eventually leads to binocular inputs with maturation.

The use of the term sensitive period has been applied to the dramatic changes that occur during the period immediately preceding adolescence or adolescence itself. Adolescence is a period when the brain strives to match environmental needs with appropriate wiring, especially in the cortex. Drs. Yang and Tseng (University of Illinois, Chicago) further our understanding of the intricate dance of maturing microcircuits during the maturation of corticolimbic circuitry. Changes in afferent inputs and their transient relationships with GABAergic interneurons are the primary focus. The fast-spiking parvalbumin GABAergic interneuron has a starring role in sculpting which synapses remain or are pruned. Relevant to the cognitive function and emotional regulation of the cortex, Yang and Tseng review the glutamate, dopamine, and GABAergic changes that occur in the cortex. Together, the chapter provides an overview of how information alters plasticity during sensitive periods.

Any new descriptions about the mechanism of sensitive periods must contain microglia to be complete. New research shows that the immune system, especially microglia, plays a key role in sculpting the immature brain. In her chapter, Drs. Dziabis and Bilbo (Duke University) describe how microglia are no longer just the "housekeepers" of the brain. Rather, microglia activity is instrumental for cell positioning, overproduction, and pruning that occur in the plastic brain. Evidence for the importance of timing of environmental information is provided for three periods: embryonic, early postnatal, and adolescence. Microglia respond to external information in the form of cell signals from other cells and the expression of "eat me" signals. Elucidating the role of microglia to shape the brain also answers long-standing questions regarding the mechanism of how the immune system can change brain development. New research and perspectives are in this exciting chapter.

Much of the earlier chapters focus on basic mechanistic research until the chapter by Drs. Luciana and Collins (University of Minnesota) which examines sensitive periods using neuroimaging. Here, Luciana and Collins review the maturation of cortical regions and associated circuits with fMRI. The authors carefully dissect the evidence for sensitive periods and raise alternative interpretations. For example, when we describe a sensitive period are we describing the endpoint of experiential effects on the brain, and not an imprinting effect? Their molar views on the human brain discuss a compacting of gray matter, rather than pruning. Whether compacting reflects the specific loss of certain neuronal types not subject to MRI analysis requires further analysis. The chapter challenges us to consider the timing of a sensitive period (is it the period leading up to observable change or when the change is evident?), its role from a psycho-social perspective, and how best to interpret imaging data.

Part II

While we know that exercise can make us feel better, little is known about how exercise influences brain development. Drs. Lubans, Leahy, Mavilidi, and Valkenborghs (University of Newcastle and the University of Wollongong) and colleagues provide a meta-analysis of the age-dependent effects of exercise. The overview includes what is known about the positive effects of exercise on behavior with an emphasis on improving executive function. Despite good news in the literature, the authors highlight that there is a paucity of data supporting benefits in other domains and in youth. Attention is paid to the differing moderating factors that make this field complicated. Finally, the authors summarize the important questions and clarifications needed to lead to greater acceptance of exercise as a strategy to moderate the brain and its functioning.

Diet is another factor that is so important to building a better brain. Drs. Cusick, Barks, and Georgieff (University of Minnesota School of Medicine) review both human and animal literature to demonstrate the importance of healthy eating. Their chapter teaches us that food does a brain good, a statement that is supported by both clinical and preclinical studies. In this chapter, the authors discuss how epidemiologic and experimental research support the positive benefits of protein, iron, iodine, and choline during sensitive periods of brain development. Negative evidence for other foods to influence brain development are critically evaluated as differences exist between normal animal studies and clinical studies in individuals who experienced malnourishment. Factors include the timing, dose, and nutritional status. The role of nutrients in building a brain is presented, when available, in a developmental context to add to the depth of this chapter.

Another positive factor, musical training, is covered by Dr. Penhune (University of Virginia). Musical training during childhood has long-term impact on the maturing brain. After reviewing the maturation of the brain regions that are involved in the complex skills of learning and performing music, the literature regarding windows of time when the brain readily learns and can play music is reviewed. Exposure to music at different ages results in brain changes that are detected with magnetic resonance imaging. Music impacts the brain regions underlying association, cognition, habit and motor, memory, reward, sensory, and the integration of function across hemispheres. Finally, Dr. Penhune includes the literature supporting a genetic predisposition for music and how the G $\times$ E interaction plays. Together, the chapter provides a compendium of all the processes needed to become musical during sensitive periods of development.

Dr. Kolb (University of Lethbridge) reviews what is known about brain injury during development across multiple mammalian species, ranging from mouse to monkey to human. Evidence presented in the chapter identifies species and age differences in sensitive periods for possible recovery of function after an insult. Ages range from prenatal, when a significant amount is known, to adolescence, when we know less. Again, the importance of both timing of the injury versus its manifestation is discussed within a sensitive period framework. The importance of understanding the inter-relationship between the stage of maturation within the individual brain region was raised. Finally, the chapter concludes with a discussion of multiple factors involved in what constitutes a sensitive period of plasticity versus unrecoverable damage.

Part III

First, a beautiful presentation of "Slic Nac" is made by Drs. Dunn, Smith, and Smith (Harvard Medical School) of a novel statistical approach to apply to epidemiology data to reveal sensitive periods. By identify key stages of development that are integral to unique brain functions, existing data can be used to optimize interventions, teaching strategies, or preventions.

Drs. Chaplo (University of North Carolina at Chapel Hill) and Fishbein (The Pennsylvania State University) direct us to use identified periods of plasticity to redirect the development of potentially harmful pathways that lead to antisocial and criminal behavior that leads to the juvenile justice system. The chapter reviews the maturation of socio-emotional circuits and their regulation with an emphasis on the pre-adolescent period. The exposure of an individual to adversity during sensitive periods of development can lead to an "alternate" phenotype associated with increased risk for further negative consequences. The underlying neurobiology involved in such changes is presented based on individuals with adverse childhood experiences (ACEs). The chapter ends with hope, guidelines, and existing (scientifically validated) interventions from prevention science to effectively intervene and re-route the trajectory and provide an equal likelihood of success for all individuals by adulthood.

My goal for this volume is that scientists and clinicians at all levels will embrace the hope and potential life-changing power that positive influences during a sensitive period may have on the cell, system, individual, and community. As prevention efforts are effective and life- and cost-saving, more research and funding toward reducing human suffering by focusing on development can have an enormous impact on society.

Belmont, MA, USA Susan L. Andersen

Contents

Part I
Mechanisms Underlying Brain Changes During Sensitive Periods

Development of Functional Properties in the Early Visual System: New Appreciations of the Roles of Lateral Geniculate Nucleus

Andrea K. Stacy and Stephen D. Van Hooser

Contents

Abstract In the years following Hubel and Wiesel's first reports on ocular dominance plasticity and amblyopia, much attention has been focused on understanding the role of cortical circuits in developmental and experience-dependent plasticity. Initial studies found few differences between retinal ganglion cells and neurons in the lateral geniculate nucleus and uncovered little evidence for an impact of altered visual experience on the functional properties of lateral geniculate nucleus neurons. In the last two decades, however, studies have revealed that the connectivity between the retina and lateral geniculate nucleus is much richer than was previously appreciated, even revealing visual plasticity – including ocular dominance plasticity – in lateral geniculate nucleus neurons. Here we review the development of the early visual system and the impact of experience with a distinct focus on recent

A. K. Stacy and S. D. Van Hooser (✉)
Department of Biology, Brandeis University, Waltham, MA, USA
e-mail: astacy@brandeis.edu; vanhoosr@brandeis.edu

Curr Topics Behav Neurosci (2022) 53: 3–36
https://doi.org/10.1007/7854_2021_297
Published Online: 3 February 2022

discoveries about lateral geniculate nucleus, its connectivity, and evidence for its plasticity and rigidity during development.

Keywords Critical period · Lateral geniculate nucleus · LGN · Plasticity · Primary visual cortex · Receptive field · Retina · Retinogeniculate · Thalamocortical · Visual development

1 Introduction

Neural connection development during critical periods in the visual system has profound effects on visual processing in the adult. In the early visual system, retinal ganglion cells send axons to the lateral geniculate nucleus (LGN) of the thalamus, which sends inputs to the primary visual cortex (V1). Proper formation of this circuit relies on normal visual experience in mammals, and the precise wiring within this system will not properly mature if vision is disrupted during the critical period. In the 1960s we learned from Hubel and Wiesel that disrupting visual experience during a critical period produces irreversible damage in visual processing (Wiesel and Hubel 1963a, b, 1965a, b). They monocularly deprived kittens of visual experience by suturing one eyelid closed during the first few months of their lives and found a dramatic change in neural responses in V1. Specifically, the distribution of ocular dominance shifted; that is, very few cells could be driven by the deprived eye. Moreover, when they performed the same experiment in adult cats, no such shift was observed (Wiesel and Hubel 1963a, b). This was the first evidence that normal sensory development depended on processes that occur during a critical period and has been the basis for examining plasticity in the visual system. Here we describe the development of the early mammalian visual system with a focus on the role of the thalamus in experience-dependent plasticity. Although the perception of the thalamus as more than a relay center is now commonly recognized, recent studies have expanded our understanding of the essential role of the thalamus in visual development. While the majority of literature concerning developmental plasticity largely focuses on primary visual cortex, we will also highlight the role of the LGN during sensitive periods of visual development. Notably, recent observations in the retinogeniculate pathway underscore the importance of earlier synapses in the visual system and this chapter will review these data within the context of what we already understand about the early visual system (Fig. 1).

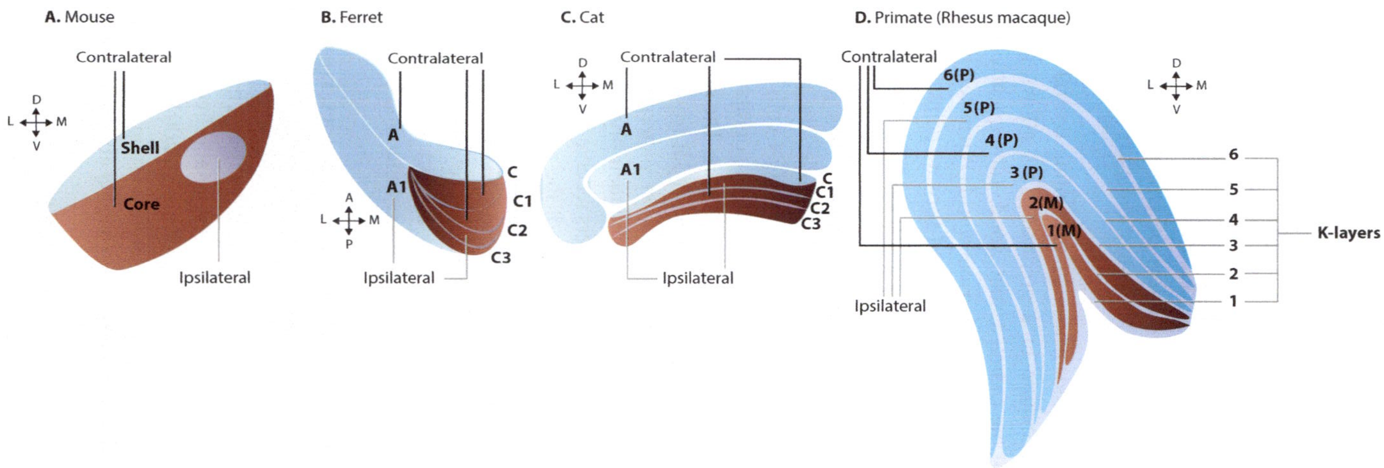

Fig. 1 Organization of the mature lateral geniculate nucleus (LGN) in mouse, ferret, cat, and primate (Rhesus macaque). (**a**) The mouse LGN is divided into the shell (blue) and core (red). The shell and core both receive inputs from the contralateral eye. A small medial portion of the core receives input from the ipsilateral eye (purple). (**b**) The ferret LGN is divided into laminae A, A1, and C (blue). The C layer is further subdivided into C, C1-C3. Laminae A and C and C1 receive projections from the contralateral eye and laminae A1 and C2 receive projections from the ipsilateral eye. Layer C3 does not receive retinal afferents. Horizontal plane is shown for visualization of laminae. (**c**) The cat LGN, like the ferret LGN, is divided into laminae A, A1, C, and C1-C3. Laminae A, C, and C2 receive projections from the contralateral eye. Laminae A1 and C1 receive input from the ipsilateral eye. Layer C3 does not receive retinal afferents. (**d**) The macaque LGN is divided into two magnocellular layers (M, layers 1–2, red), four parvocellular layers (P, layers 3–6, blue), and six koniocellular layers (purple). Mouse, cat, and macaque LGN are shown in the coronal plane. References: Mouse: (Godement et al. 1984; Grubb and Thompson 2004); Ferret: (Zahs and Stryker 1985); Cat: (Hickey and Guillery 1974); Macaque: (Malpeli and Baker 1975; Connolly and Van Essen 1984); *D* dorsal, *V* ventral, *M* medial, *L* lateral, *A* anterior, *P* posterior

2 Gross Organization and Development of LGN

The visual input from the brain arises from retinal ganglion cells, the only input from the retina to the brain. These retinal ganglion cells are divided into separate classes that tile the retinal surface and convey different types of information to the brain (Sherman and Spear 1982; Livingstone and Hubel 1987; DeYoe and Van Essen 1988; Maunsell 1992; Hendry and Reid 2000; Kawasaki et al. 2004). Modern genetic methods have identified 46 distinct classes of retinal ganglion cells in mouse (Tran et al. 2019), 18 such classes in macaque monkey (Peng et al. 2019), and 12 classes in human (Yan et al. 2020).

The LGN begins as separate progenitor neurons representing central vision and peripheral vision and divide and mature in a gradient. This allows for earlier development of cells representing the central visual world, subsequently followed by cells carrying information from the peripheral visual world (Wiencken-barger and Casagrande 2002). As the visual system develops, a feed-forward pathway is formed from the retina to the lateral geniculate nucleus (LGN) to the primary visual cortex (V1). This pathway consists of two well-defined synapses. The first synapse, the retinogeniculate synapse, describes the connection between retinal ganglion cells (RGCs) and the thalamocortical cells of the LGN, while the second, the thalamocortical synapse, refers to the connection between LGN cells and cells in V1. RGCs encode visual information and send their axons topographically to the LGN, which is located in the thalamus adjacent to the optic tract. The optic tract is a thick bundle of axons that provides the major feed-forward input from the retina.

In many mammals, the LGN exhibits clear lamination patterns that can be observed with Nissl staining, but the number of layers and their organization varies from species to species. In primates, most individual layers receive retinal input from a single eye (Kaas et al. 1972). For example, the LGN of the macaque monkey is divided into six layers. Layers 1 and 2, the magnocellular layers, and layers 3–6, the parvocellular layers, can be further characterized by the eye-specific inputs they receive. Layers 1, 4, and 6 receive input from the contralateral eye, while layers 2, 3, and 5 receive input from the ipsilateral eye. A third distinct set of layers, the koniocellular layers, are interspersed between the magnocellular and parvocellular layers.

Cells in each LGN layer often exhibit similar functional properties. For example, macaque magnocellular layers can be characterized by their transient responses, shorter response latencies, and lack of chromatic selectivity (Sherman et al. 1976; Kaplan and Shapley 1982; Yeh et al. 1995; Usrey and Reid 2000; Levitt et al. 2001). In contrast, parvocellular layers are distinguished by their smaller receptive fields, sustained responses, and chromatic contrast sensitivity (Xu et al. 2001). Much less is known about the koniocellular layers but their cells feature a broad set of characteristics, suggesting that there are multiple classes of cells (Irvin et al. 1986; Hendry and Reid 2000; Xu et al. 2001).

In carnivores, such as the cat and ferret, the LGN is divided into the A, A1, and C layers (Guillery 1969; Linden et al. 1981) (Fig. 1). These layers are innervated by retinal ganglion cell classes called X, Y, and W cells (Shapley 1984). Cells in LGN exhibit similar receptive field properties as these retinal ganglion cell classes and the LGN neurons are also divided into classes referred to as X, Y, and W (Howland 1984). Retinal ganglion X-cells drastically outnumber Y-cells and send eye-specific inputs to layer A or A1 of the LGN (Lennie 1980). Retinal ganglion Y-cells send their projections to all three layers, with those from the contralateral eye sending synapses to layer C or A, and Y-cells from the ipsilateral eye sending synapses to layer A1 (Hickey and Guillery 1974) (Fig. 1). W-cells send inputs to the C layer, and, like koniocellular cells, are a heterogeneous superclass that is comprised of many cell types, including cells with non-concentric receptive fields, longer latencies, and many other types (Stone and Fukuda 1974; Shapley 1984; Hendry and Reid 2000). The relationships between X-cells and Y-cells of carnivores and the parvocellular and magnocellular cells of primates are unclear, with some investigators citing the linearity of X and parvocellular cells and non-linearity of Y cells and magnocellular cells as evidence of homology, while others cite the common contrast gain of X, Y, and magnocellular cells and suggest that the color-sensitive parvocellular cells are a class unique to primates (Shapley and Hugh Perry 1986; Van Hooser et al. 2003).

In other species, such as rats and mice, the division of cells in LGN is more subtle. The mouse LGN is divided into the shell and core (Reese 1988). The shell and core both receive inputs from the contralateral eye while only a small population of ipsilateral RGC afferents project to the core (Fig. 1). The shell is innervated by a heterogeneous population of RGCs, including cells that are orientation selective or direction selective, while the core receives inputs from α RGCs. α RGCs can be classified based on their large soma and dendritic field size and are found among a wide range of mammalian species (Peichl et al. 1987; Peichl 1991; Sun et al. 2002).

3 Innervation from the Retina: Molecular Cues

The circuitry of the visual system is remarkably precise. Its anatomical and functional organization begins to emerge prior to the onset of visual experience and develops in a feed-forward manner. Before eye-opening, visual system development begins with an initial phase during which axon mapping and rearrangement of different classes of retinal ganglion cells (RGCs) depend on molecular guidance cues and spontaneous activity. Molecular cues act to establish the initial retinotopic mapping of the LGN, forming a basic framework for vision.

RGCs are first guided by the ephrin/Eph family of axon guidance molecules (Penn et al. 1998). Ephrins and Eph receptors, expressed in the retina and its central visual targets, act as graded guidance cues for sending RGC axons to their appropriate locations (McLaughlin and O'Leary 2005; Flanagan 2006). As evidence for the essential role of ephrins in formation of topographic maps, it has been shown that loss or gain of EphA/ephrin-A function in mice produces defects in the placement of layers within the dLGN (Huberman et al. 2005a, b; Pfeiffenberger et al. 2005). Nevertheless, division of axonal projections from the two eyes into layers persists, suggesting that another process must be responsible for this laminar segregation, as we discuss next.

4 Innervation from the Retina: Spontaneous Activity and Laminar Plasticity

Although RGCs guided by molecular cues are able to send axons to their correct targets, without coordinated waves of activity, RGC axons fail to form dense terminations (Grubb et al. 2003; McLaughlin and O'Leary 2005). Instead, they form diffuse axon arborizations that fail to convey a complete representation of the visual world. These coordinated waves, known as retinal waves, begin prior to visual experience and can be described as coordinated, spontaneous waves of action potentials among neighboring RGCs. Retinal waves are initiated randomly and propagate across the retina, stimulating correlated LGN neuron spiking activity during this process (Mooney et al. 1996). Although they are generated randomly, waves of spontaneous activity retain a high degree of structure. For example, studies have shown that activity from neurons in same-eye LGN layers and same-sign (On or Off) sublayers is more correlated than activity in opposite eye layers in opposite-sign sublayers (Weliky and Katz 1999; Huberman et al. 2008) (For more in-depth review, see Feller (1999)). Further, in mice it has been shown that retinal waves are biased to travel in a temporal to nasal direction, which mimics the front-to-back motion that the animal would experience if it were moving forward in its environment (Ge et al. 2021).

Across species, retinal waves can be grouped into three stages during which distinct mechanisms result in different activity patterns that provide diverse functions. Studies have shown dramatic effects of blocking spontaneous activity during Stage II and Stage III waves. In ferrets, for example, when stage II retinal waves are blocked with epibatidine in both eyes during the normal period of eye-specific segregation (P1-P10), eye-specific segregation does not occur. However, blocking only one eye's wave activity still allows for retinogeniculate segregation, but this manipulation results in the normal eye expanding its thalamic axonal territory, while the blocked eye has a dramatically reduced projection territory (Penn et al. 1998). If retinal waves are completely inhibited in mice, using ephrin-A2/A3/A5 triple knockouts, there is a disruption in eye-specific segregation and layer placement in the LGN (Pfeiffenberger et al. 2005). This and other studies demonstrate that, together with molecular guidance cues, this correlated activity is necessary for the development of retinotopic maps and eye-specific segregation of LGN.

In primates, retinogeniculate eye-specific segregation of LGN layers begins before the onset of visual experience. Eye-specific connections form prenatally and undergo accelerated growth during the last two trimesters (Rakic 1976), beginning segregation during the second trimester in layers 5 and 6 of the LGN. By the third trimester segregation of the LGN is complete, and the magnocellular and parvocellular layers can be differentiated by their characteristic response properties by 1 week postnatal (Huberman et al. 2005a, b) (Fig. 2).

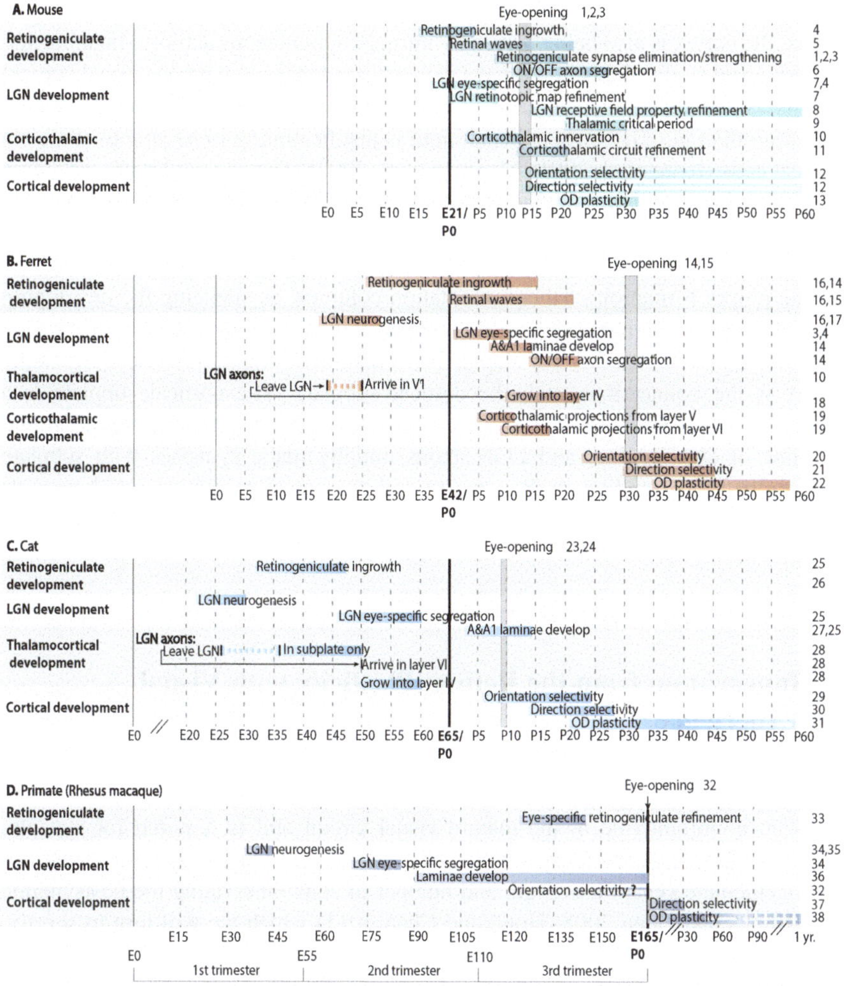

Fig. 2 Timeline of developmental events in the retino-geniculo-cortical pathway of the (**a**) mouse, (**b**) ferret, (**c**) cat, and (**d**) primate (Rhesus macaque) visual systems. (**a**) *Mouse* (teal). References: 1. (Chen and Regehr 2000); 2. (Hong et al. 2014); 3. (Jaubert-Miazza et al. 2005) 4. (Godement et al. 1984); 5. (Demas et al. 2003); 6. (Tian and Copenhagen 2003); 7. (Pfeiffenberger et al. 2005); 8. (Tschetter et al. 2018); 9. (Liang and Chen 2020); 10. (Seabrook et al. 2013); 11. (Thompson et al. 2016); 12. (Rochefort et al. 2011); 13. (Gordon and Stryker 1996); (**b**) *Ferret* (pink). References: 14. (Linden et al. 1981); 15. (Wong et al. 1993); 16. (Johnson and Casagrande 1993); 17. (Jackson et al. 1989); 18. (Herrmann et al. 1994); 19. (Clascá et al. 2012); 20. (Chapman and Stryker 1993); 21. (Li et al. 2006); 22. (Issa et al. 1999); (**c**) *Cat* (blue). References: 23. (Wiesel and Hubel 1963a, b); 24. (Hubel and Wiesel 1964); 25. (Allendoerfer and Shatz 1994); 26. (Hickey and Hitchcock 1984); 27. (Kalil 1978); 28. (Ghosh and Shatz 1992); 29. (Albus and Wolf 1984); 30. (Albus and Fries 1980); 31. (Hubel and Wiesel 1970); (**d**) *Primate* (Rhesus macaque) (purple). References: 32. (Wiesel and Hubel 1974); 33. (Rakic 1976); 34. (Huberman et al. 2005a, b); 35. (Rakic 1977); 36. (Rakic et al. 1977); 37. (Hatta et al. 1998); 38. (LeVay et al. 1980); Approximate ages of eye-opening for each species are indicated above timelines. Conception and birth are labeled as embryonic (E0) and postnatal day zero (P0). Outlined bars indicate period of less developmental plasticity and dashed line (purple) indicates minimal plasticity. *LGN* lateral geniculate nucleus, *OD* ocular dominance

While the retina is producing spontaneous activity that propagates to LGN and cortex, the cortex is also being driven by intrinsic spontaneous activity. Imaging and physiological studies in the ferret have demonstrated that spontaneous visual cortical activity is already highly modular 10 days before eye-opening (Chiu and Weliky 2001; Smith et al. 2018) when visual input is just beginning to be able to stimulate cortical cells (Krug et al. 2001; Akerman et al. 2002). These modules are the future positions of orientation columns in the cortex (Smith et al. 2018), indicating that considerable functional network connectivity in cortex has been established even before the onset of signals from the LGN or retina. Activity in the cortex is necessary for the proper formation of these orientation columns, as silencing the developing cortex during this period prevents the emergence of orientation selectivity (Chapman and Stryker 1993). Another key step in the formation of orientation selectivity in the cortex is the connections with the cortical subplate. The cortical subplate is a transient cortical structure that exists early in development and disappears around the time of eye-opening, and LGN axons initially make synapses with subplate neurons before extending their axons to make connections with layer 4 neurons (Allendoerfer and Shatz 1994). If the subplate is ablated, orientation selectivity never forms in visual cortex (Kanold et al. 2003).

5 Innervation from the Retina: Plasticity with Visual Experience

While eye-specific segregation occurs before eye-opening, early visual experience plays an important role in the mature visual circuit and is required for synaptic plasticity. Visual deprivation during early vision leads to a profound change in retinogeniculate synaptic strength and number of cells innervating the postsynaptic cell (Hooks and Chen 2008; Hong and Chen 2011) which we will turn to shortly. Cell-class-specific changes can also be observed in LGN lamina that lack visual experience. Monocular deprivation by lid suture (MDLS) is a classical, reversable method for understanding experience-dependent changes during early critical periods. With MDLS, there is a shift in OD toward the non-deprived eye, and associated structural changes in synapse development, in addition to alteration of properties of cells themselves, suggesting substantial network rewiring (Wiesel and Hubel 1963a, b). For example, in studies of monocular deprivation in primate, magnocellular cells of the deprived eye were found to exhibit somewhat faster latencies to optic chiasm stimulation, slightly stronger receptive field surrounds, and lower responsivities (Levitt et al. 2001). Magnocellular and parvocellular cells of the deprived eye also exhibited lower nonlinearities.

In an anatomical study of MDLS monkeys, laminae corresponding to the deprived eye exhibited cell shrinkage compared to non-deprived laminae with early lid closure (Vital-Durand et al. 1978; von Noorden and Crawford 1978; Headon et al. 1979; Sherman and Spear 1982; Tigges et al. 1984; Levitt et al.

2001), and cells were shown to have elevated synaptic densities overall at the soma of deprived eye (Lachica et al. 1990; Wilson and Forestner 1995; Levitt et al. 2001). In MDLS cats, there is a specific loss of Y cells, and an X-cell-specific reduction in spatial resolution in the deprived laminae (Sherman et al. 1972; Lehmkuhle et al. 1980; Sherman and Spear 1982).

As an interesting aside, the choice of manipulation for monocular deprivation has profound effects on the changes in neural activity that are induced in LGN and cortex. A study of visual deprivation via eyelid closure or tetrodotoxin (TTX) inactivation of the retina at the peak of the visual critical period in mice (~P28) showed that while both manipulations eliminated visually evoked activity, they did not affect spontaneous activity and average firing rate of LGN neurons. Instead, it seems that these two different manipulations had two different effects. Eyelid closure led to a decrease in correlative firing between simultaneously recorded cells, while TTX inactivation resulted in an increase in thalamic bursting activity (Linden et al. 2009). Therefore, it is important to think of MDLS or MD via inactivation as manipulations that do not remove activity but instead substantially alter its quality, and in different ways.

6 Synapse Development and Convergence Between Retina and LGN

After an initial experience-independent phase of retinotopic mapping, retinogeniculate cells undergo fine-scale refinement. During this phase of development, RGC synapse elimination and strengthening relies on spontaneous activity and visually evoked activity (Hong and Chen 2011; Hooks and Chen 2006). At this critical period of heightened plasticity, the onset of visual experience drives the elimination of weak retinogeniculate inputs so that, at maturity, an LGN neuron only receives strong input from a few RGCs (Chen and Regehr 2000; Jaubert-Miazza et al. 2005; Hooks and Chen 2006) (For more in-depth review, see Huberman (2007)) (Fig. 3). This describes the characteristic functional convergence of RGCs. However, functional convergence/divergence and morphological convergence/divergence are considered to be distinct properties of the retinogeniculate synapse. Functional convergence/divergence describes the number of RGCs that synapse onto a single LGN neuron and the number of LGN neurons that receive synaptic input from an RGC, while morphological convergence/divergence refers to the number of terminal boutons or axon arbors. Thus, a single LGN neuron may receive input from many different boutons or arbors, but only a few may functionally drive that cell (Tavazoie and Reid 2000). Following eye-opening, the LGN begins to develop in an experience-dependent manner.

While past studies have found low convergence of RGCs on single thalamic cells (Cleland et al. 1971a, b, Mastronarde 1992; Chen and Regehr 2000; Ziburkus and

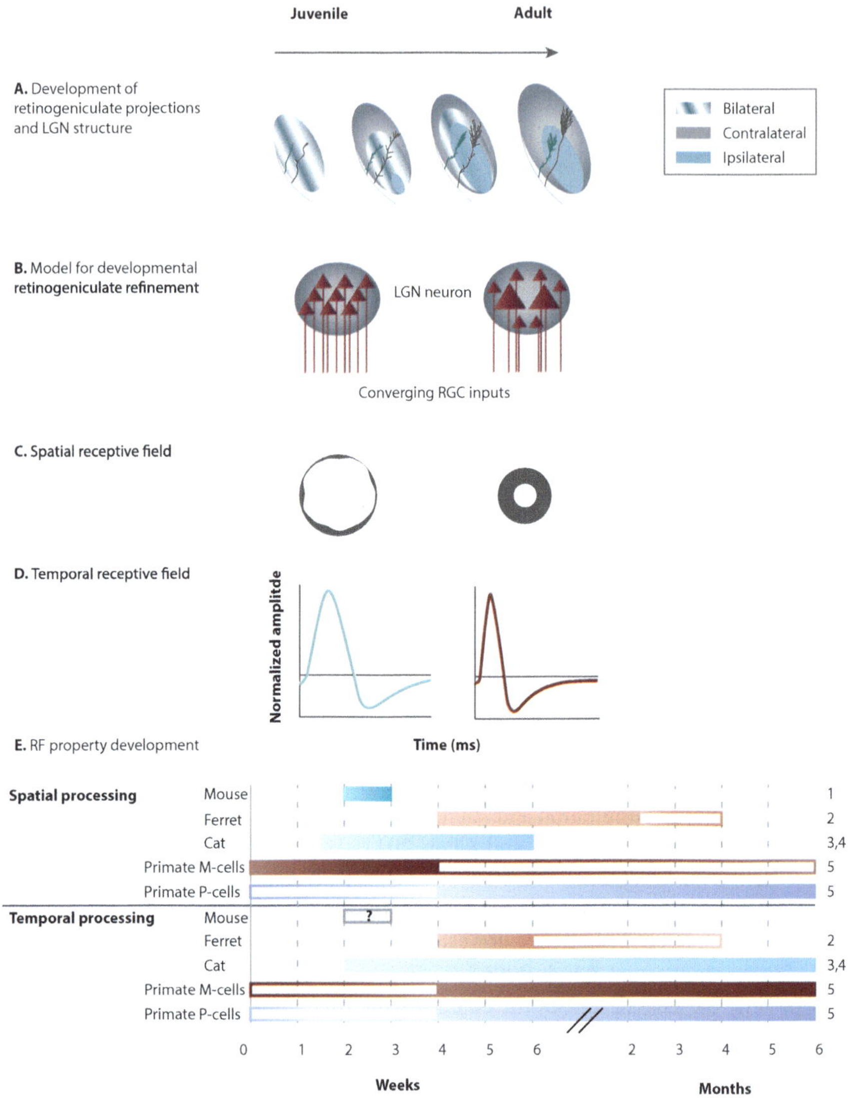

Fig. 3 Development of the lateral geniculate nucleus (LGN). (**a**) Development of retinogeniculate projections and LGN structure from juvenile to adult. Initially retinal afferents are overlapping. Contralateral and ipsilateral projection targets and axons develop while some afferents are still overlapping. Finally, the mature LGN segregates into eye-specific layers. For simplicity, LGN diagram features a single ipsilateral and contralateral layer. (**b**) Model for developmental retinogeniculate refinement. An LGN neuron initially receives many weak retinal ganglion cell (RGC) inputs. With development, inputs are weakened or eliminated and only a few inputs are strengthened. (**c**) Spatial receptive field development. With development, an unstructured spatial receptive field structure decreases in size, acquires a concentric center-surround antagonistic organization and the ratio of amplitudes of the center and surround components increases. (**d**) Temporal receptive field development. With development, an LGN temporal response peak latency and duration decreases. (**e**) Timeline of peak and overall developmental periods of spatial and temporal receptive field development in mouse, ferret, cat, and primate (Rhesus macaque) LGN.

Guido 2006; Hammer et al. 2015), there is recent evidence for a higher number of retinogeniculate inputs than previously thought. A number of physiological and ultrastructural studies of retinogeniculate inputs in mice have all revealed evidence of numerous RGC inputs to single LGN cells (Hammer et al. 2015; Morgan et al. 2016; Weyand 2016; Rompani et al. 2017; Liang and Chen 2020). In addition, a study in cat that transiently blocked ON retinal ganglion cells with 2-amino-4-phosphonobutyric acid found that ON-center LGN cells suddenly became sensitive to OFF inputs in the center of their receptive fields, revealing a hidden input that was not appreciated earlier (Moore et al. 2011). It is now estimated that there is an average of ten functional RGC inputs converging onto a single thalamocortical cell but only a few of these connections provide strong functional input (Chen and Regehr 2000; Jaubert-Miazza et al. 2005; Hooks and Chen 2006).

What is the function of these additional, weaker inputs? Retinogeniculate convergence is thought to serve a number of purposes and in mice has been described as conveying information in three modes (Rompani et al. 2017). In relay mode, different RGC boutons converging on a single LGN neuron share similar functions. That is, they share the same tuning properties. Relay mode represents the classic description of LGN cells that was appreciated in the time of Hubel and Wiesel's collaboration. In relay mode, thalamocortical cells may increase or suppress signals conveyed by the retina and relay them to cortex, but they do not greatly modify the information being sent. They effectively reflect the sum of their inputs. However, in combination mode, converging axons exhibit the same preference for one visual feature but have a diverse range of other tuning properties. In this case a single LGN neuron could be selective for more complex stimulus features. The third mode describes binocular integration. Like combination mode, a single LGN neuron can combine inputs from a number of RGCs but in this mode the afferents originate from both eyes (Rompani et al. 2017). Regardless, it seems that by combining inputs tuned for similar features, an LGN neuron can reduce noise and trial-to-trial variability from individual inputs, while increasing the signal-to-noise ratio and robustness of its response. In this way, weaker inputs that may not individually generate a response in the LGN neuron can be combined, resulting in activation of the postsynaptic cell (Litvina and Chen 2017). Moreover, the combination of temporally correlated inputs can increase transmission efficiency (Liang and Chen 2020).

The two phases of retinogeniculate development described so far detail an initial phase that requires no visual experience and a second phase that requires visually-

Fig. 3 (continued) Primate cell development is divided into magnocellular (M) cells and parvocellular (P) cells. Saturated bars indicate peak developmental time periods and outlined bars indicate period of initial plasticity or continued development. References: 1. (Tschetter et al. 2018); 2. (Tavazoie and Reid 2000); 3. (Cai et al. 1997); 4. (Kiley and Usrey 2017); 5. (Movshon et al. 2005). Additional: (C J Shatz and Sretavan 1986); (Hooks and Chen 2006). *LGN* lateral geniculate nucleus, *RGC* retinal ganglion cell, *M* magnocellular, *P* parvocellular

evoked activity for fine-scale refinement of the retinogeniculate synapse. However, there is also a recently observed final phase of retinogeniculate refinement that depends on normal visual experience. During this final critical period in visual development, visual manipulations have a greater potential to remodel retinogeniculate synapses (Hooks and Chen 2008). After eye-opening there is continued growth of axon arborization size and complexity (Liang and Chen 2020) and this persistence of development leaves open a window for plasticity. Late dark-rearing of mice during this period (P20-P35) after the onset of visual experience (~P12) has revealed unexpected changes in retinogeniculate synaptic refinement. In normally-reared mice, although size and branching complexity remain stable (Kim et al. 2010) at this time, bouton size and distribution along the arbor change (Hong et al. 2014). The initial broad distribution of boutons along the terminal arbor transforms into distinct clusters. However, if animals are dark-reared during this later time period, there is a corresponding increase in retinogeniculate inputs accompanied by a decrease in RGC input strength (Hooks and Chen 2006) effectively leading to a regression of visual development to its state prior to eye-opening. Moreover, manipulations before or after this critical period do not have the same robust effect on plasticity. Chronic visual deprivation (dark-rearing mice from birth) leads to modest changes in synaptic maturation without significantly affecting retinogeniculate input number or strength. Additionally, short-term deprivation (<4 days), or visual deprivation beginning at P16 or P25 does not have the same effect on synapse development as late dark-rearing (Hooks and Chen 2008), indicating a distinct period for experience-dependent plasticity.

7 Receptive Field Development and Plasticity

A receptive field is characterized by the spatiotemporal aspects of a visual stimulus that causes a cell to fire action potentials. Across species, a period of receptive field refinement begins before the onset of visual experience and RF properties continue to develop after eye-opening. Accordingly, the immature state of the receptive field looks different than that of the mature adult (Fig. 3). However, visual processing persists throughout this process, demonstrating that there must be certain developmental changes that occur to allow for visual information to be conveyed at any point during this sensitive period of refinement.

8 Spatial Processing in LGN

Some aspects of LGN receptive fields are firmly established at the onset of visual experience, while other aspects undergo large-scale refinement. Because some developmental changes are already complete before eye-opening, such as the segregation of eye-specific retinogeniculate inputs, some spatial receptive field properties are also fully developed at eye-opening and do not seem to require visual experience. This includes the degree of On/Off specificity which can be seen across species, in mice (Tschetter et al. 2018) and ferrets (Linden et al. 1981; Hahm et al. 1991; 1999; Akerman et al. 2002). On the other hand, receptive field size and shape undergoes a substantial change, converting broad and imprecise receptive fields into ones that are smaller, sharpened, or more temporally precise in a number of species, including cat (Wiesel and Hubel 1963a, b, Daniels et al. 1978; Tootle and Friedlander 1989; Gary-Bobo et al. 1995; Cai et al. 1997; Suematsu et al. 2013) ferret (Tavazoie and Reid 2000; Akerman et al. 2002; Davis et al. 2015), mouse (Tschetter et al. 2018), and primate (Blakemore and Vital-Durand 1986) (Fig. 3).

The decrease in receptive field size develops gradually with visual experience (Daniels et al. 1978). In cats, for example, this process begins at 1 week postnatally and can be observed progressively at 2 weeks, 1 month and 6 weeks postnatal (Kiley and Usrey 2017) (Fig. 3). During this period of refinement LGN receptive fields of X-cells and Y-cells may undergo changes if the animal is visually deprived, such as a reduction in spatial resolution (Lehmkuhle et al. 1978; Kratz et al. 1979; Mower et al. 1982; Sherman and Spear 1982; Yin et al. 2006). In mouse, a decrease in receptive field size occurs during the first week after eye-opening (Tschetter et al. 2018) (Fig. 3). Spatial receptive field shape modification in LGN during development by the convergence of RGC inputs has been classically demonstrated in monosynaptic paired recordings of RGCs and LGN cells. These recordings of RGC-LGN cell receptive field pairs have demonstrated remarkable similarity in receptive field sign, size, and spatial position (Levick et al. 1972; Mastronarde 1987; Usrey et al. 1999). Functionally, LGN cells' receptive fields therefore largely reflect their retinal inputs. With a decrease in receptive field size there is also an increase in stimulus size suppression that develops early and an increase in peak spatial frequency. These changes are indicative of the development of visual acuity.

9 Temporal Processing in LGN

Temporal properties of individual neurons play an important role in the successful transmission of signals across synapses. RGC spike timing can determine whether there is a postsynaptic thalamocortical action potential. Additionally, timing of thalamocortical spikes can affect the synaptic conveyance of visual information to cortex (Mastronarde 1987; Usrey et al. 1999; Usrey 2002). Paired extracellular recordings of RGCs and LGN neurons have shown that even if there is a strong synaptic connection, a single RGC spike may not be sufficient to cause the postsynaptic cell to fire (Cleland and Lee 1985; Usrey et al. 1999; Usrey 2002). Instead, it seems that local activity, such as modulatory inputs and membrane potential, determines the response mode of LGN cells (whether they fire rapid bursts or tonic trains of spikes) (McCormick and Bal 1994; Usrey et al. 1998; Sherman and Guillery 2002; Wang et al. 2007; Litvina and Chen 2017).

The transformation of temporal processing over development may also be attributed to retinogeniculate convergence. By combining RGC inputs with similar tuning for visual stimulus features, a single LGN neuron will receive temporally correlated synaptic inputs that summate and jointly increase the efficacy of transmission of signals, causing the postsynaptic neuron to fire. It is well documented that paired-stimulus enhancement occurs when two spikes are fired by an RGC within 30 ms of each other, so that the second spike is much more likely to drive a response in the LGN cell (Mastronarde 1987; Usrey et al. 1998). Synchrony of timing early in development may provide the initial patterning of RGC synapses in LGN and synchrony in the adult may be important for efficient transfer of information across synapses. The importance of initial mapping of RGC synapses can be understood by the mature state of sustained and transient responses already at eye-opening (Tschetter et al. 2018).

The developmental time course of some temporal response properties extends for much longer, particularly in comparison with spatial response properties (Suematsu et al. 2013). One seemingly ubiquitous receptive field property seen in immature LGN cells across species is long-latency responses (Fig. 3). In primates, initial long-latency responses and a reduced range of temporal frequencies that elicit responses can be observed in infants at 5 weeks old (Blakemore and Vital-Durand 1986; Hawken et al. 1997; Movshon et al. 2005). Longer latency responses in young animals may be necessary in order to extend the time window for temporal summation of spontaneous and/or immature visually-evoked activity. This might allow for Hebbian mechanisms to strengthen correlated responses early in development. The decrease in response latency is thought to be due, in part, to the development of retinal axon myelination (Elgeti et al. 1976; Moore et al. 1976) and biophysical membrane properties (Monyer et al. 1994; Dunah et al. 1996; Wenzel et al. 1996; Ramoa and Prusky 1997; Liu and Chen 2008). For this reason, temporal response properties continue to develop with visual experience during the first few months of life (Stavros and Kiorpes 2008) (Fig. 3). In primates, at 1-month postnatal, temporal resolution is still less than one half the response of adult values. However, contrast gain and peak response rates for optimal stimuli have reached two-thirds of adult values, indicating modest maturation during early development. By adulthood (24 weeks), primates have fully developed responses to these stimuli (Movshon et al. 2005). In the ferret, LGN cell latencies exhibit a gradual reduction during the first 2 weeks of visual experience and continue to become slightly faster into adulthood (Tavazoie and Reid 2000) (Fig. 3).

10 Orientation and Direction Selectivity

Orientation and direction selectivity are well-documented receptive field properties. Universal among examined mammalian species, neurons in the primary visual cortex exhibit a preference for the orientation of a visual stimulus (Van Hooser 2007). That is, when an object in the visual field is oriented at a specific angle, a cell will increase its firing rate. In carnivores and primates cortical orientation selectivity is organized in columns, spanning the surface of the primary visual cortex, such that adjacent columns have similar orientation preferences that are slightly shifted (Hubel and Wiesel 1962; Thompson et al. 1983; Weliky et al. 1996).

In some species, cortical neurons exhibit selectivity for stimulus direction in addition to stimulus orientation. Direction-selective cells are found in certain layers of the primate visual cortex (Orban et al. 1986), throughout all layers of visual cortex in cats (Gilbert 1977; Ohki et al. 2005) and ferrets (Weliky et al. 1996; Li et al. 2008), and in some layers of the mouse (Niell and Stryker 2008; Rochefort et al. 2011; Hoy and Niell 2015) and rabbit visual cortex (Zhuang et al. 2013, 2014). In other species, such as tree shrew and squirrel, there are few direction-selective cells in visual cortex (Heimel et al. 2005; Van Hooser et al. 2013). Many mammals, such as mice (Kim et al. 2008; Huberman et al. 2009), rabbits (Barlow and Levick 1965), and squirrels (McCourt and Jacobs 1984), have large populations of retinal ganglion cells that are direction-selective, but these direction-selective retinal cells are apparently very rare or absent in the carnivore (Cleland and Levick 1974) and primate (Dhande et al. 2019) retina.

In carnivores, cortical orientation selectivity begins developing before eye-opening but requires visual experience for full maturation (Chapman and Stryker 1993; Li et al. 2006). Experiments in ferrets have demonstrated that when animals are dark-reared and visual experience is entirely eliminated, they still develop orientation-selective responses still develop, but not to the same level as that of a normally-reared animal. Binocular lid suture before the onset of visual experience, which affects the pattern of visually-evoked activity, results in dramatic defects and orientation selectivity is largely eliminated. While binocular deprivation via lid suture still allows for light activation of the retina, only low spatial and temporal frequencies are conveyed to the visual system. Thus it seems that the pattern of visual experience is crucial to the development of orientation selectivity (White et al. 2001). Similarly, direction selectivity requires patterned visual activity for development and matures during the first 2 weeks of visual experience in carnivores (Humphrey and Saul 1998; Li et al. 2006) (Fig. 2). However, if the animal is dark-reared, direction maps do not form and V1 neurons do not develop direction selectivity. Moreover, reintroducing normal visual experience at P45 (~2 weeks after eye-opening) allows for the return of some V1 response properties, such as orientation selectivity and contrast sensitivity, but direction tuning cannot be rescued (Li et al. 2006). Although direction selectivity has not been examined in dark-reared primates, there is also evidence in primates that orientation selectivity is present at birth and that direction selectivity requires visual experience. Direction selectivity matures over the first 4 weeks postnatal in the macaque (Hatta et al. 1998) (Fig. 2). Thus, there seems to be a precise critical period during which visual experience is necessary to cortical direction selectivity development.

While there has been considerable research into the circuit origins of orientation selectivity (see Ferster and Miller (2000) and Priebe (2016) for reviews), the precise mechanisms underlying direction tuning in V1 are still an active topic of research. One set of hypotheses for this circuit involves direct inheritance of direction selectivity from feed-forward inputs from individual LGN cells. We now have evidence that both orientation and direction selectivity can be encoded in individual LGN cells in a number of species, particularly in mice (Marshel et al. 2012; Piscopo et al. 2013; Scholl et al. 2013; Zhao et al. 2013), rabbit (Levick et al. 1969; Swadlow and Weyand 1985; Hei et al. 2014), and more weakly in squirrel (Zaltsman et al. 2015), cat (Hubel and Wiesel 1961; Daniels et al. 1977; Levick and Thibos 1980; Vidyasagar and Urbas 1982; Soodak et al. 1987; Shou and Leventhal 1989; Thompson et al. 1994), and primate (Lee et al. 1979; Smith et al. 1990; Cheong et al. 2013). In addition, orientation and direction-selective retinal ganglion cells (DSGCs), among other complex properties, can also be found in rodents and rabbits (Barlow and Levick 1965; Gollisch and Meister 2010). Additionally, it has been shown in cat (Vidyasagar and Urbas 1982) and mouse (Scholl et al. 2013; Zhao et al. 2013) that inactivating V1 has no significant effect on orientation selectivity in LGN cells, suggesting that this feature selectivity does not rely on a corticothalamic feedback mechanism and that V1 could potentially inherit such properties from LGN cells. Genetic tools have also revealed in mice that these DSGCs do indeed send monosynaptic inputs to LGN (Huberman et al. 2009; Kay et al. 2011; Rivlin-Etzion et al. 2011) and that these signals are then conveyed to superficial layers of V1 (Huberman et al. 2009). These data suggest that, in mouse, some V1 cells may inherit direction selectivity from the retina (Marshel et al. 2012; Cruz-Martin et al. 2014; Sun et al. 2015; Hillier et al. 2017).

Another family of hypotheses suggests that direction selectivity may arise from the patterns of convergence of LGN cells onto V1 cells. One possibility is that "lagged" and "nonlagged" LGN cells contribute to the spatiotemporal offsets (Cleland et al. 1971a, b, Marrocco 1976; Adelson and Bergen 1985; Mastronarde 1987; Saul and Humphrey 1990; Moore et al. 2005; Piscopo et al. 2013). Lagged and nonlagged cells differ in timing at low temporal frequencies but not high temporal frequencies. Therefore, as temporal frequency responses develop, there are a wider array of response times available (Saul and Feidler 2002). It is also possible that sustained and transient properties contribute to the development of direction selectivity (Cleland et al. 1971a, b, Marr and Ullman 1981; Lien and Scanziani 2018). Early sustained and transient responses driven by retinal waves may also inform direction selectivity. Sustained units active at low spatial phases with transient units at high spatial phases with different decay constants could combine to produce the spatiotemporal offset necessary for direction selectivity (Lien and Scanziani 2018). It could also be that temporal dynamics of ON and OFF cells confer direction selectivity. If ON and OFF cells exhibited different impulse responses, then their early maturation modified by temporal processing maturation could result in a direction-selective V1 cell (Chariker et al. 2021). Finally, direction selectivity could be conferred by inputs within the cortex. In the retina, null direction inhibition is a key mechanism for providing direction selectivity (Barlow and Levick 1965; Euler et al. 2002; Briggman et al. 2011), and there is some recent evidence that the inhibition to excitation ratio of signals arriving at a V1 neuron is slightly greater from the null side than the preferred side (Wilson et al. 2018; Rossi et al. 2020) (but see Priebe and Ferster (2005)). There are models of intracortical direction selectivity that suggest that direction selectivity can be produced via receptive field asymmetries and intracortical interactions (Goodwin et al. 1975; Emerson and Gerstein 1977; Sillito 1977; Ganz and Felder 1984; Suarez et al. 1995; Maex and Orban 1996;

Livingstone 1998; Li et al. 2014). For example, time delays within a cortical cell's receptive field could be generated in cortex, or intracortical inhibition between cells with different spatiotemporal receptive fields could produce cortical cells selective for opposite directions.

So what is the definitive mechanism and how does experience play a role? It's complicated. A critical period of early visual experience appears to be required for the development of direction selectivity in carnivores and primates (Hatta et al. 1998). However, there is conflicting evidence describing experience-dependent plasticity of direction selectivity in rodents and rabbits. Some studies show that visual experience is not required for the development of direction selectivity in the retina or cortex and that, at eye-opening, DSGCs are already present and V1 cells are already tuned for direction (Chan and Chiao 2008; Elstrott et al. 2008; Chen et al. 2009, 2014; Sun et al. 2011; Wei et al. 2011). However, there is also evidence in these species for the requirement of normal visual experience in the development of direction tuning (Pearson et al. 1981). Direction selectivity in mouse layer 4 increases with visual experience, suggesting an experience-dependent influence (Hoy and Niell 2015). Research has shown that visual experience is required for the clustering of cells along the cardinal directions in mice DSGCs (Bos et al. 2016). Within one study, there was evidence that there are distinct retina-independent and retina-dependent computations for cortical direction selectivity (Hillier et al. 2017), which leaves open the possibility that some cortical direction selectivity is computed from retinal direction-selective sources and other cortical direction selectivity is computed via inputs that are not themselves direction-selective.

11 Clues from the Rapid Development of Direction Selectivity

Developmental changes can sometimes be used to tease apart circuit mechanisms. In the case of the ferret, cortical direction selectivity can be rapidly induced in naïve animals by providing 3–9 h of experience with a moving visual stimulus (Li et al. 2008; Van Hooser et al. 2012; Ritter et al. 2017; Roy et al. 2020; Li et al. 2008; Van Hooser et al. 2012; Ritter et al. 2017. Stacy et al. (2021) examined what changes occurred in LGN receptive fields during the rapid induction of direction selectivity and found that LGN cell latencies, sustainedness/transience, and orientation and direction selectivity were unchanged after 6 h of visual experience with a stimulus that causes a profound increase in direction selectivity in visual cortex. These results indicated that cortical direction selectivity can precede the maturation of LGN cell latencies (Tavazoie and Reid 2000), so it is unlikely that absolute LGN cell latencies are critical for direction selectivity in cortex. This lack of influence of short-term experience on LGN cells is consistent with another study that examined the impact of long-term dark-rearing on orientation or direction tuning of cat LGN cells and found no influence (Zhou et al. 1995). Consistent with these results, another experiment where optogenetic stimulation was provided to the cortex in lieu of visual experience showed that non-specific cortical activity was sufficient to cause an emergence of cortical direction selectivity (Roy et al. 2016), suggesting that cortical mechanisms are likely to be the main drivers of the development of direction selectivity in ferret V1.

12 Ocular Dominance and Its Plasticity

If the view through one eye is blurred during early development or if the alignment of the two eyes is compromised by strabismus, the impact of stimulation from that eye on the cortex will be greatly reduced, in a condition referred to as amblyopia (Hensch and Quinlan 2018). This condition is often modeled in animals by providing monocular deprivation through artificial lid suture (Wiesel and Hubel 1963a, b). The brain is only sensitive to this discrepancy in the quality of the input from the two eyes for a limited time after the onset of visual experience. This sensitive period has been measured carefully in animals such as the mouse (Gordon and Stryker 1996), ferret (Issa et al. 1999), and cat (Hubel and Wiesel 1970), and generally opens 1–2 weeks after the onset of visual experience and closes some weeks later, although some plasticity is possible in adult mice. In monkeys and humans, this critical period extends for several months or years (Lewis and Maurer 2005). Very recently, it has been shown that temporarily providing a complete inactivation of the dominant eye with tetrodotoxin can restore visual acuity through the weak eye even after the critical period has passed in both cats and mice (Fong et al. 2021).

Several cortical mechanisms have been identified that contribute to this phenomenon. After MDLS, there is a substantial weakening of the thalamic inputs from the deprived eye (Heynen et al. 2003). This weakening affects both feed-forward excitatory and inhibitory inputs, and does not, by itself, account for the reduced net drive of the deprived eye in cortical circuits (Miska et al. 2018). Intracortical inhibition is also increased, which serves to provide a net reduction of activity in response to input from the deprived eye (Maffei et al. 2006; Miska et al. 2018). These functional changes in feed-forward input reflect synaptic changes but occur well before the gross rearrangement of thalamocortical axons is a result of critical period plasticity (Antonini and Stryker 1996; Antonini et al. 1999; Silver and Stryker 1999; Trachtenberg et al. 2000; Trachtenberg and Stryker 2001). In a second step of plasticity, the overall drop in activity during MDLS activates homeostatic synaptic mechanisms and homeostatic increases in cortical cell excitability, which serve to potentiate overall responses from either eye (Kaneko et al. 2008; Hengen et al. 2013; Lambo and Turrigiano 2013). The net effect of these changes after several days is a net decrease in drive from the deprived eye, and a net increase in drive from the open eye.

Until very recently, ocular dominance plasticity was thought to be mediated exclusively through cortical methods and not through any changes in the lateral geniculate nucleus. When Hubel and Wiesel first described ocular dominance, their MDLS experiments in kittens provided the common notion that long-term visual deprivation leads to a decrease in thalamocortical axon size (Antonini and Stryker 1993; Antonini et al. 1999) and changes in LGN cell size were recapitulated in monkey LGN (Movshon and Dürsteler 1977), but that the functional properties of LGN neurons remained the same (Wiesel and Hubel 1963a, b). Further, because OD plasticity was assumed to result from a competitive process between the inputs from the two eyes, it was thought that OD plasticity was a distinctly cortical phenomenon, due to clear eye-specific segregation among LGN layers and what they thought were exclusively monocularly driven LGN cells (Wiesel and Hubel 1963a, b; Gilbert and Wiesel 1992). However, new findings in mice have challenged these ideas, both structurally and functionally.

In brief, the LGN of the mouse is divided into a shell and core (Clascá et al. 2012), which both receive inputs from the contralateral eye, and it was thought that only a small portion of the core received input from the ipsilateral eye (Fig. 1). However, newer data from adults and juveniles have shown that there may not be any cells that respond exclusively to ipsilateral eye stimulation. In addition, a large percentage of the LGN cell population is binocularly driven (Howarth et al. 2014; Rompani et al. 2017; Sommeijer et al. 2017), indicating that a fraction of eye-specific inputs are integrated prior to sending input to cortex. Furthermore, this evidence for binocular convergence was recapitulated in the primate, demonstrating that this finding is not a distinct feature of the mouse retinogeniculate synapse but may be common in mammals (Zeater et al. 2015).

While it has long been established that there are eye-specific laminae in the primate LGN, recent work characterizing the koniocellular layers has illuminated more complexity of individual cell responses than was originally appreciated. It was shown in marmoset that a relatively large percentage of the koniocellular layers (~30% of cells) are binocularly driven (Zeater et al. 2015). The presence of binocularly driven cells in the primate LGN has large implications for signal integration. These findings indicate that binocular signals may be integrated earlier in the visual system than previously thought. In cats there is also evidence for binocular interactions in the LGN, although the majority of binocular cells are suppressive in nature (Sanderson et al. 1971). Additionally, similar to the monkey, the total percentage of the LGN cell population thought to be binocularly driven is low (3% in monkey; 2–10% in cat) (Erulkar and Fillenz 1960; Bishop et al. 1962; Kinston et al. 1969; Dacey 1994; Cheong et al. 2013; Zeater et al. 2015; Dougherty et al. 2019). However, these data are relevant for understanding where in the visual pathway signals from two eyes converge, which is essential for understanding binocular vision.

New findings have also elucidated the potential for distinct mechanisms regulating thalamic plasticity and cortical plasticity. Data showed that, after a week of visual deprivation, LGN boutons that were monocularly driven exhibited a reduced or lack of response to the deprived eye, and often began to respond to the non-deprived eye (Jaepel et al. 2017; Huh et al. 2020). These responses were not propagated backward to the LGN from cortex, as silencing of cortex did not alter the changes in LGN axon plasticity with MDLS (Jaepel et al. 2017). In juvenile mice, early visual deprivation resulted in a change in LGN neuron response properties via depression of the deprived eye only. However, when they looked at cortex, they found that OD plasticity was the result of strengthening of LGN responses from the non-deprived eye (Rose et al. 2016; Sommeijer et al. 2017). These results suggest that changes in neurons in the LGN directly contribute to ocular dominance phenotypes in amblyopia.

13 Corticothalamic Development

There is a high likelihood that an interaction between retinogeniculate and corticothalamic refinement during development shapes mature LGN neurons. Although retinogeniculate inputs are the primary source of excitatory drive of thalamocortical cells, corticothalamic neurons make up a much higher percentage of synapses in the LGN compared to retinogeniculate afferents (Guillery 1969; Erişir et al. 1997a, b). In the mammalian visual system LGN cells project primarily to layer 4 and less densely to layer 6 of V1 (Hubel and Wiesel 1972; Gilbert and Kelly 1975; Hendrickson et al. 1978; Blasdel and Lund 1983; Ferster and Lindström 1983; Thompson et al. 2017). Layer 6 of the visual cortex in turn provides the larger part of corticothalamic feedback to the LGN (Gilbert and Kelly 1975; Katz 1987; Fitzpatrick et al. 1994; Briggs et al. 2016). Corticothalamic cells of the feedback pathway provide excitatory synaptic inputs to LGN directly in a retinotopic manner, in addition to sending synapses to other cortical layers that LGN projects to.

While much less is understood about the role of corticothalamic development, it is thought that corticothalamic feedback acts to modulate LGN activity, sharpening receptive fields of LGN cells in addition to enhancing signal transmission through LGN (Briggs and Usrey 2008). Like retinogeniculate development, corticothalamic development requires visual experience (Fagiolini et al. 1994; Gordon and Stryker 1996; Kang et al. 2013). However, corticothalamic projections develop later than retinogeniculate inputs and complete innervation after the onset of visual experience (Shatz and Rakic 1981; Jacobs et al. 2007; Brooks et al. 2013). Evidence from studies of layer 6 corticothalamic suppression suggests that during an intermediate window of maturation in the early visual pathway, corticothalamic innervation continues to influence the fine-tuning of the visual circuit (Thompson et al. 2016; Liang and Chen 2020). The enduring plasticity of retinal inputs to LGN at this time likely allows for their continued refinement via feedback from V1 and optimization of connectivity and complex feature tuning (Thompson et al. 2016). Thus, the timing of their developmental periods is essential for proper circuit formation.

14 Conclusions

Recent interest in characterization of the thalamic contribution to visual development and information encoding continues to evolve our understanding of the important role of the LGN in visual processing. In addition, growing evidence for signal integration and more complex feature selectivity of individual LGN neurons continues to alter not only our methods for understanding the retino-geniculo-cortical pathway, but also our interpretations of past findings, including what we thought we knew about ocular dominance plasticity (Figs. 2 and 3). Moreover, the advancements in technology development continue to allow for a more comprehensive depiction of the inputs and outputs of the LGN and the role of experience-independent and experience-dependent refinement. We describe a number of sensitive periods in visual development that importantly contribute to the refinement of the visual system and normal visual processing. Despite the anatomical differences in the organization of LGN across species, including mouse, ferret, cat, and primate, the changes in receptive fields of LGN neurons during development have several elements in common. Thus, findings from other species during their respective sensitive periods can aid in our understanding of general visual circuit wiring. Findings from typical development and experimental manipulations during these periods will continue to elucidate necessary properties of working neural circuits and differences that have evolved to maintain such robust organization across structures and species.

References

Adelson EH, Bergen JR (1985) Spatiotemporal energy models for the perception of motion. J Opt Soc Am A 2(2):284–299

Akerman CJ, Smyth D, Thompson ID (2002) Visual experience before eye-opening and the development of the retinogeniculate pathway. Neuron 36(5):869–879

Albus K, Fries W (1980) Inhibitory sidebands of complex receptive fields in the cat's striate cortex. Vision Res 20(4):369–372

Albus K, Wolf W (1984) Early post-natal development of neuronal function in the kitten's visual cortex: a laminar analysis. J Physiol 348(1):153–185

Allendoerfer KL, Shatz CJ (1994) The subplate, a transient neocortical structure: its role in the development of connections between thalamus and cortex. Annu Rev Neurosci 17:185–218

Antonini A, Stryker MP (1993) Rapid remodeling of axonal arbors in the visual cortex. Science 260(5115):1819–1821

Antonini A, Stryker MP (1996) Plasticity of geniculocortical afferents following brief or prolonged monocular occlusion in the cat. J Comp Neurol 369(1):64–82

Antonini A, Fagiolini M, Stryker MP (1999) Anatomical correlates of functional plasticity in mouse visual cortex. J Neurosci 19(11):4388–4406

Barlow HB, Levick WR (1965) The mechanism of directionally selective units in rabbit's retina. J Physiol 178(3):477–504

Bishop PO, Burke W, Davis R (1962) The interpretation of the extracellular response of single lateral geniculate cells. J Physiol 162(3):451–472

Blakemore C, Vital-Durand F (1986) Organization and post-natal development of the monkey's lateral geniculate nucleus. J Physiol 380:453–491
Blasdel G, Lund J (1983) Termination of afferent axons in macaque striate cortex. J Neurosci 3(7): 1389–1413
Bos R, Gainer C, Feller MB (2016) Role for visual experience in the development of direction-selective circuits. Curr Biol 26(10):1367–1375
Briggman KL, Helmstaedter M, Denk W (2011) Wiring specificity in the direction-selectivity circuit of the retina. Nature 471(7337):183–188
Briggs F, Usrey WM (2008) Emerging views of corticothalamic function. Curr Opin Neurobiol 18(4):403–407
Briggs F, Kiley CW, Callaway EM, Usrey WM (2016) Morphological substrates for parallel streams of corticogeniculate feedback originating in both V1 and V2 of the macaque monkey. Neuron 90(2):388–399
Brooks JM, Su J, Levy C, Wang JS, Seabrook TA, Guido W, Fox MA (2013) A molecular mechanism regulating the timing of corticogeniculate innervation. Cell Rep 5(3):573–581
Cai D, Deangelis GC, Freeman RD (1997) Spatiotemporal receptive field organization in the lateral geniculate nucleus of cats and kittens. J Neurophysiol 78(2):1045–1061
Chan Y-C, Chiao C-C (2008) Effect of visual experience on the maturation of ON–OFF direction selective ganglion cells in the rabbit retina. Vision Res 48(23):2466–2475
Chapman B, Stryker M (1993) Development of orientation selectivity in ferret visual cortex and effects of deprivation. J Neurosci 13(12):5251–5262
Chariker L, Shapley R, Hawken M, Young L-S (2021) A theory of direction selectivity for macaque primary visual cortex. Proc Natl Acad Sci 118(32):e2105062118
Chen C, Regehr WG (2000) Developmental remodeling of the retinogeniculate synapse. Neuron 28(3):955–966
Chen M, Weng S, Deng Q, Xu Z, He S (2009) Physiological properties of direction-selective ganglion cells in early postnatal and adult mouse retina. J Physiol 587(Pt 4):819–828
Chen H, Liu X, Tian N (2014) Subtype-dependent postnatal development of direction- and orientation-selective retinal ganglion cells in mice. J Neurophysiol 112(9):2092–2101
Cheong SK, Tailby C, Solomon SG, Martin PR (2013) Cortical-like receptive fields in the lateral geniculate nucleus of marmoset monkeys. J Neurosci 33(16):6864–6876
Chiu C, Weliky M (2001) Spontaneous activity in developing ferret visual cortex in vivo. J Neurosci 21(22):8906–8914
Clascá F, Rubio-Garrido P, Jabaudon D (2012) Unveiling the diversity of thalamocortical neuron subtypes. Eur J Neurosci 35(10):1524–1532
Cleland BG, Lee BB (1985) A comparison of visual responses of cat lateral geniculate nucleus neurones with those of ganglion cells afferent to them. J Physiol 369:249–268
Cleland BG, Levick WR (1974) Properties of rarely encountered types of ganglion cells in the cat's retina and an overall classification. J Physiol 240(2):457–492
Cleland BG, Dubin MW, Levick WR (1971a) Simultaneous recording of input and output of lateral geniculate neurones. Nat New Biol 231(23):191–192
Cleland BG, Dubin MW, Levick WR (1971b) Sustained and transient neurones in the cat's retina and lateral geniculate nucleus. J Physiol 217(2):473–496
Connolly M, Van Essen D (1984) The representation of the visual field in parvicellular and magnocellular layers of the lateral geniculate nucleus in the macaque monkey. J Comp Neurol 226(4):544–564
Cruz-Martin A, El-Danaf RN, Osakada F, Sriram B, Dhande OS, Nguyen PL, Callaway EM, Ghosh A, Huberman AD (2014) A dedicated circuit links direction-selective retinal ganglion cells to the primary visual cortex. Nature 507(7492):358–361
Dacey DM (1994) Physiology, morphology and spatial densities of identified ganglion cell types in primate retina. Ciba Found Symp 184:12–28. discussion 28–34, 63–70
Daniels JD, Norman JL, Pettigrew JD (1977) Biases for oriented moving bars in lateral geniculate nucleus neurons of normal and stripe-reared cats. Exp Brain Res 29(2):155–172

Daniels JD, Pettigrew JD, Norman JL (1978) Development of single-neuron responses in kitten's lateral geniculate nucleus. J Neurophysiol 41(6):1373–1393

Davis ZW, Chapman B, Cheng HJ (2015) Increasing spontaneous retinal activity before eye opening accelerates the development of geniculate receptive fields. J Neurosci 35(43): 14612–14623

Demas J, Eglen SJ, Wong ROL (2003) Developmental loss of synchronous spontaneous activity in the mouse retina is independent of visual experience. J Neurosci 23(7):2851–2860

DeYoe EA, Van Essen DC (1988) Concurrent processing streams in monkey visual cortex. Trends Neurosci 11(5):219–226

Dhande OS, Stafford BK, Franke K, El-Danaf R, Percival KA, Phan AH, Li P, Hansen BJ, Nguyen PL, Berens P, Taylor WR, Callaway E, Euler T, Huberman AD (2019) Molecular fingerprinting of on-off direction-selective retinal ganglion cells across species and relevance to primate visual circuits. J Neurosci 39(1):78–95

Dougherty K, Schmid MC, Maier A (2019) Binocular response modulation in the lateral geniculate nucleus. J Comp Neurol 527(3):522–534

Dunah AW, Yasuda RP, Wang YH, Luo J, Dávila-García M, Gbadegesin M, Vicini S, Wolfe BB (1996) Regional and ontogenic expression of the NMDA receptor subunit NR2D protein in rat brain using a subunit-specific antibody. J Neurochem 67(6):2335–2345

Elgeti H, Elgeti R, Fleischhauer K (1976) Postnatal growth of the dorsal lateral geniculate nucleus of the cat. Anat Embryol 149(1):1–13

Elstrott J, Anishchenko A, Greschner M, Sher A, Litke AM, Chichilnisky EJ, Feller MB (2008) Direction selectivity in the retina is established independent of visual experience and cholinergic retinal waves. Neuron 58(4):499–506

Emerson RC, Gerstein GL (1977) Simple striate neurons in the cat. I. Comparison of responses to moving and stationary stimuli. J Neurophysiol 40(1):119–135

Erişir A, Van Horn SC, Bickford ME, Sherman SM (1997a) Immunocytochemistry and distribution of parabrachial terminals in the lateral geniculate nucleus of the cat: a comparison with corticogeniculate terminals. J Comp Neurol 377(4):535–549

Erişir A, Van Horn SC, Sherman SM (1997b) Relative numbers of cortical and brainstem inputs to the lateral geniculate nucleus. Proc Natl Acad Sci U S A 94(4):1517–1520

Erulkar SD, Fillenz M (1960) Single-unit activity in the lateral geniculate body of the cat. J Physiol 154(1):206–218

Euler T, Detwiler PB, Denk W (2002) Directionally selective calcium signals in dendrites of starburst amacrine cells. Nature 418(6900):845–852

Fagiolini M, Pizzorusso T, Berardi N, Domenici L, Maffei L (1994) Functional postnatal development of the rat primary visual cortex and the role of visual experience: dark rearing and monocular deprivation. Vision Res 34(6):709–720

Feller MB (1999) Spontaneous correlated activity in developing neural circuits. Neuron 22(4): 653–656

Ferster D, Lindström S (1983) An intracellular analysis of geniculo-cortical connectivity in area 17 of the cat. J Physiol 342(1):181–215

Ferster D, Miller KD (2000) Neural mechanisms of orientation selectivity in the visual cortex. Annu Rev Neurosci 23:441–471

Fitzpatrick D, Usrey WM, Schofield BR, Einstein G (1994) The sublaminar organization of corticogeniculate neurons in layer 6 of macaque striate cortex. Vis Neurosci 11(2):307–315

Flanagan JG (2006) Neural map specification by gradients. Curr Opin Neurobiol 16(1):59–66

Fong MF, Duffy KR, Leet MP, Candler CT, Bear MF (2021) Correction of amblyopia in cats and mice after the critical period. Elife 10

Ganz L, Felder R (1984) Mechanism of directional selectivity in simple neurons of the cat's visual cortex analyzed with stationary flash sequences. J Neurophysiol 51(2):294–324

Gary-Bobo E, Przybyslawski J, Saillour P (1995) Experience-dependent maturation of the spatial and temporal characteristics of the cell receptive fields in the kitten visual cortex. Neurosci Lett 189(3):147–150

Ge X, Zhang K, Gribizis A, Hamodi AS, Sabino AM, Crair MC (2021) Retinal waves prime visual motion detection by simulating future optic flow. Science 373(6553)
Ghosh A, Shatz CJ (1992) Involvement of subplate neurons in the formation of ocular dominance columns. Science 255(5050):1441–1443
Gilbert CD (1977) Laminar differences in receptive field properties of cells in cat primary visual cortex. J Physiol 268(2):391–421
Gilbert CD, Kelly JP (1975) The projections of cells in different layers of the cat's visual cortex. J Comp Neurol 163(1):81–105
Gilbert CD, Wiesel TN (1992) Receptive field dynamics in adult primary visual cortex. Nature 356(6365):150–152
Godement P, Salaün J, Imbert M (1984) Prenatal and postnatal development of retinogeniculate and retinocollicular projections in the mouse. J Comp Neurol 230(4):552–575
Gollisch T, Meister M (2010) Eye smarter than scientists believed: neural computations in circuits of the retina. Neuron 65(2):150–164
Goodwin AW, Henry GH, Bishop PO (1975) Direction selectivity of simple striate cells: properties and mechanism. J Neurophysiol 38(6):1500–1523
Gordon JA, Stryker MP (1996) Experience-dependent plasticity of binocular responses in the primary visual cortex of the mouse. J Neurosci 16(10):3274–3286
Grubb MS, Thompson ID (2004) Biochemical and anatomical subdivision of the dorsal lateral geniculate nucleus in normal mice and in mice lacking the β2 subunit of the nicotinic acetylcholine receptor. Vision Res 44(28):3365–3376
Grubb MS, Rossi FM, Changeux JP, Thompson ID (2003) Abnormal functional organization in the dorsal lateral geniculate nucleus of mice lacking the beta 2 subunit of the nicotinic acetylcholine receptor. Neuron 40(6):1161–1172
Guillery RW (1969) The organization of synaptic interconnections in the laminae of the dorsal lateral geniculate nucleus of the cat. Z Zellforsch Mikrosk Anat 96(1):1–38
Hahm J-O, Langdon RB, Sur M (1991) Disruption of retinogeniculate afferent segregation by antagonists to NMDA receptors. Nature 351(6327):568–570
Hahm J-O, Cramer KS, Sur M (1999) Pattern formation by retinal afferents in the ferret lateral geniculate nucleus: developmental segregation and the role of N-methyl-D-aspartate receptors. J Comp Neurol 411(2):327–345
Hammer S, Monavarfeshani A, Lemon T, Su J, Fox MA (2015) Multiple retinal axons converge onto relay cells in the adult mouse thalamus. Cell Rep 12(10):1575–1583
Hatta S, Kumagami T, Qian J, Thornton M, Smith EL 3rd, Chino YM (1998) Nasotemporal directional bias of V1 neurons in young infant monkeys. Invest Ophthalmol Vis Sci 39(12): 2259–2267
Hawken MJ, Blakemore C, Morley JW (1997) Development of contrast sensitivity and temporal-frequency selectivity in primate lateral geniculate nucleus. Exp Brain Res 114(1):86–98
Headon MP, Sloper JJ, Hiorns RW, Powell TPS (1979) Cell size changes in undeprived laminae of monkey lateral geniculate nucleus after monocular closure. Nature 281(5732):572–574
Hei X, Stoelzel CR, Zhuang J, Bereshpolova Y, Huff JM, Alonso JM, Swadlow HA (2014) Directional selective neurons in the awake LGN: response properties and modulation by brain state. J Neurophysiol 112(2):362–373
Heimel JA, Van Hooser SD, Nelson SB (2005) Laminar organization of response properties in primary visual cortex of the gray squirrel (Sciurus carolinensis). J Neurophysiol 94(5): 3538–3554
Hendrickson AE, Wilson JR, Ogren MP (1978) The neuroanatomical organization of pathways between the dorsal lateral geniculate nucleus and visual cortex in Old World and New World primates. J Comp Neurol 182(1):123–136
Hendry SH, Reid RC (2000) The koniocellular pathway in primate vision. Annu Rev Neurosci 23: 127–153
Hengen KB, Lambo ME, Van Hooser SD, Katz DB, Turrigiano GG (2013) Firing rate homeostasis in visual cortex of freely behaving rodents. Neuron 80(2):335–342

Hensch TK, Quinlan EM (2018) Critical periods in amblyopia. Vis Neurosci 35:E014–E014
Herrmann K, Antonini A, Shatz CJ (1994) Ultrastructural evidence for synaptic interactions between thalamocortical axons and subplate neurons. Eur J Neurosci 6(11):1729–1742
Heynen AJ, Yoon BJ, Liu CH, Chung HJ, Huganir RL, Bear MF (2003) Molecular mechanism for loss of visual cortical responsiveness following brief monocular deprivation. Nat Neurosci 6(8): 854–862
Hickey TL, Guillery RW (1974) An autoradiographic study of retinogeniculate pathways in the cat and the fox. J Comp Neurol 156(2):239–253
Hickey TL, Hitchcock PF (1984) Genesis of neurons in the dorsal lateral geniculate nucleus of the cat. J Comp Neurol 228(2):186–199
Hillier D, Fiscella M, Drinnenberg A, Trenholm S, Rompani SB, Raics Z, Katona G, Juettner J, Hierlemann A, Rozsa B, Roska B (2017) Causal evidence for retina-dependent and -independent visual motion computations in mouse cortex. Nat Neurosci 20(7):960–968
Hong YK, Chen C (2011) Wiring and rewiring of the retinogeniculate synapse. Curr Opin Neurobiol 21(2):228–237
Hong YK, Park S, Litvina EY, Morales J, Sanes JR, Chen C (2014) Refinement of the retinogeniculate synapse by bouton clustering. Neuron 84(2):332–339
Hooks BM, Chen C (2006) Distinct roles for spontaneous and visual activity in remodeling of the retinogeniculate synapse. Neuron 52(2):281–291
Hooks BM, Chen C (2008) Vision triggers an experience-dependent sensitive period at the retinogeniculate synapse. J Neurosci 28(18):4807–4817
Howarth M, Walmsley L, Brown TM (2014) Binocular integration in the mouse lateral geniculate nuclei. Curr Biol 24(11):1241–1247
Howland HC (1984) Parallel processing in the visual system. The classification of retinal ganglion cells and its impact on the neurobiology of vision. Jonathan Stone. Q Rev Biol 59(4):502–503
Hoy JL, Niell CM (2015) Layer-specific refinement of visual cortex function after eye opening in the awake mouse. J Neurosci 35(8):3370–3383
Hubel DH, Wiesel TN (1961) Integrative action in the cat's lateral geniculate body. J Physiol 155(2):385–398
Hubel DH, Wiesel TN (1962) Receptive fields, binocular interaction and functional architecture in the cat's visual cortex. J Physiol 160(1):106–154
Hubel DH, Wiesel TN (1964) Effects of monocular deprivation in kittens. Naunyn Schmiedebergs Arch Exp Pathol Pharmakol 248(6):492–497
Hubel DH, Wiesel TN (1970) The period of susceptibility to the physiological effects of unilateral eye closure in kittens. J Physiol 206(2):419–436
Hubel DH, Wiesel TN (1972) Laminar and columnar distribution of geniculo-cortical fibers in the macaque monkey. J Comp Neurol 146(4):421–450
Huberman AD (2007) Mechanisms of eye-specific visual circuit development. Curr Opin Neurobiol 17(1):73–80
Huberman AD, Dehay C, Berland M, Chalupa LM, Kennedy H (2005a) Early and rapid targeting of eye-specific axonal projections to the dorsal lateral geniculate nucleus in the fetal macaque. J Neurosci 25(16):4014–4023
Huberman AD, Murray KD, Warland DK, Feldheim DA, Chapman B (2005b) Ephrin-As mediate targeting of eye-specific projections to the lateral geniculate nucleus. Nat Neurosci 8(8): 1013–1021
Huberman AD, Feller MB, Chapman B (2008) Mechanisms underlying development of visual maps and receptive fields. Ann Rev Neurosci 31:479–509
Huberman AD, Wei W, Elstrott J, Stafford BK, Feller MB, Barres BA (2009) Genetic identification of an on-off direction-selective retinal ganglion cell subtype reveals a layer-specific subcortical map of posterior motion. Neuron 62(3):327–334
Huh CYL, Abdelaal K, Salinas KJ, Gu D, Zeitoun J, Figueroa Velez DX, Peach JP, Fowlkes CC, Gandhi SP (2020) Long-term monocular deprivation during juvenile critical period disrupts binocular integration in mouse visual thalamus. J Neurosci 40(3):585–604

Humphrey AL, Saul AB (1998) Strobe rearing reduces direction selectivity in area 17 by altering spatiotemporal receptive-field structure. J Neurophysiol 80(6):2991–3004

Irvin GE, Norton TT, Sesma MA, Casagrande VA (1986) W-like response properties of interlaminar zone cells in the lateral geniculate nucleus of a primate (Galago crassicaudatus). Brain Res 362(2):254–270

Issa NP, Trachtenberg JT, Chapman B, Zahs KR, Stryker MP (1999) The critical period for ocular dominance plasticity in the Ferret's visual cortex. J Neurosci 19(16):6965–6978

Jackson CA, Peduzzi J, Hickey T (1989) Visual cortex development in the ferret. I. Genesis and migration of visual cortex neurons. J Neurosci 9:1242–1253

Jacobs EC, Campagnoni C, Kampf K, Reyes SD, Kalra V, Handley V, Xie YY, Hong-Hu Y, Spreur V, Fisher RS, Campagnoni AT (2007) Visualization of corticofugal projections during early cortical development in a tau-GFP-transgenic mouse. Eur J Neurosci 25(1):17–30

Jaepel J, Hübener M, Bonhoeffer T, Rose T (2017) Lateral geniculate neurons projecting to primary visual cortex show ocular dominance plasticity in adult mice. Nat Neurosci 20(12):1708–1714

Jaubert-Miazza L, Green E, Lo FS, Bui K, Mills J, Guido W (2005) Structural and functional composition of the developing retinogeniculate pathway in the mouse. Vis Neurosci 22(5): 661–676

Johnson JK, Casagrande V (1993) Prenatal development of axon outgrowth and connectivity in the ferret visual system. Vis Neurosci 10(1):117–130

Kaas JH, Guillery RW, Allman JM (1972) Some principles of organization in the dorsal lateral geniculate nucleus. Brain Behav Evol 6(1-6):253–299

Kalil R (1978) Dark rearing in the cat: effects on visuomotor behavior and cell growth in the dorsal lateral geniculate nucleus. J Comp Neurol 178(3):451–467

Kaneko M, Stellwagen D, Malenka RC, Stryker MP (2008) Tumor necrosis factor-alpha mediates one component of competitive, experience-dependent plasticity in developing visual cortex. Neuron 58(5):673–680

Kang E, Durand S, LeBlanc JJ, Hensch TK, Chen C, Fagiolini M (2013) Visual acuity development and plasticity in the absence of sensory experience. J Neurosci 33(45):17789–17796

Kanold PO, Kara P, Reid RC, Shatz CJ (2003) Role of subplate neurons in functional maturation of visual cortical columns. Science 301(5632):521–525

Kaplan E, Shapley RM (1982) X and Y cells in the lateral geniculate nucleus of macaque monkeys. J Physiol 330:125–143

Katz L (1987) Local circuitry of identified projection neurons in cat visual cortex brain slices. J Neurosci 7(4):1223–1249

Kawasaki H, Crowley JC, Livesey FJ, Katz LC (2004) Molecular organization of the ferret visual thalamus. J Neurosci 24(44):9962–9970

Kay JN, De la Huerta I, Kim I-J, Zhang Y, Yamagata M, Chu MW, Meister M, Sanes JR (2011) Retinal ganglion cells with distinct directional preferences differ in molecular identity, structure, and central projections. J Neurosci 31(21):7753–7762

Kiley CW, Usrey WM (2017) Orientation tuning of correlated activity in the developing lateral geniculate nucleus. J Neurosci 37(48):11549–11558

Kim IJ, Zhang Y, Yamagata M, Meister M, Sanes JR (2008) Molecular identification of a retinal cell type that responds to upward motion. Nature 452(7186):478–482

Kim IJ, Zhang Y, Meister M, Sanes JR (2010) Laminar restriction of retinal ganglion cell dendrites and axons: subtype-specific developmental patterns revealed with transgenic markers. J Neurosci 30(4):1452–1462

Kinston WJ, Vadas MA, Bishop PO (1969) Multiple projection of the visual field to the medical portion of the dorsal lateral geniculate nucleus and the adjacent nuclei of the thalamus of the cat. J Comp Neurol 136(3):295–315

Kratz K, Sherman S, Kalil R (1979) Lateral geniculate nucleus in dark-reared cats: loss of Y cells without changes in cell size. Science 203(4387):1353–1355

Krug K, Akerman CJ, Thompson ID (2001) Responses of neurons in neonatal cortex and thalamus to patterned visual stimulation through the naturally closed lids. J Neurophysiol 85(4): 1436–1443

Lachica EA, Crooks MW, Casagrande VA (1990) Effects of monocular deprivation on the morphology of retinogeniculate axon arbors in a primate. J Comp Neurol 296(2):303–323

Lambo ME, Turrigiano GG (2013) Synaptic and intrinsic homeostatic mechanisms cooperate to increase L2/3 pyramidal neuron excitability during a late phase of critical period plasticity. J Neurosci 33(20):8810–8819

Lee BB, Creutzfeldt OD, Elepfandt A (1979) The responses of magno- and parvocellular cells of the monkey's lateral geniculate body to moving stimuli. Exp Brain Res 35(3):547–557

Lehmkuhle S, Kratz KE, Mangel SC, Sherman SM (1978) An effect of early monocular lid suture upon the development of X-cells in the cat's lateral geniculate nucleus. Brain Res 157(2): 346–350

Lehmkuhle S, Kratz KE, Mangel SC, Sherman SM (1980) Effects of early monocular lid suture on spatial and temporal sensitivity of neurons in dorsal lateral geniculate nucleus of the cat. J Neurophysiol 43(2):542–556

Lennie P (1980) Parallel visual pathways: a review. Vision Res 20(7):561–594

LeVay S, Wiesel TN, Hubel DH (1980) The development of ocular dominance columns in normal and visually deprived monkeys. J Comp Neurol 191(1):1–51

Levick WR, Thibos LN (1980) Orientation bias of cat retinal ganglion cells. Nature 286(5771): 389–390

Levick WR, Oyster CW, Takahashi E (1969) Rabbit lateral geniculate nucleus: sharpener of directional information. Science 165(3894):712–714

Levick WR, Cleland BG, Dubin MW (1972) Lateral geniculate neurons of cat: retinal inputs and physiology. Invest Ophthalmol 11(5):302–311

Levitt JB, Schumer RA, Sherman SM, Spear PD, Movshon JA (2001) Visual response properties of neurons in the LGN of normally reared and visually deprived macaque monkeys. J Neurophysiol 85(5):2111–2129

Lewis TL, Maurer D (2005) Multiple sensitive periods in human visual development: evidence from visually deprived children. Dev Psychobiol 46(3):163–183

Li Y, Fitzpatrick D, White LE (2006) The development of direction selectivity in ferret visual cortex requires early visual experience. Nat Neurosci 9(5):676–681

Li Y, Van Hooser SD, Mazurek M, White LE, Fitzpatrick D (2008) Experience with moving visual stimuli drives the early development of cortical direction selectivity. Nature 456(7224):952–956

Li Y-T, Liu B-H, Chou X-L, Zhang LI, Tao HW (2014) Strengthening of direction selectivity by broadly tuned and spatiotemporally slightly offset inhibition in mouse visual cortex. Cereb Cortex 25(9):2466–2477

Liang L, Chen C (2020) Organization, function, and development of the mouse retinogeniculate synapse. Annu Rev Vis Sci 6:261–285

Lien AD, Scanziani M (2018) Cortical direction selectivity emerges at convergence of thalamic synapses. Nature 558(7708):80–86

Linden DC, Guillery RW, Cucchiaro J (1981) The dorsal lateral geniculate nucleus of the normal ferret and its postnatal development. J Comp Neurol 203(2):189–211

Linden ML, Heynen AJ, Haslinger RH, Bear MF (2009) Thalamic activity that drives visual cortical plasticity. Nat Neurosci 12(4):390–392

Litvina EY, Chen C (2017) Functional convergence at the retinogeniculate synapse. Neuron 96(2): 330–338.e335

Liu X, Chen C (2008) Different roles for AMPA and NMDA receptors in transmission at the immature retinogeniculate synapse. J Neurophysiol 99(2):629–643

Livingstone MS (1998) Mechanisms of direction selectivity in macaque V1. Neuron 20(3):509–526

Livingstone M, Hubel D (1987) Psychophysical evidence for separate channels for the perception of form, color, movement, and depth. J Neurosci 7(11):3416–3468

Maex R, Orban GA (1996) Model circuit of spiking neurons generating directional selectivity in simple cells. J Neurophysiol 75(4):1515–1545
Maffei A, Nataraj K, Nelson SB, Turrigiano GG (2006) Potentiation of cortical inhibition by visual deprivation. Nature 443(7107):81–84
Malpeli JG, Baker FH (1975) The representation of the visual field in the lateral geniculate nucleus of Macaca mulatta. J Comp Neurol 161(4):569–594
Marr D, Ullman S (1981) Directional selectivity and its use in early visual processing. Proc R Soc Lond B Biol Sci 211(1183):151–180
Marrocco RT (1976) Sustained and transient cells in monkey lateral geniculate nucleus: conduction velocites and response properties. J Neurophysiol 39(2):340–353
Marshel JH, Kaye AP, Nauhaus I, Callaway EM (2012) Anterior-posterior direction opponency in the superficial mouse lateral geniculate nucleus. Neuron 76(4):713–720
Mastronarde DN (1987) Two classes of single-input X-cells in cat lateral geniculate nucleus. II. Retinal inputs and the generation of receptive-field properties. J Neurophysiol 57(2):381–413
Mastronarde DN (1992) Nonlagged relay cells and interneurons in the cat lateral geniculate nucleus: receptive-field properties and retinal inputs. Vis Neurosci 8(5):407–441
Maunsell JH (1992) Functional visual streams. Curr Opin Neurobiol 2(4):506–510
McCormick DA, Bal T (1994) Sensory gating mechanisms of the thalamus. Curr Opin Neurobiol 4(4):550–556
McCourt ME, Jacobs GH (1984) Directional filter characteristics of optic nerve fibers in California ground squirrel (Spermophilus beecheyi). J Neurophysiol 52(6):1200–1212
McLaughlin T, O'Leary DD (2005) Molecular gradients and development of retinotopic maps. Annu Rev Neurosci 28:327–355
Miska NJ, Richter LM, Cary BA, Gjorgjieva J, Turrigiano GG (2018) Sensory experience inversely regulates feedforward and feedback excitation-inhibition ratio in rodent visual cortex. Elife 7
Monyer H, Burnashev N, Laurie DJ, Sakmann B, Seeburg PH (1994) Developmental and regional expression in the rat brain and functional properties of four NMDA receptors. Neuron 12(3): 529–540
Mooney R, Penn AA, Gallego R, Shatz CJ (1996) Thalamic relay of spontaneous retinal activity prior to vision. Neuron 17(5):863–874
Moore CL, Kalil R, Richards W (1976) Development of myelination in optic tract of the cat. J Comp Neurol 165(2):125–136
Moore BDT, Alitto HJ, Usrey WM (2005) Orientation tuning, but not direction selectivity, is invariant to temporal frequency in primary visual cortex. J Neurophysiol 94(2):1336–1345
Moore BDT, Kiley CW, Sun C, Usrey WM (2011) Rapid plasticity of visual responses in the adult lateral geniculate nucleus. Neuron 71(5):812–819
Morgan JL, Berger DR, Wetzel AW, Lichtman JW (2016) The fuzzy logic of network connectivity in mouse visual thalamus. Cell 165(1):192–206
Movshon JA, Dürsteler MR (1977) Effects of brief periods of unilateral eye closure on the kitten's visual system. J Neurophysiol 40(6):1255–1265
Movshon JA, Kiorpes L, Hawken MJ, Cavanaugh JR (2005) Functional maturation of the macaque's lateral geniculate nucleus. J Neurosci 25(10):2712–2722
Mower GD, Burchfiel JL, Duffy FH (1982) Animal models of strabismic amblyopia: physiological studies of visual cortex and the lateral geniculate nucleus. Brain Res 281(3):311–327
Niell CM, Stryker MP (2008) Highly selective receptive fields in mouse visual cortex. J Neurosci 28(30):7520–7536
Ohki K, Chung S, Ch'ng YH, Kara P, Reid RC (2005) Functional imaging with cellular resolution reveals precise micro-architecture in visual cortex. Nature 433(7026):597–603
Orban GA, Kennedy H, Bullier J (1986) Velocity sensitivity and direction selectivity of neurons in areas V1 and V2 of the monkey: influence of eccentricity. J Neurophysiol 56(2):462–480
Pearson HE, Berman N, Murphy EH (1981) Stroboscopic rearing reduces direction selectivity in rabbit visual cortex. Dev Brain Res 1(1):127–131

Peichl L (1991) Alpha ganglion cells in mammalian retinae: Common properties, species differences, and some comments on other ganglion cells. Vis Neurosci 7(1-2):155–169

Peichl L, Ott H, Boycott BB (1987) Alpha ganglion cells in mammalian retinae. Proc Royal Soc Lond Ser B Biol Sci 231(1263):169–197

Peng YR, Shekhar K, Yan W, Herrmann D, Sappington A, Bryman GS, van Zyl T, Do MTH, Regev A, Sanes JR (2019) Molecular classification and comparative taxonomics of foveal and peripheral cells in primate retina. Cell 176(5):1222–1237.e1222

Penn AA, Riquelme PA, Feller MB, Shatz CJ (1998) Competition in retinogeniculate patterning driven by spontaneous activity. Science 279(5359):2108–2112

Pfeiffenberger C, Cutforth T, Woods G, Yamada J, Rentería RC, Copenhagen DR, Flanagan JG, Feldheim DA (2005) Ephrin-As and neural activity are required for eye-specific patterning during retinogeniculate mapping. Nat Neurosci 8(8):1022–1027

Piscopo DM, El-Danaf RN, Huberman AD, Niell CM (2013) Diverse visual features encoded in mouse lateral geniculate nucleus. J Neurosci 33(11):4642–4656

Priebe NJ (2016) Mechanisms of orientation selectivity in the primary visual cortex. Annu Rev Vis Sci 2:85–107

Priebe NJ, Ferster D (2005) Direction selectivity of excitation and inhibition in simple cells of the cat primary visual cortex. Neuron 45(1):133–145

Rakic P (1976) Prenatal genesis of connections subserving ocular dominance in the rhesus monkey. Nature 261(5560):467–471

Rakic P (1977) Genesis of the dorsal lateral geniculate nucleus in the rhesus monkey: site and time of origin, kinetics of proliferation, routes of migration and pattern of distribution of neurons. J Comp Neurol 176(1):23–52

Rakic P, Barlow HB, Gaze RM (1977) Prenatal development of the visual system in rhesus monkey. Philos Trans R Soc Lond B Biol Sci 278(961):245–260

Ramoa AS, Prusky G (1997) Retinal activity regulates developmental switches in functional properties and ifenprodil sensitivity of NMDA receptors in the lateral geniculate nucleus. Brain Res Dev Brain Res 101(1-2):165–175

Reese BE (1988) 'Hidden lamination' in the dorsal lateral geniculate nucleus: the functional organization of this thalamic region in the rat. Brain Res Rev 13(2):119–137

Ritter NJ, Anderson NM, Van Hooser SD (2017) Visual stimulus speed does not influence the rapid emergence of direction selectivity in ferret visual cortex. J Neurosci 37(6):1557–1567

Rivlin-Etzion M, Zhou K, Wei W, Elstrott J, Nguyen PL, Barres BA, Huberman AD, Feller MB (2011) Transgenic mice reveal unexpected diversity of on-off direction-selective retinal ganglion cell subtypes and brain structures involved in motion processing. J Neurosci 31(24): 8760–8769

Rochefort NL, Narushima M, Grienberger C, Marandi N, Hill DN, Konnerth A (2011) Development of direction selectivity in mouse cortical neurons. Neuron 71(3):425–432

Rompani SB, Müllner FE, Wanner A, Zhang C, Roth CN, Yonehara K, Roska B (2017) Different modes of visual integration in the lateral geniculate nucleus revealed by single-cell-initiated transsynaptic tracing. Neuron 93(4):767–776.e766

Rose T, Jaepel J, Hübener M, Bonhoeffer T (2016) Cell-specific restoration of stimulus preference after monocular deprivation in the visual cortex. Science 352(6291):1319–1322

Rossi LF, Harris KD, Carandini M (2020) Spatial connectivity matches direction selectivity in visual cortex. Nature 588(7839):648–652

Roy A, Osik JJ, Ritter NJ, Wang S, Shaw JT, Fiser J, Van Hooser SD (2016) Optogenetic spatial and temporal control of cortical circuits on a columnar scale. J Neurophysiol 115(2):1043–1062

Roy A, Wang S, Meschede-Krasa B, Breffle J, Van Hooser SD (2020) An early phase of instructive plasticity before the typical onset of sensory experience. Nat Commun 11(1):11

Sanderson KJ, Bishop PO, Darian-Smith I (1971) The properties of the binocular receptive fields of lateral geniculate neurons. Exp Brain Res 13(2):178–207

Saul AB, Feidler JC (2002) Development of response timing and direction selectivity in cat visual thalamus and cortex. J Neurosci 22(7):2945–2955

Saul AB, Humphrey AL (1990) Spatial and temporal response properties of lagged and nonlagged cells in cat lateral geniculate nucleus. J Neurophysiol 64(1):206–224
Scholl B, Tan AY, Corey J, Priebe NJ (2013) Emergence of orientation selectivity in the Mammalian visual pathway. J Neurosci 33(26):10616–10624
Seabrook TA, El-Danaf RN, Krahe TE, Fox MA, Guido W (2013) Retinal input regulates the timing of corticogeniculate innervation. J Neurosci 33(24):10085–10097
Shapley R (1984) The neurobiology of vision: parallel processing in the visual system. The classification of retinal ganglion cells and its impact on the neurobiology of vision. Jonathan Stone. Plenum, New York, 1983. xvi, 438 pp., illus. $55. Perspectives in vision research. Science 223(4643):1403–1404
Shapley R, Hugh Perry V (1986) Cat and monkey retinal ganglion cells and their visual functional roles. Trends Neurosci 9:229–235
Shatz CJ, Rakic P (1981) The genesis of efferent connections from the visual cortex of the fetal rhesus monkey. J Comp Neurol 196(2):287–307
Shatz CJ, Sretavan DW (1986) Interactions between retinal ganglion cells during the development of the mammalian visual system. Annu Rev Neurosci 9(1):171–207
Sherman SM, Guillery RW (2002) The role of the thalamus in the flow of information to the cortex. Philos Trans R Soc Lond B Biol Sci 357(1428):1695–1708
Sherman SM, Spear PD (1982) Organization of visual pathways in normal and visually deprived cats. Physiol Rev 62(2):738–855
Sherman SM, Hoffmann KP, Stone J (1972) Loss of a specific cell type from dorsal lateral geniculate nucleus in visually deprived cats. J Neurophysiol 35(4):532–541
Sherman SM, Wilson JR, Kaas JH, Webb SV (1976) X- and Y-cells in the dorsal lateral geniculate nucleus of the owl monkey (Aotus trivirgatus). Science 192(4238):475–477
Shou TD, Leventhal AG (1989) Organized arrangement of orientation-sensitive relay cells in the cat's dorsal lateral geniculate nucleus. J Neurosci 9(12):4287–4302
Sillito AM (1977) Inhibitory processes underlying the directional specificity of simple, complex and hypercomplex cells in the cat's visual cortex. J Physiol 271(3):699–720
Silver MA, Stryker MP (1999) Synaptic density in geniculocortical afferents remains constant after monocular deprivation in the cat. J Neurosci 19(24):10829–10842
Smith EL 3rd, Chino YM, Ridder WH 3rd, Kitagawa K, Langston A (1990) Orientation bias of neurons in the lateral geniculate nucleus of macaque monkeys. Vis Neurosci 5(6):525–545
Smith GB, Hein B, Whitney DE, Fitzpatrick D, Kaschube M (2018) Distributed network interactions and their emergence in developing neocortex. Nat Neurosci 21(11):1600–1608
Sommeijer JP, Ahmadlou M, Saiepour MH, Seignette K, Min R, Heimel JA, Levelt CN (2017) Thalamic inhibition regulates critical-period plasticity in visual cortex and thalamus. Nat Neurosci 20(12):1715–1721
Soodak RE, Shapley RM, Kaplan E (1987) Linear mechanism of orientation tuning in the retina and lateral geniculate nucleus of the cat. J Neurophysiol 58(2):267–275
Stacy AK, Schneider NA, Gilman NK, Van Hooser SD (2021) Impact of acute visual experience on development of LGN receptive fields in the ferret. bioRxiv. https://doi.org/10.1101/2021.07.16.452697
Stavros KA, Kiorpes L (2008) Behavioral measurement of temporal contrast sensitivity development in macaque monkeys (Macaca nemestrina). Vision Res 48(11):1335–1344
Stone J, Fukuda Y (1974) Properties of cat retinal ganglion cells: a comparison of W-cells with X- and Y-cells. J Neurophysiol 37(4):722–748
Suarez H, Koch C, Douglas R (1995) Modeling direction selectivity of simple cells in striate visual cortex within the framework of the canonical microcircuit. J Neurosci 15(10):6700–6719
Suematsu N, Naito T, Miyoshi T, Sawai H, Sato H (2013) Spatiotemporal receptive field structures in retinogeniculate connections of cat. Front Syst Neurosci 7(103)
Sun W, Li N, He S (2002) Large-scale morphological survey of mouse retinal ganglion cells. J Comp Neurol 451(2):115–126

Sun L, Han X, He S (2011) Direction-selective circuitry in rat retina develops independently of GABAergic, cholinergic and action potential activity. PLoS One 6(5):e19477

Sun W, Tan Z, Mensh B, Ji N (2015) Thalamus provides layer 4 of primary visual cortex with orientation- and direction-tuned inputs. Nat Neurosci 19

Swadlow HA, Weyand TG (1985) Receptive-field and axonal properties of neurons in the dorsal lateral geniculate nucleus of awake unparalyzed rabbits. J Neurophysiol 54(1):168–183

Tavazoie SF, Reid RC (2000) Diverse receptive fields in the lateral geniculate nucleus during thalamocortical development. Nat Neurosci 3(6):608–616

Thompson ID, Kossut M, Blakemore C (1983) Development of orientation columns in cat striate cortex revealed by 2-deoxyglucose autoradiography. Nature 301(5902):712–715

Thompson KG, Zhou Y, Leventhal AG (1994) Direction-sensitive X and Y cells within the A laminae of the cat's LGNd. Vis Neurosci 11(5):927–938

Thompson AD, Picard N, Min L, Fagiolini M, Chen C (2016) Cortical feedback regulates feedforward retinogeniculate refinement. Neuron 91(5):1021–1033

Thompson A, Gribizis A, Chen C, Crair MC (2017) Activity-dependent development of visual receptive fields. Curr Opin Neurobiol 42:136–143

Tian N, Copenhagen DR (2003) Visual stimulation is required for refinement of on and off pathways in postnatal retina. Neuron 39(1):85–96

Tigges M, Hendrickson AE, Tigges J (1984) Anatomical consequences of long-term monocular eyelid closure on lateral geniculate nucleus and striate cortex in squirrel monkey. J Comp Neurol 227(1):1–13

Tootle JS, Friedlander MJ (1989) Postnatal development of the spatial contrast sensitivity of X- and Y-cells in the kitten retinogeniculate pathway. J Neurosci 9(4):1325–1340

Trachtenberg JT, Stryker MP (2001) Rapid anatomical plasticity of horizontal connections in the developing visual cortex. J Neurosci 21(10):3476–3482

Trachtenberg JT, Trepel C, Stryker MP (2000) Rapid extragranular plasticity in the absence of thalamocortical plasticity in the developing primary visual cortex. Science 287(5460): 2029–2032

Tran NM, Shekhar K, Whitney IE, Jacobi A, Benhar I, Hong G, Yan W, Adiconis X, Arnold ME, Lee JM, Levin JZ, Lin D, Wang C, Lieber CM, Regev A, He Z, Sanes JR (2019) Single-cell profiles of retinal ganglion cells differing in resilience to injury reveal neuroprotective genes. Neuron 104(6):1039–1055.e1012

Tschetter WW, Govindaiah G, Etherington IM, Guido W, Niell CM (2018) Refinement of Spatial Receptive Fields in the Developing Mouse Lateral Geniculate Nucleus Is Coordinated with Excitatory and Inhibitory Remodeling. J Neurosci 38(19):4531–4542

Usrey WM (2002) Spike timing and visual processing in the retinogeniculocortical pathway. Philos Trans R Soc Lond B Biol Sci 357(1428):1729–1737

Usrey WM, Reid RC (2000) Visual physiology of the lateral geniculate nucleus in two species of new world monkey: Saimiri sciureus and Aotus trivirgatis. J Physiol 523(Pt 3):755–769

Usrey WM, Reppas JB, Reid RC (1998) Paired-spike interactions and synaptic efficacy of retinal inputs to the thalamus. Nature 395(6700):384–387

Usrey WM, Reppas JB, Reid RC (1999) Specificity and strength of retinogeniculate connections. J Neurophysiol 82(6):3527–3540

Van Hooser SD (2007) Similarity and diversity in visual cortex: is there a unifying theory of cortical computation? Neuroscientist 13(6):639–656

Van Hooser SD, Heimel JA, Nelson SB (2003) Receptive field properties and laminar organization of lateral geniculate nucleus in the gray squirrel (Sciurus carolinensis). J Neurophysiol 90(5): 3398–3418

Van Hooser SD, Li Y, Christensson M, Smith GB, White LE, Fitzpatrick D (2012) Initial neighborhood biases and the quality of motion stimulation jointly influence the rapid emergence of direction preference in visual cortex. J Neurosci 32(21):7258–7266

Van Hooser SD, Roy A, Rhodes HJ, Culp JH, Fitzpatrick D (2013) Transformation of receptive field properties from lateral geniculate nucleus to superficial V1 in the tree shrew. J Neurosci 33(28):11494–11505
Vidyasagar TR, Urbas JV (1982) Orientation sensitivity of cat LGN neurones with and without inputs from visual cortical areas 17 and 18. Exp Brain Res 46(2):157–169
Vital-Durand F, Garey LJ, Blakemore C (1978) Monocular and binocular deprivation in the monkey: morphological effects and reversibility. Brain Res 158(1):45–64
von Noorden GK, Crawford ML (1978) Morphological and physiological changes in the monkey visual system after short-term lid suture. Invest Ophthalmol Vis Sci 17(8):762–768
Wang X, Wei Y, Vaingankar V, Wang Q, Koepsell K, Sommer FT, Hirsch JA (2007) Feedforward excitation and inhibition evoke dual modes of firing in the cat's visual thalamus during naturalistic viewing. Neuron 55(3):465–478
Wei W, Hamby AM, Zhou K, Feller MB (2011) Development of asymmetric inhibition underlying direction selectivity in the retina. Nature 469(7330):402–406
Weliky M, Katz LC (1999) Correlational structure of spontaneous neuronal activity in the developing lateral geniculate nucleus in vivo. Science 285(5427):599–604
Weliky M, Bosking WH, Fitzpatrick D (1996) A systematic map of direction preference in primary visual cortex. Nature 379(6567):725–728
Wenzel A, Villa M, Mohler H, Benke D (1996) Developmental and regional expression of NMDA receptor subtypes containing the NR2D subunit in rat brain. J Neurochem 66(3):1240–1248
Weyand TG (2016) The multifunctional lateral geniculate nucleus. Rev Neurosci 27(2):135–157
White LE, Coppola DM, Fitzpatrick D (2001) The contribution of sensory experience to the maturation of orientation selectivity in ferret visual cortex. Nature 411(6841):1049–1052
Wiencken-barger AE, Casagrande VA (2002) Visual system development and neural activity. In: Ramachandran VS (ed) Encyclopedia of the human brain. Academic Press, New York, pp 791–804
Wiesel TN, Hubel DH (1963a) Effects of visual deprivation on morphology and physiology of cells in the cat's lateral geniculate body. J Neurophysiol 26(6):978–993
Wiesel TN, Hubel DH (1963b) Single-cell responses in striate cortex of kittens deprived of vision in one eye. J Neurophysiol 26(6):1003–1017
Wiesel TN, Hubel DH (1965a) Comparison of the effects of unilateral and bilateral eye closure on cortical unit responses in kittens. J Neurophysiol 28(6):1029–1040
Wiesel TN, Hubel DH (1965b) Extent of recovery from the effects of visual deprivation in kittens. J Neurophysiol 28(6):1060–1072
Wiesel TN, Hubel DH (1974) Ordered arrangement of orientation columns in monkeys lacking visual experience. J Comp Neurol 158(3):307–318
Wilson JR, Forestner DM (1995) Synaptic inputs to single neurons in the lateral geniculate nuclei of normal and monocularly deprived squirrel monkeys. J Comp Neurol 362(4):468–488
Wilson DE, Scholl B, Fitzpatrick D (2018) Differential tuning of excitation and inhibition shapes direction selectivity in ferret visual cortex. Nature 560(7716):97–101
Wong RO, Meister M, Shatz CJ (1993) Transient period of correlated bursting activity during development of the mammalian retina. Neuron 11(5):923–938
Xu X, Ichida JM, Allison JD, Boyd JD, Bonds AB, Casagrande VA (2001) A comparison of koniocellular, magnocellular and parvocellular receptive field properties in the lateral geniculate nucleus of the owl monkey (Aotus trivirgatus). J Physiol 531(Pt 1):203–218
Yan W, Peng Y-R, van Zyl T, Regev A, Shekhar K, Juric D, Sanes JR (2020) Cell atlas of the human fovea and peripheral retina. Sci Rep 10(1):9802
Yeh T, Lee BB, Kremers J, Cowing JA, Hunt DM, Martin PR, Troy JB (1995) Visual responses in the lateral geniculate nucleus of dichromatic and trichromatic marmosets (Callithrix jacchus). J Neurosci 15(12):7892–7904
Yin ZQ, Crewther SG, Wang C, Crewther DP (2006) Pre- and post-critical period induced reduction of Cat-301 immunoreactivity in the lateral geniculate nucleus and visual cortex of cats Y-blocked as adults or made strabismic as kittens. Mol Vis 12:858–866

Zahs KR, Stryker MP (1985) The projection of the visual field onto the lateral geniculate nucleus of the ferret. J Comp Neurol 241(2):210–224
Zaltsman JB, Heimel JA, Hooser SDV (2015) Weak orientation and direction selectivity in lateral geniculate nucleus representing central vision in the gray squirrel Sciurus carolinensis. J Neurophysiol 113(7):2987–2997
Zeater N, Cheong SK, Solomon SG, Dreher B, Martin PR (2015) Binocular visual responses in the primate lateral geniculate nucleus. Curr Biol 25(24):3190–3195
Zhao X, Chen H, Liu X, Cang J (2013) Orientation-selective responses in the mouse lateral geniculate nucleus. J Neurosci 33(31):12751–12763
Zhou Y, Leventhal AG, Thompson KG (1995) Visual deprivation does not affect the orientation and direction sensitivity of relay cells in the lateral geniculate nucleus of the cat. J Neurosci 15(1 Pt 2):689–698
Zhuang J, Stoelzel CR, Bereshpolova Y, Huff JM, Hei X, Alonso JM, Swadlow HA (2013) Layer 4 in primary visual cortex of the awake rabbit: contrasting properties of simple cells and putative feedforward inhibitory interneurons. J Neurosci 33(28):11372–11389
Zhuang J, Bereshpolova Y, Stoelzel CR, Huff JM, Hei X, Alonso JM, Swadlow HA (2014) Brain state effects on layer 4 of the awake visual cortex. J Neurosci 34(11):3888–3900
Ziburkus J, Guido W (2006) Loss of binocular responses and reduced retinal convergence during the period of retinogeniculate axon segregation. J Neurophysiol 96(5):2775–2784

Maturation of Corticolimbic Functional Connectivity During Sensitive Periods of Brain Development

Shaolin Yang and Kuei Y. Tseng

Contents

Abstract The maturation of key corticolimbic structures and the prefrontal cortex during sensitive periods of brain development from early life through adolescence is crucial for the acquisition of a variety of cognitive and affective processes associated with adult behavior. In this chapter, we first review how key cellular and circuit level changes during adolescence dictate the development of the prefrontal cortex and its capacity to integrate contextual and emotional information from the ventral hippocampus and the amygdala. We further discuss how afferent transmission from ventral hippocampal and amygdala inputs displays unique age-dependent trajectories that directly impact prefrontal functional maturation through adolescence. We conclude by proposing that time-sensitive strengthening of specific corticolimbic synapses is a critical contributing factor for the protracted maturation of cognitive and emotional regulation by the prefrontal cortex.

S. Yang
Department of Psychiatry, University of Pittsburgh, Pittsburgh, PA, USA

K. Y. Tseng (✉)
Department of Anatomy and Cell Biology, University of Illinois at Chicago – College of Medicine, Chicago, IL, USA
e-mail: tsengky@uic.edu

Curr Topics Behav Neurosci (2022) 53: 37–54
https://doi.org/10.1007/7854_2021_239
Published Online: 13 August 2021

Keywords Adolescence · Amygdala · Connectivity · Development · Dopamine · GABA · Hippocampus · LTD · LTP · NMDA · Prefrontal cortex

1 Introduction

Brain maturation and its control of complex behaviors are dictated by the dynamic structural and functional remodeling of neuronal circuitry during sensitive periods of development. Of particular interest is the impact of early life experiences through adolescence and the maturation of functional connectivity among key interconnected corticolimbic brain regions implicated in cognitive and affective regulation (i.e., prefrontal cortex, ventral hippocampus, amygdala). Even though the exact span of adolescence varies across the different species (Andersen 2003; Tseng et al. 2009; Brenhouse and Andersen 2011; Caballero and Tseng 2016), it is typically defined as a transitional period between childhood and adulthood during which the functional maturation of neuronal circuitry involved in the control of cognition and affect emerges (Spear 2000). Thus, understanding the precise neurobiology underlying these maturational processes requires the integration of psychological and imaging data from developing human subjects and the implementation of animal models that ultimately test the causal contribution of specific neural circuit mechanisms for the acquisition of adult behavior (Caballero et al. 2016; Caballero and Tseng 2016).

2 Functional Maturation of the Prefrontal Cortex (PFC) During Adolescence

It is during the peri-adolescent transitional period when the maturation of specific functional domains within the PFC, which are required for proper cognitive control of affect, begins to occur. Key developmental changes that impact PFC maturation during adolescence include the gain of local neuromodulatory capacity of synaptic activity (i.e., dopamine, GABA, glutamate) that ultimately enhances the capacity of prefrontal integration of contextual and emotional information from the ventral hippocampus and the amygdala (Caballero et al. 2016; Caballero and Tseng 2016). The following sections will review how dopamine modulation of PFC activity changes during adolescence, particularly due to the maturation of both glutamatergic and GABAergic neuronal responses to D1 and D2 receptor stimulation. We will also review major findings highlighting adolescence as a critical period for the functional remodeling of local excitatory and inhibitory synaptic transmission as both systems display unique age-dependent changes that directly impact PFC output function and plasticity.

2.1 Maturation of Dopaminergic Control of PFC Activity

The development and acquisition of key cognitive processes such as working memory, inhibitory control, and attention are thought to be regulated by the maturation of dopamine actions in the PFC (Goldman-Rakic et al. 2000; Horvitz 2000; Jay 2003; O'Donnell 2003). While prefrontal dopamine innervation can be detected early after birth, it continues to increase until postnatal day (P) 60 (Kalsbeek et al. 1988). Despite the different patterns of innervation observed in the PFC across the different species (Lewis et al. 1987; Goldman-Rakic et al. 1989; Rosenberg and Lewis 1994; Lewis et al. 1998; Raghanti et al. 2008), dopamine terminals in both rodent and primates are remarkably similar at the ultrastructural level (Seguela et al. 1988; Goldman-Rakic et al. 1989). The distribution of dopamine receptors also follows closely the pattern of innervation (Gaspar et al. 1995; Muly et al. 1998) with most pyramidal neurons and GABAergic interneurons expressing both classes of D1 and D2 receptors (Bouthenet et al. 1987; Gaspar et al. 1995; Muly et al. 1998; Santana et al. 2009).

Functionally, the late adolescent acquisition of dopamine-dependent control of excitatory and inhibitory transmission in the PFC is thought to provide a critical neurobiological step to fine-tuning prefrontal output activity responsible for the development and maturation of adult cognitive abilities (Tseng et al. 2009; O'Donnell and Tseng 2010). It has been proposed that a D1 receptor-mediated facilitation of NMDA receptor response during adolescence (Fig. 1a) (Tseng and O'Donnell 2005; Heng et al. 2011; Flores-Barrera et al. 2014) contributes to the maturation of PFC-dependent cognitive functions (Tseng et al. 2009; O'Donnell and Tseng 2010) given the fact that proper levels of D1 receptor function in the PFC are needed to improve memory retrieval and working memory (Seamans et al. 1998; Floresco and Phillips 2001), and for appetitive instrumental learning in adult rats (Baldwin et al. 2002). Thus, any disruption that altered the acquisition of adult levels of D1 (Leslie et al. 1991; Williams 1993; Williams et al. 1993; Monyer et al. 1994; Tarazi et al. 1999; Tarazi and Baldessarini 2000; Flores-Barrera et al. 2014) is likely to impair the functional maturation of the PFC and the subsequent development of cognitive abilities associated with adult behaviors (Brenhouse et al. 2008; Sonntag et al. 2014).

Dopamine-mediated inhibitory control of PFC output activity is also developmentally regulated, mainly through the maturation of D2 receptor signaling onto pyramidal neurons (Tseng and O'Donnell 2004), and indirectly through the facilitation of local GABAergic transmission during adolescence (Fig. 1b) (Pirot et al. 1992; Gulledge and Jaffe 1998; Gorelova et al. 2002; Tseng and O'Donnell 2004; Tseng et al. 2006). In fact, GABAergic interneurons in the PFC do express both D1 and D2 receptors (Vincent et al. 1993; Smiley et al. 1994; Gaspar et al. 1995; Mrzljak et al. 1996; Muly et al. 1998) and dopamine contacts onto GABAergic cells also increase during the transition to young adulthood (P60) (Benes et al. 1996). While the facilitation of interneurons' excitability by dopamine in the juvenile PFC (P15–35) is strictly D1 receptor-mediated (Gorelova et al. 2002; Tseng et al. 2006), a

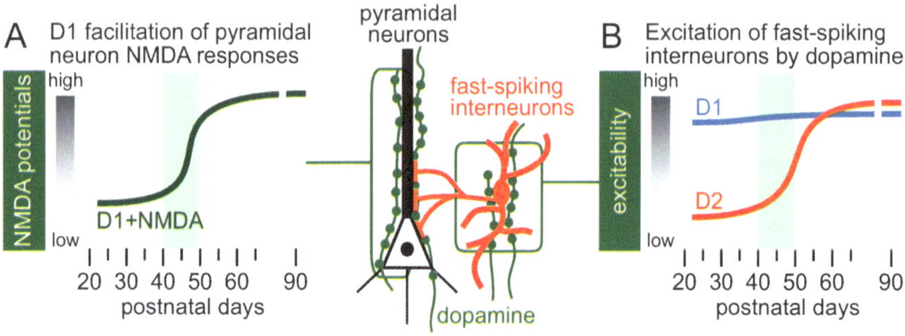

Fig. 1 Developmental trajectory of dopamine actions in the prefrontal cortex. (**a**) D1 receptor-mediated facilitation of NMDA receptor responses begin to increase at P45 through P90 (adulthood). (**b**) Dopamine facilitation of fast-spiking interneurons activity also increases after P45 through adulthood, due to the acquisition of a non-canonical D2 receptor-mediated excitation

powerful excitatory action by D2 receptor signaling emerges in the PFC after P50 to drive local GABAergic activity (Fig. 1b) (Tseng et al. 2006; Tseng and O'Donnell 2007) through a mechanism that appears to involve beta-arrestin signaling (Urs et al. 2016). Thus, the developmental recruitment of PFC GABAergic activity by dopamine through D1 and D2-dependent mechanisms (Tseng et al. 2006; Tseng and O'Donnell 2007) is thought to provide a powerful inhibitory control of prefrontal output input selectivity and its computational capacity to enable network stability for sustaining working memory (Lew and Tseng 2014).

2.2 *Maturation of Glutamatergic Control of PFC Activity*

In addition to the maturation of dopamine's action (Fig. 1), changes in L-type Ca^{2+} function and post-synaptic PKA signaling after P40 (Fig. 2a) are critical to enhance PFC output excitability during adolescence (Heng et al. 2011). At the synaptic level, proper coordination of afferent excitatory transmission to the PFC also contributes to the maturation of prefrontal output function during adolescence (Maroun and Richter-Levin 2003; Tseng et al. 2009; Caballero et al. 2016). Of particular interest is the subunit composition of NMDA receptors, which has been shown to directly impact cortical circuit plasticity (Zhao et al. 2005) over the course of postnatal development by increasing the GluN2A contribution of synaptic transmission (Dumas 2005; Wang et al. 2008). In the PFC, however, it is the gain of GluN2B (not GluN2A) transmission during adolescence that strengthens ventral hippocampal inputs through adulthood (Flores-Barrera et al. 2014). Of note, such acquisition of GluN2B transmission after P50 is restricted to the apical dendrite of layer V pyramidal neurons (Fig. 2b) (Flores-Barrera et al. 2014) approximately 1 week after the apical compartment becomes functionally coupled to the soma (i.e., P42)

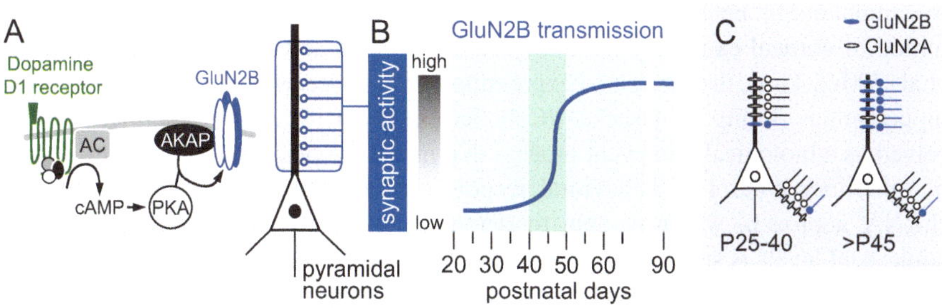

Fig. 2 Developmental trajectory of glutamatergic transmission in the prefrontal cortex. (**a**) Diagram illustrating the mechanism underlying the late-adolescent facilitation of NMDA-GluN2B transmission by D1 receptor signaling. (**b**) Trajectory of the late-adolescent onset of GluN2B-mediated potentiation of synaptic activity onto layer V pyramidal neurons. (**c**) The late-adolescent facilitation of GluN2B function occurs primarily in the apical dendrite of layer V pyramidal neurons

and major structural remodeling of its dendritic complexity has been completed (Zhu 2000; Heng et al. 2011). Due to the slow kinetic of GluN2B-containing NMDA receptors (Vicini et al. 1998), the late adolescent incorporation of GluN2B transmission into a functionally mature apical dendrite could provide a computational advantage for input integration (Wang 1999) and selectively amplify afferent information from the ventral hippocampus (Flores-Barrera et al. 2014). Accordingly, proper levels of GluN2B-mediated transmission in the PFC are key for a variety of PFC-dependent functions including working memory processes (Wang et al. 2008; Wang et al. 2013) and trace fear conditioning (Gilmartin et al. 2013; Miguelez Fernandez et al. 2021). It is conceivable that any developmental disruption that compromises the recruitment of GluN2B transmission during adolescence will negatively impact PFC maturation by limiting the input-specific capacity of afferent integration and processing.

2.3 *Maturation of GABAergic Control of PFC Activity*

The functionality of GABAergic interneurons in the PFC also undergoes major remodeling during adolescence as revealed by the levels of expression of calcium binding proteins parvalbumin (PV) and calretinin (CR) (Erickson and Lewis 2002; Fung et al. 2010; Caballero et al. 2014a). By means of biochemical and immunohistochemical measures, we found that these two populations of interneurons show opposite developmental trajectories, such that PV expression increases whereas CR decreases sharply during the adolescent transition to adulthood (Caballero et al. 2014a). As a "slow calcium buffer" (Chard et al. 1993; Lee et al. 2000; Collin et al. 2005), the presence of PV enables cortical GABAergic interneurons to respond to high frequency inputs. Accordingly, PV expression is typically found in a subset of GABAergic cells known as fast-spiking interneurons, which limit the activity of pyramidal output neurons through a feedforward inhibitory mechanism and the fine-tuning of cortical excitatory-inhibitory balance (Celio and Heizmann 1981; Bartos et al. 2007). Thus, the protracted expression of PV observed in the PFC and ventral hippocampus during adolescence (Caballero et al. 2013, 2014a) is currently perceived as a biologically relevant process that must take place to attain mature levels of inhibitory control of behavioral responses (Caballero et al. 2020). Using an shRNA approach, we were able to demonstrate that downregulation of PV to adolescent levels is sufficient to disrupt the functional trajectory of two biologically relevant events of synaptic maturation in the PFC (Fig. 3). Both the normal facilitation of excitatory synaptic activity onto PV-positive fast-spiking interneurons (Fig. 3a) and the gain of GABAergic transmission onto layers V-VI pyramidal neurons (Fig. 3b) are no longer enabled when PV expression levels failed to reach adult levels (Caballero et al. 2020). These findings provide a novel, developmentally relevant mechanism for the refinement of prefrontal GABAergic function regulated by the trajectory of PV upregulation during adolescence.

Parallel to the increased PV expression in the PFC during adolescence is the facilitation of excitatory synaptic activity onto PV-positive fast-spiking interneurons (Fig. 3a) (Caballero et al. 2014a). Interestingly, such developmental event failed to occur when NMDA receptor transmission is transiently blocked during adolescence (Flores-Barrera et al. 2020), which resembles the effect observed following PV downregulation (Caballero et al. 2020). As PV expression is strongly dependent on glutamatergic signaling (Behrens et al. 2007), it is possible that NMDA receptor transmission from multiple long-range inputs (Flores-Barrera et al. 2014; Bogart and O'Donnell 2018) is needed to enhance the functionality of PV-positive, fast-spiking interneurons in the PFC during adolescence to enable the transition into an "adult" profile (Caballero et al. 2016; Caballero and Tseng 2016). Accordingly, transient disruption of NMDA receptor signaling during adolescence also prevented the gain of inhibitory synaptic activity onto PFC output neurons that normally emerges during this period (Fig. 3b) (Flores-Barrera et al. 2020). In this regard, a failure to reach sufficient level of afferent-mediated glutamatergic transmission during

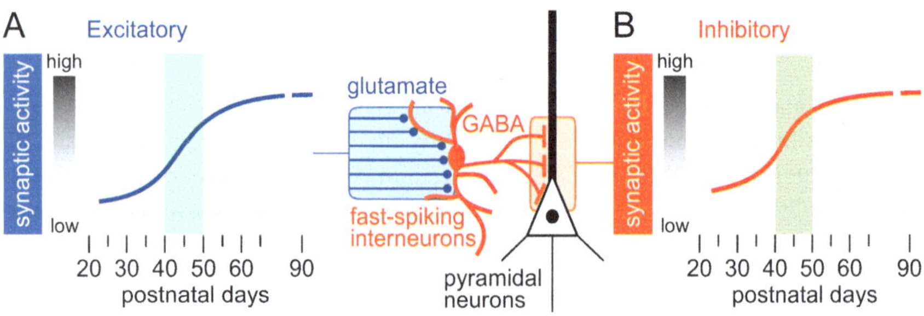

Fig. 3 Developmental trajectory of GABAergic function in the prefrontal cortex. (**a**) Typical developmental facilitation of excitatory synaptic activity (as revealed by the frequency of excitatory postsynaptic currents) onto fast-spiking interneurons during adolescence. (**b**) Concurrently, basal inhibitory synaptic activity (as determined by the frequency of inhibitory postsynaptic currents) increases sharply after P40–45

adolescence will arrest the functional maturation of fast-spiking interneurons in the PFC at a developmental stage of insufficient inhibitory control of prefrontal output.

3 Functional Maturation of PFC Connectivity During Adolescence

At the neural circuitry level, it becomes evident that the maturation of cognitive (Satterthwaite et al. 2013) and emotional regulation (Gee et al. 2013; Swartz et al. 2014) are paralleled by significant changes in connectivity between key corticolimbic structures (Sowell and Jernigan 1998) and the PFC (Paus et al. 1999) during adolescence. Of particular interest is the integration of ventral hippocampal and amygdalar inputs carrying contextual and emotional information by the PFC (Floresco et al. 1997; Ishikawa and Nakamura 2003; Tse et al. 2015) as remodeling of several anatomical features within the hippocampal-PFC and amygdalar-PFC connectivity continues through adolescence (Benes 1989; Cunningham et al. 2002; Cressman et al. 2010). In fact, disruptions of these pathways are sufficient to negatively impact the acquisition of a wide range of cognitive abilities associated with adult behavior (Casey et al. 2000; Tseng et al. 2009; Best and Miller 2010). Functionally, it has long been recognized that synchronized pre- and post-synaptic events are required for strengthening specific corticolimbic-PFC pathways during adolescence to drive the protracted maturation of cognitive and emotional regulation (Tseng et al. 2009). The following sections will review major findings highlighting how the development trajectory of activity-dependent plasticity through adolescence directly impacts the remodeling and maturation of hippocampal-PFC and amygdalar-PFC functional connectivity.

3.1 Maturation of Hippocampal-PFC Functional Connectivity

The glutamatergic ventral hippocampal projections to the PFC innervate predominantly pyramidal output neurons (Carr and Sesack 1996) and a subclass of GABAergic interneurons expressing parvalbumin (Gabbott et al. 2002). The integrity of this pathway is critical for working memory processes and associated cognitive functions (Friedman and Goldman-Rakic 1988; Floresco et al. 1997; Wang and Cai 2006), which begin to reach maturity during the adolescent transition to young adulthood (Luna et al. 2004). Although there is evidence from human studies showing increased hippocampal volume during postnatal development through late adolescence and young adulthood (Suzuki et al. 2005; Gogtay et al. 2006; Goddings et al. 2014), data on the precise timing and anatomical trajectory by which ventral hippocampal innervation to the PFC reach maturity are lacking with the exception of the studies by the Benes group (Benes 1989; Benes et al. 1994). A recent longitudinal study characterizing developmental changes from 8 to 32 years of age revealed significant increases in hippocampal-PFC functional connectivity at rest, particularly with the ventromedial (not dorsolateral) region of the PFC (Calabro et al. 2020). Importantly, the trajectory of the hippocampal-ventromedial PFC connectivity can predict the cognitive ability to problem solving and future planning supporting the view that maturation of high-level cognition requires increased functional integration across the hippocampal-ventromedial PFC circuitry through adolescence (Murty et al. 2016; Calabro et al. 2020).

Mechanistically, the functional monosynaptic ventral hippocampal-PFC connectivity has been studied for some time in adult animals (Jay 2003). Despite the well-established role of NMDA receptor transmission and D1 receptor signaling in modulating the gain of PFC response to hippocampal inputs (Gurden et al. 2000), it took over a decade to uncover the extent of functional remodeling occurring within the hippocampal-PFC pathway that continues through late adolescence (Fig. 4) distinct from the developmental trajectory of the amygdala-PFC connectivity (see Sect. 3.2, Fig. 5).

The pattern of PFC response to ventral hippocampal inputs is dictated by the frequency of afferent drive and the functional state of local GABAergic and glutamatergic synapses. Both inhibitory and excitatory components of the hippocampal-PFC pathway undergo functional maturation during adolescence, in part due to the recruitment of fast-spiking interneurons (Fig. 3) and the gain of GluN2B transmission onto pyramidal neurons (Fig. 2). Typically, proper levels of $GABA_AR$ signaling are needed to enable the acquisition of PFC inhibitory control of hippocampal afferent transmission at 20 and 40 Hz (Cass et al. 2013; Thomases et al. 2013). Such maturation of GABAergic function begins to occur after P45 (Fig. 3) ventral hippocampal recruitment of PFC parvalbumin-positive fast-spiking interneurons becomes online (Caballero et al. 2020). At higher frequencies (e.g., 100 Hz) ventral hippocampal inputs can potentiate GABAergic transmission in the

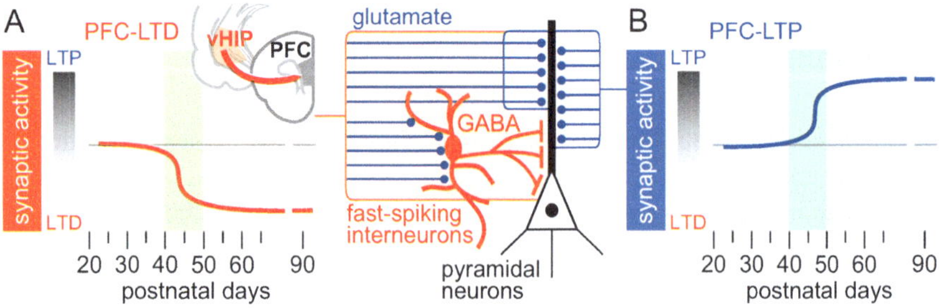

Fig. 4 Developmental trajectory of the ventral hippocampal-prefrontal cortex connectivity. (**a**) Characteristic developmental facilitation of hippocampal-induced LTD in the prefrontal cortex that begins to emerge after P45. Note the recruitment of fast-spiking interneurons' activity by ventral hippocampal excitatory inputs and subsequent potentiation of GABAergic synapses onto pyramidal neurons. (**b**) Parallel to the potentiated GABAergic transmission in the prefrontal cortex, there is a GluN2B-mediated LTP response at layer V pyramidal neurons' apical dendrite

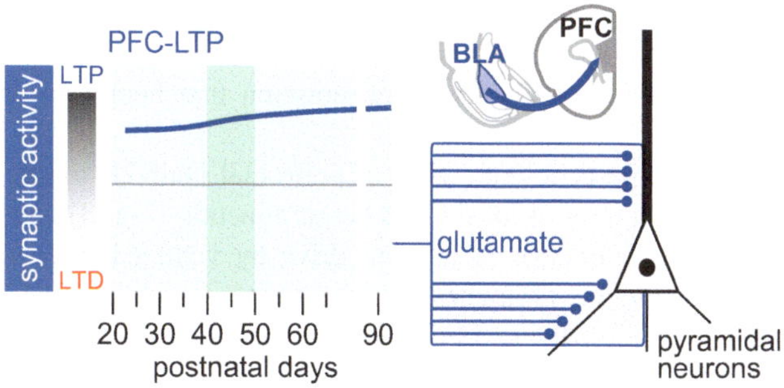

Fig. 5 Developmental trajectory of the amygdalar-prefrontal cortex connectivity. Contrary to the trajectory of ventral hippocampal-prefrontal LTP, the amygdalar counterpart is already present by P30. However, the pattern of LTP at P30 is significantly less potentiated than the response observed after P50

PFC (Cass et al. 2013; Caballero et al. 2014b; Thomases et al. 2014) and subsequently induce a form of LTD that also emerges after P45 (Fig. 4a). Recent studies from our laboratory indicate that preventing the normal upregulation of parvalbumin expression in the PFC is sufficient to arrest the recruitment of local GABAergic transmission needed to enable the inhibitory control of afferent integration by the ventral hippocampus and its regulation of extinction learning (Caballero et al. 2020). Furthermore, transient disturbances in dopamine, cannabinoid, and NMDA receptor signaling during early adolescence elicit a similar state of prefrontal GABAergic deficit and impaired ventral hippocampal-PFC functional connectivity (Cass et al. 2013, 2014; Thomases et al. 2013, 2014) that can be traced to a deficient glutamatergic drive onto fast-spiking interneurons (Caballero et al. 2020; Flores-Barrera et al. 2020). The input-specific impact of PV downregulation suggests that prior GABAergic maturation is needed for the maturation of PFC processing of hippocampal inputs. Although not mutually exclusive, it is also possible that strengthening of specific inputs like the ventral hippocampus is required to enhance the gain of fast-spiking GABAergic transmission in the PFC.

In addition to the recruitment of PFC GABAergic circuit is the late adolescent strengthening of ventral hippocampal glutamatergic inputs onto prefrontal output pyramidal neurons (Caballero et al. 2014b; Flores-Barrera et al. 2014) by the acquisition of GluN2B transmission (Fig. 2b, c) that provides a mechanism for dopamine regulation of the hippocampal-PFC functional connectivity in an input-specific manner (Flores-Barrera et al. 2014). Relative to the potentiated response elicited by the amygdala that is already enabled by P30 (Fig. 5), the prefrontal LTP observed following ventral hippocampal stimulation does not emerge until P50 (Fig. 4b) (Caballero et al. 2014b; Flores-Barrera et al. 2014). Notably, both forms of LTP are mediated by NMDA receptor signaling, yet an intact GluN2B transmission (through a D1 and PKA-mediated mechanism; Fig. 4a) is needed to sustain the potentiated hippocampal response (Flores-Barrera et al. 2014) and its regulation of behavior (Miguelez Fernandez et al. 2021). Thus, any developmental disruption that compromises the gain of GluN2B function during adolescence will arrest the functional maturation of the ventral hippocampal-PFC connectivity by limiting the input-specific strengthening of afferent transmission that begins to emerge after P40–45 (Fig. 4b).

In sum, the late-adolescent expression of prefrontal GluN2B plasticity (Fig. 2) together with the facilitation of local GABAergic function (Fig. 3) and the recruitment of PFC inhibitory control of afferent drive by ventral hippocampal inputs (Fig. 4) could explain the protracted development of cognitive and emotional regulation subserved by different PFC domains. Of particular relevance for such functional implication is the acquisition of PFC control over amygdalar-dependent processes during adolescence (e.g., after P40) (Thomases et al. 2014) when input-specific remodeling of pre- and post-synaptic mechanisms contributing to strengthening the ventral hippocampal-PFC connectivity begins to emerge (Caballero and Tseng 2016; Caballero et al. 2016, 2020; Flores-Barrera et al. 2020; Miguelez Fernandez et al. 2021).

3.2 Maturation of Amygdalar-PFC Functional Connectivity

Monosynaptic glutamatergic inputs originated from the amygdala can be observed in the PFC (Krettek and Price 1977; Bacon et al. 1996; McDonald 1996; Verwer et al. 1996) shortly after birth with a bilaminar pattern arising between P12 and P16 (Verwer et al. 1996; Cunningham et al. 2002), which continues to increase through adolescence until young adulthood (P65) (Cunningham et al. 2002). Although amygdalar inputs contacting both pyramidal neurons and GABAergic interneurons can be found in the PFC (Cunningham et al. 2002, 2008; Gabbott et al. 2006), the increasing density of amygdalar projections observed during adolescence reflects preferentially the innervation onto pyramidal neurons (Cunningham et al. 2002). Functionally, amygdalar inputs can effectively recruit local interneurons in the adult PFC to exert feedforward inhibition onto pyramidal neurons (Dilgen et al. 2013). However, the developmental impact of amygdalar inputs to drive LTP in the PFC is already enabled by P30 and does not involve local GABAergic transmission (Caballero et al. 2014b). Consistent with the increasing amygdalar innervation, the pattern of LTP elicited in the P30–40 PFC is significantly less potentiated than the response observed in adults (Fig. 5). Thus, it is reasonable to conclude that the gain of amygdalar-prefrontal transmission during adolescence is likely to impact PFC output function through its action on pyramidal neurons, and to less extent on local GABAergic interneurons.

Early life disruption of the amygdalar-PFC connectivity has been correlated to some extent with the severity of affective disorders (Burghy et al. 2012). This is not surprising since PFC processing of amygdalar inputs is required for the modulation of amygdala-dependent behaviors (Garcia et al. 1999; Hariri et al. 2003) including emotional regulation and the consolidation of learning and memory (Davis and Whalen 2001). Recent findings using task-based functional magnetic resonance imaging indicate that the amygdala-PFC functional connectivity progressively shifts from positive to negative correlation during the transition from childhood to young adulthood (Gee et al. 2013; Wu et al. 2016). Importantly, this trajectory of functional connectivity is no longer detectable in children and adults experiencing generalized anxiety disorder (Kujawa et al. 2016). At the neurochemical level, recent research in human subjects suggests that changes in GABA levels predict the variability of functional connectivity across brain networks (Duncan et al. 2014; Delli Pizzi et al. 2017a, b). In addition, increased prefrontal glutamate concentration was observed in young adults, but not adolescents with high anxiety responses (Cortese and Phan 2005; Strawn et al. 2013), suggesting that fine-tuning of excitatory-inhibitory balance in the PFC is required for the developmental switch of the immature amygdala-PFC functional connectivity to a mature state.

4 Summary and Conclusions

The neurobiology underlying the protracted remodeling of the corticolimbic functional connectivity challenges the view that not all mechanisms of plasticity contributing to the maturation of cortical circuits are acquired within the first 21–28 days of postnatal development in rodents as traditionally proposed (Dan and Poo 2006). Furthermore, the relative early onset of amygdalar-driven plasticity (Fig. 5) compared to the hippocampal counterpart (Fig. 4) indicates that timely recruitment of distinct synaptic mechanisms through the peri-adolescent development is critical for the maturation of PFC functional connectivity in an input-specific manner. Certainly, any early developmental events capable of altering the functional trajectory of PFC maturation could trigger abnormal assembly of hippocampal- and amygdalar-PFC connectivity that would not become apparent until late adolescence or young adulthood when prefrontal integration of input-specific processes are enabled.

Acknowledgements *Grant Sponsor:* National Institute of Mental Health grants MH086507, MH105488, and MH123147 to KYT, and MH116475 to SY and KYT. The funding source had no role in the study design, data analysis, or writing of this report.

References

Andersen SL (2003) Trajectories of brain development: point of vulnerability or window of opportunity? Neurosci Biobehav Rev 27:3–18

Bacon SJ, Headlam AJ, Gabbott PL, Smith AD (1996) Amygdala input to medial prefrontal cortex (mPFC) in the rat: a light and electron microscope study. Brain Res 720:211–219

Baldwin AE, Sadeghian K, Kelley AE (2002) Appetitive instrumental learning requires coincident activation of NMDA and dopamine D1 receptors within the medial prefrontal cortex. J Neurosci 22:1063–1071

Bartos M, Vida I, Jonas P (2007) Synaptic mechanisms of synchronized gamma oscillations in inhibitory interneuron networks. Nat Rev Neurosci 8:45–56
Behrens MM, Ali SS, Dao DN, Lucero J, Shekhtman G, Quick KL, Dugan LL (2007) Ketamine-induced loss of phenotype of fast-spiking interneurons is mediated by NADPH-oxidase. Science 318:1645–1647
Benes FM (1989) Myelination of cortical-hippocampal relays during late adolescence. Schizophr Bull 15:585–593
Benes FM, Turtle M, Khan Y, Farol P (1994) Myelination of a key relay zone in the hippocampal formation occurs in the human brain during childhood, adolescence, and adulthood. Arch Gen Psychiatry 51:477–484
Benes FM, Vincent SL, Molloy R, Khan Y (1996) Increased interaction of dopamine-immunoreactive varicosities with GABA neurons of rat medial prefrontal cortex occurs during the postweanling period. Synapse 23:237–245
Best JR, Miller PH (2010) A developmental perspective on executive function. Child Dev 81:1641–1660
Bogart LJ, O'Donnell P (2018) Multiple long-range inputs evoke NMDA currents in prefrontal cortex fast-spiking interneurons. Neuropsychopharmacology 43:2101–2108
Bouthenet ML, Martres MP, Sales N, Schwartz JC (1987) A detailed mapping of dopamine D-2 receptors in rat central nervous system by autoradiography with [125I]iodosulpride. Neuroscience 20:117–155
Brenhouse HC, Andersen SL (2011) Developmental trajectories during adolescence in males and females: a cross-species understanding of underlying brain changes. Neurosci Biobehav Rev 35:1687–1703
Brenhouse HC, Sonntag KC, Andersen SL (2008) Transient D1 dopamine receptor expression on prefrontal cortex projection neurons: relationship to enhanced motivational salience of drug cues in adolescence. J Neurosci 28:2375–2382
Burghy CA, Stodola DE, Ruttle PL, Molloy EK, Armstrong JM, Oler JA, Fox ME, Hayes AS, Kalin NH, Essex MJ, Davidson RJ, Birn RM (2012) Developmental pathways to amygdala-prefrontal function and internalizing symptoms in adolescence. Nat Neurosci 15:1736–1741
Caballero A, Tseng KY (2016) GABAergic function as a limiting factor for prefrontal maturation during adolescence. Trends Neurosci 39:441–448
Caballero A, Diah KC, Tseng KY (2013) Region-specific upregulation of parvalbumin-, but not calretinin-positive cells in the ventral hippocampus during adolescence. Hippocampus 23:1331–1336
Caballero A, Flores-Barrera E, Cass DK, Tseng KY (2014a) Differential regulation of parvalbumin and calretinin interneurons in the prefrontal cortex during adolescence. Brain Struct Funct 219:395–406
Caballero A, Thomases DR, Flores-Barrera E, Cass DK, Tseng KY (2014b) Emergence of GABAergic-dependent regulation of input-specific plasticity in the adult rat prefrontal cortex during adolescence. Psychopharmacology (Berl) 231:1789–1796
Caballero A, Granberg R, Tseng KY (2016) Mechanisms contributing to prefrontal cortex maturation during adolescence. Neurosci Biobehav Rev 70:4–12
Caballero A, Flores-Barrera E, Thomases DR, Tseng KY (2020) Downregulation of parvalbumin expression in the prefrontal cortex during adolescence causes enduring prefrontal disinhibition in adulthood. Neuropsychopharmacology 45:1527–1535
Calabro FJ, Murty VP, Jalbrzikowski M, Tervo-Clemmens B, Luna B (2020) Development of hippocampal-prefrontal cortex interactions through adolescence. Cereb Cortex 30:1548–1558
Carr DB, Sesack SR (1996) Hippocampal afferents to the rat prefrontal cortex: synaptic targets and relation to dopamine terminals. J Comp Neurol 369:1–15
Casey BJ, Giedd JN, Thomas KM (2000) Structural and functional brain development and its relation to cognitive development. Biol Psychol 54:241–257

Cass DK, Thomases DR, Caballero A, Tseng KY (2013) Developmental disruption of gamma-aminobutyric acid function in the medial prefrontal cortex by noncontingent cocaine exposure during early adolescence. Biol Psychiatry 74:490–501

Cass DK, Flores-Barrera E, Thomases DR, Vital WF, Caballero A, Tseng KY (2014) CB1 cannabinoid receptor stimulation during adolescence impairs the maturation of GABA function in the adult rat prefrontal cortex. Mol Psychiatry 19:536–543

Celio MR, Heizmann CW (1981) Calcium-binding protein parvalbumin as a neuronal marker. Nature 293:300–302

Chard PS, Bleakman D, Christakos S, Fullmer CS, Miller RJ (1993) Calcium buffering properties of calbindin D28k and parvalbumin in rat sensory neurones. J Physiol 472:341–357

Collin T, Chat M, Lucas MG, Moreno H, Racay P, Schwaller B, Marty A, Llano I (2005) Developmental changes in parvalbumin regulate presynaptic Ca2+ signaling. J Neurosci 25:96–107

Cortese BM, Phan KL (2005) The role of glutamate in anxiety and related disorders. CNS Spectr 10:820–830

Cressman VL, Balaban J, Steinfeld S, Shemyakin A, Graham P, Parisot N, Moore H (2010) Prefrontal cortical inputs to the basal amygdala undergo pruning during late adolescence in the rat. J Comp Neurol 518:2693–2709

Cunningham MG, Bhattacharyya S, Benes FM (2002) Amygdalo-cortical sprouting continues into early adulthood: implications for the development of normal and abnormal function during adolescence. J Comp Neurol 453:116–130

Cunningham MG, Bhattacharyya S, Benes FM (2008) Increasing interaction of amygdalar afferents with GABAergic interneurons between birth and adulthood. Cereb Cortex 18:1529–1535

Dan Y, Poo MM (2006) Spike timing-dependent plasticity: from synapse to perception. Physiol Rev 86:1033–1048

Davis M, Whalen PJ (2001) The amygdala: vigilance and emotion. Mol Psychiatry 6:13–34

Delli Pizzi S, Chiacchiaretta P, Mantini D, Bubbico G, Edden RA, Onofrj M, Ferretti A, Bonanni L (2017a) GABA content within medial prefrontal cortex predicts the variability of fronto-limbic effective connectivity. Brain Struct Funct 222:3217–3229

Delli Pizzi S, Chiacchiaretta P, Mantini D, Bubbico G, Ferretti A, Edden RA, Di Giulio C, Onofrj M, Bonanni L (2017b) Functional and neurochemical interactions within the amygdala-medial prefrontal cortex circuit and their relevance to emotional processing. Brain Struct Funct 222:1267–1279

Dilgen J, Tejeda HA, O'Donnell P (2013) Amygdala inputs drive feedforward inhibition in the medial prefrontal cortex. J Neurophysiol 110:221–229

Dumas TC (2005) Late postnatal maturation of excitatory synaptic transmission permits adult-like expression of hippocampal-dependent behaviors. Hippocampus 15:562–578

Duncan NW, Wiebking C, Northoff G (2014) Associations of regional GABA and glutamate with intrinsic and extrinsic neural activity in humans-a review of multimodal imaging studies. Neurosci Biobehav Rev 47:36–52

Erickson SL, Lewis DA (2002) Postnatal development of parvalbumin- and GABA transporter-immunoreactive axon terminals in monkey prefrontal cortex. J Comp Neurol 448:186–202

Flores-Barrera E, Thomases DR, Heng LJ, Cass DK, Caballero A, Tseng KY (2014) Late adolescent expression of GluN2B transmission in the prefrontal cortex is input-specific and requires postsynaptic protein kinase A and D1 dopamine receptor signaling. Biol Psychiatry 75:508–516

Flores-Barrera E, Thomases DR, Tseng KY (2020) MK-801 exposure during adolescence elicits enduring disruption of prefrontal E-I balance and its control of fear extinction behavior. J Neurosci 40:4881–4887

Floresco SB, Phillips AG (2001) Delay-dependent modulation of memory retrieval by infusion of a dopamine D1 agonist into the rat medial prefrontal cortex. Behav Neurosci 115:934–939

Floresco SB, Seamans JK, Phillips AG (1997) Selective roles for hippocampal, prefrontal cortical, and ventral striatal circuits in radial-arm maze tasks with or without a delay. J Neurosci 17:1880–1890

Friedman HR, Goldman-Rakic PS (1988) Activation of the hippocampus and dentate gyrus by working-memory: a 2-deoxyglucose study of behaving rhesus monkeys. J Neurosci 8:4693–4706
Fung SJ, Webster MJ, Sivagnanasundaram S, Duncan C, Elashoff M, Weickert CS (2010) Expression of interneuron markers in the dorsolateral prefrontal cortex of the developing human and in schizophrenia. Am J Psychiatry 167:1479–1488
Gabbott P, Headlam A, Busby S (2002) Morphological evidence that CA1 hippocampal afferents monosynaptically innervate PV-containing neurons and NADPH-diaphorase reactive cells in the medial prefrontal cortex (areas 25/32) of the rat. Brain Res 946:314–322
Gabbott PL, Warner TA, Busby SJ (2006) Amygdala input monosynaptically innervates parvalbumin immunoreactive local circuit neurons in rat medial prefrontal cortex. Neuroscience 139:1039–1048
Garcia R, Vouimba RM, Baudry M, Thompson RF (1999) The amygdala modulates prefrontal cortex activity relative to conditioned fear. Nature 402:294–296
Gaspar P, Bloch B, Le Moine C (1995) D1 and D2 receptor gene expression in the rat frontal cortex: cellular localization in different classes of efferent neurons. Eur J Neurosci 7:1050–1063
Gee DG, Humphreys KL, Flannery J, Goff B, Telzer EH, Shapiro M, Hare TA, Bookheimer SY, Tottenham N (2013) A developmental shift from positive to negative connectivity in human amygdala-prefrontal circuitry. J Neurosci 33:4584–4593
Gilmartin MR, Kwapis JL, Helmstetter FJ (2013) NR2A- and NR2B-containing NMDA receptors in the prelimbic medial prefrontal cortex differentially mediate trace, delay, and contextual fear conditioning. Learn Mem 20(6):290–294
Goddings AL, Mills KL, Clasen LS, Giedd JN, Viner RM, Blakemore SJ (2014) The influence of puberty on subcortical brain development. Neuroimage 88:242–251
Gogtay N, Nugent TF 3rd, Herman DH, Ordonez A, Greenstein D, Hayashi KM, Clasen L, Toga AW, Giedd JN, Rapoport JL, Thompson PM (2006) Dynamic mapping of normal human hippocampal development. Hippocampus 16:664–672
Goldman-Rakic PS, Leranth C, Williams SM, Mons N, Geffard M (1989) Dopamine synaptic complex with pyramidal neurons in primate cerebral cortex. Proc Natl Acad Sci U S A 86:9015–9019
Goldman-Rakic PS, Muly EC 3rd, Williams GV (2000) D(1) receptors in prefrontal cells and circuits. Brain Res Brain Res Rev 31:295–301
Gorelova N, Seamans JK, Yang CR (2002) Mechanisms of dopamine activation of fast-spiking interneurons that exert inhibition in rat prefrontal cortex. J Neurophysiol 88:3150–3166
Gulledge AT, Jaffe DB (1998) Dopamine decreases the excitability of layer V pyramidal cells in the rat prefrontal cortex. J Neurosci 18:9139–9151
Gurden H, Takita M, Jay TM (2000) Essential role of D1 but not D2 receptors in the NMDA receptor-dependent long-term potentiation at hippocampal-prefrontal cortex synapses in vivo. J Neurosci 20:RC106
Hariri AR, Mattay VS, Tessitore A, Fera F, Weinberger DR (2003) Neocortical modulation of the amygdala response to fearful stimuli. Biol Psychiatry 53:494–501
Heng LJ, Markham JA, Hu XT, Tseng KY (2011) Concurrent upregulation of postsynaptic L-type Ca(2+) channel function and protein kinase A signaling is required for the periadolescent facilitation of Ca(2+) plateau potentials and dopamine D1 receptor modulation in the prefrontal cortex. Neuropharmacology 60:953–962
Horvitz JC (2000) Mesolimbocortical and nigrostriatal dopamine responses to salient non-reward events. Neuroscience 96:651–656
Ishikawa A, Nakamura S (2003) Convergence and interaction of hippocampal and amygdalar projections within the prefrontal cortex in the rat. J Neurosci 23:9987–9995
Jay TM (2003) Dopamine: a potential substrate for synaptic plasticity and memory mechanisms. Prog Neurobiol 69:375–390
Kalsbeek A, Voorn P, Buijs RM, Pool CW, Uylings HB (1988) Development of the dopaminergic innervation in the prefrontal cortex of the rat. J Comp Neurol 269:58–72

Krettek JE, Price JL (1977) Projections from the amygdaloid complex to the cerebral cortex and thalamus in the rat and cat. J Comp Neurol 172:687–722

Kujawa A, Wu M, Klumpp H, Pine DS, Swain JE, Fitzgerald KD, Monk CS, Phan KL (2016) Altered development of amygdala-anterior cingulate cortex connectivity in anxious youth and young adults. Biol Psychiatry Cogn Neurosci Neuroimaging 1:345–352

Lee SH, Schwaller B, Neher E (2000) Kinetics of Ca2+ binding to parvalbumin in bovine chromaffin cells: implications for [Ca2+] transients of neuronal dendrites. J Physiol 525 (Pt 2):419–432

Leslie CA, Robertson MW, Cutler AJ, Bennett JP Jr (1991) Postnatal development of D1 dopamine receptors in the medial prefrontal cortex, striatum and nucleus accumbens of normal and neonatal 6-hydroxydopamine treated rats: a quantitative autoradiographic analysis. Brain Res Dev Brain Res 62:109–114

Lew SE, Tseng KY (2014) Dopamine modulation of GABAergic function enables network stability and input selectivity for sustaining working memory in a computational model of the prefrontal cortex. Neuropsychopharmacology 39:3067–3076

Lewis DA, Campbell MJ, Foote SL, Goldstein M, Morrison JH (1987) The distribution of tyrosine hydroxylase-immunoreactive fibers in primate neocortex is widespread but regionally specific. J Neurosci 7:279–290

Lewis DA, Sesack SR, Levey AI, Rosenberg DR (1998) Dopamine axons in primate prefrontal cortex: specificity of distribution, synaptic targets, and development. Adv Pharmacol 42:703–706

Luna B, Garver KE, Urban TA, Lazar NA, Sweeney JA (2004) Maturation of cognitive processes from late childhood to adulthood. Child Dev 75:1357–1372

Maroun M, Richter-Levin G (2003) Exposure to acute stress blocks the induction of long-term potentiation of the amygdala-prefrontal cortex pathway in vivo. J Neurosci 23:4406–4409

McDonald AJ (1996) Glutamate and aspartate immunoreactive neurons of the rat basolateral amygdala: colocalization of excitatory amino acids and projections to the limbic circuit. J Comp Neurol 365:367–379

Miguelez Fernandez AMM, Molla HM, Thomases DR, Tseng KY (2021) Prefrontal alpha7nAChR signaling differentially modulates afferent drive and trace fear conditioning behavior in adolescent and adult rats. J Neurosci 41(9):1908–1916

Monyer H, Burnashev N, Laurie DJ, Sakmann B, Seeburg PH (1994) Developmental and regional expression in the rat brain and functional properties of four NMDA receptors. Neuron 12:529–540

Mrzljak L, Bergson C, Pappy M, Huff R, Levenson R, Goldman-Rakic PS (1996) Localization of dopamine D4 receptors in GABAergic neurons of the primate brain. Nature 381:245–248

Muly EC 3rd, Szigeti K, Goldman-Rakic PS (1998) D1 receptor in interneurons of macaque prefrontal cortex: distribution and subcellular localization. J Neurosci 18:10553–10565

Murty VP, Calabro F, Luna B (2016) The role of experience in adolescent cognitive development: integration of executive, memory, and mesolimbic systems. Neurosci Biobehav Rev 70:46–58

O'Donnell P, Tseng KY (2010) Postnatal maturation of dopamine actions in the prefrontal cortex. In: Iversen L, Iversen S, Dunnett S, Bjorklund A (eds) Dopamine handbook. Oxford University Press, Oxford

O'Donnell P (2003) Dopamine gating of forebrain neural ensembles. Eur J Neurosci 17:429–435

Paus T, Zijdenbos A, Worsley K, Collins DL, Blumenthal J, Giedd JN, Rapoport JL, Evans AC (1999) Structural maturation of neural pathways in children and adolescents: in vivo study. Science 283:1908–1911

Pirot S, Godbout R, Mantz J, Tassin JP, Glowinski J, Thierry AM (1992) Inhibitory effects of ventral tegmental area stimulation on the activity of prefrontal cortical neurons: evidence for the involvement of both dopaminergic and GABAergic components. Neuroscience 49:857–865

Raghanti MA, Stimpson CD, Marcinkiewicz JL, Erwin JM, Hof PR, Sherwood CC (2008) Cortical dopaminergic innervation among humans, chimpanzees, and macaque monkeys: a comparative study. Neuroscience 155:203–220

Rosenberg DR, Lewis DA (1994) Changes in the dopaminergic innervation of monkey prefrontal cortex during late postnatal development: a tyrosine hydroxylase immunohistochemical study. Biol Psychiatry 36:272–277
Santana N, Mengod G, Artigas F (2009) Quantitative analysis of the expression of dopamine D1 and D2 receptors in pyramidal and GABAergic neurons of the rat prefrontal cortex. Cereb Cortex 19:849–860
Satterthwaite TD, Wolf DH, Erus G, Ruparel K, Elliott MA, Gennatas ED, Hopson R, Jackson C, Prabhakaran K, Bilker WB, Calkins ME, Loughead J, Smith A, Roalf DR, Hakonarson H, Verma R, Davatzikos C, Gur RC, Gur RE (2013) Functional maturation of the executive system during adolescence. J Neurosci 33:16249–16261
Seamans JK, Floresco SB, Phillips AG (1998) D1 receptor modulation of hippocampal-prefrontal cortical circuits integrating spatial memory with executive functions in the rat. J Neurosci 18:1613–1621
Seguela P, Watkins KC, Descarries L (1988) Ultrastructural features of dopamine axon terminals in the anteromedial and the suprarhinal cortex of adult rat. Brain Res 442:11–22
Smiley JF, Levey AI, Ciliax BJ, Goldman-Rakic PS (1994) D1 dopamine receptor immunoreactivity in human and monkey cerebral cortex: predominant and extrasynaptic localization in dendritic spines. Proc Natl Acad Sci U S A 91:5720–5724
Sonntag KC, Brenhouse HC, Freund N, Thompson BS, Puhl M, Andersen SL (2014) Viral over-expression of D1 dopamine receptors in the prefrontal cortex increase high-risk behaviors in adults: comparison with adolescents. Psychopharmacology (Berl) 231:1615–1626
Sowell ER, Jernigan TL (1998) Further MRI evidence of late brain maturation: limbic volume increases and changing asymmetries during childhood and adolescence. Dev Neuropsychol 14:599–617
Spear LP (2000) The adolescent brain and age-related behavioral manifestations. Neurosci Biobehav Rev 24:417–463
Strawn JR, Chu WJ, Whitsel RM, Weber WA, Norris MM, Adler CM, Eliassen JC, Phan KL, Strakowski SM, DelBello MP (2013) A pilot study of anterior cingulate cortex neurochemistry in adolescents with generalized anxiety disorder. Neuropsychobiology 67:224–229
Suzuki M, Hagino H, Nohara S, Zhou SY, Kawasaki Y, Takahashi T, Matsui M, Seto H, Ono T, Kurachi M (2005) Male-specific volume expansion of the human hippocampus during adolescence. Cereb Cortex 15:187–193
Swartz JR, Carrasco M, Wiggins JL, Thomason ME, Monk CS (2014) Age-related changes in the structure and function of prefrontal cortex-amygdala circuitry in children and adolescents: a multi-modal imaging approach. Neuroimage 86:212–220
Tarazi FI, Baldessarini RJ (2000) Comparative postnatal development of dopamine D(1), D(2) and D(4) receptors in rat forebrain. Int J Dev Neurosci 18:29–37
Tarazi FI, Tomasini EC, Baldessarini RJ (1999) Postnatal development of dopamine D1-like receptors in rat cortical and striatolimbic brain regions: an autoradiographic study. Dev Neurosci 21:43–49
Thomases DR, Cass DK, Tseng KY (2013) Periadolescent exposure to the NMDA receptor antagonist MK-801 impairs the functional maturation of local GABAergic circuits in the adult prefrontal cortex. J Neurosci 33:26–34
Thomases DR, Cass DK, Meyer JD, Caballero A, Tseng KY (2014) Early adolescent MK-801 exposure impairs the maturation of ventral hippocampal control of basolateral amygdala drive in the adult prefrontal cortex. J Neurosci 34:9059–9066
Tse MT, Piantadosi PT, Floresco SB (2015) Prefrontal cortical gamma-aminobutyric acid transmission and cognitive function: drawing links to schizophrenia from preclinical research. Biol Psychiatry 77:929–939
Tseng KY, O'Donnell P (2004) Dopamine-glutamate interactions controlling prefrontal cortical pyramidal cell excitability involve multiple signaling mechanisms. J Neurosci 24:5131–5139
Tseng KY, O'Donnell P (2005) Post-pubertal emergence of prefrontal cortical up states induced by D1-NMDA co-activation. Cereb Cortex 15:49–57

Tseng KY, O'Donnell P (2007) Dopamine modulation of prefrontal cortical interneurons changes during adolescence. Cereb Cortex 17:1235–1240

Tseng KY, Mallet N, Toreson KL, Le Moine C, Gonon F, O'Donnell P (2006) Excitatory response of prefrontal cortical fast-spiking interneurons to ventral tegmental area stimulation in vivo. Synapse 59:412–417

Tseng KY, Chambers RA, Lipska BK (2009) The neonatal ventral hippocampal lesion as a heuristic neurodevelopmental model of schizophrenia. Behav Brain Res 204:295–305

Urs NM, Gee SM, Pack TF, McCorvy JD, Evron T, Snyder JC, Yang X, Rodriguiz RM, Borrelli E, Wetsel WC, Jin J, Roth BL, O'Donnell P, Caron MG (2016) Distinct cortical and striatal actions of a beta-arrestin-biased dopamine D2 receptor ligand reveal unique antipsychotic-like properties. Proc Natl Acad Sci U S A 113:E8178–E8186

Verwer RW, Van Vulpen EH, Van Uum JF (1996) Postnatal development of amygdaloid projections to the prefrontal cortex in the rat studied with retrograde and anterograde tracers. J Comp Neurol 376:75–96

Vicini S, Wang JF, Li JH, Zhu WJ, Wang YH, Luo JH, Wolfe BB, Grayson DR (1998) Functional and pharmacological differences between recombinant N-methyl-D-aspartate receptors. J Neurophysiol 79:555–566

Vincent SL, Khan Y, Benes FM (1993) Cellular distribution of dopamine D1 and D2 receptors in rat medial prefrontal cortex. J Neurosci 13:2551–2564

Wang XJ (1999) Synaptic basis of cortical persistent activity: the importance of NMDA receptors to working memory. J Neurosci 19:9587–9603

Wang GW, Cai JX (2006) Disconnection of the hippocampal-prefrontal cortical circuits impairs spatial working memory performance in rats. Behav Brain Res 175:329–336

Wang H, Stradtman GG 3rd, Wang XJ, Gao WJ (2008) A specialized NMDA receptor function in layer 5 recurrent microcircuitry of the adult rat prefrontal cortex. Proc Natl Acad Sci U S A 105:16791–16796

Wang M, Yang Y, Wang CJ, Gamo NJ, Jin LE, Mazer JA, Morrison JH, Wang XJ, Arnsten AF (2013) NMDA receptors subserve persistent neuronal firing during working memory in dorsolateral prefrontal cortex. Neuron 77:736–749

Williams K (1993) Ifenprodil discriminates subtypes of the N-methyl-D-aspartate receptor: selectivity and mechanisms at recombinant heteromeric receptors. Mol Pharmacol 44:851–859

Williams K, Russell SL, Shen YM, Molinoff PB (1993) Developmental switch in the expression of NMDA receptors occurs in vivo and in vitro. Neuron 10:267–278

Wu M, Kujawa A, Lu LH, Fitzgerald DA, Klumpp H, Fitzgerald KD, Monk CS, Phan KL (2016) Age-related changes in amygdala-frontal connectivity during emotional face processing from childhood into young adulthood. Hum Brain Mapp 37:1684–1695

Zhao MG, Toyoda H, Lee YS, Wu LJ, Ko SW, Zhang XH, Jia Y, Shum F, Xu H, Li BM, Kaang BK, Zhuo M (2005) Roles of NMDA NR2B subtype receptor in prefrontal long-term potentiation and contextual fear memory. Neuron 47:859–872

Zhu JJ (2000) Maturation of layer 5 neocortical pyramidal neurons: amplifying salient layer 1 and layer 4 inputs by Ca2+ action potentials in adult rat tuft dendrites. J Physiol 526(Pt 3):571–587

Microglia and Sensitive Periods in Brain Development

Julia E. Dziabis and Staci D. Bilbo

Contents

Abstract From embryonic neuronal migration to adolescent circuit refinement, the immune system plays an essential role throughout central nervous system (CNS) development. Immune signaling molecules serve as a common language between the immune system and CNS, allowing them to work together to modulate brain function both in health and disease. As the resident CNS macrophage, microglia comprise the majority of immune cells in the brain. Much like their peripheral counterparts, microglia survey their environment for pathology, clean up debris, and propagate inflammatory responses when necessary. Beyond this, recent studies have highlighted that microglia perform a number of complex tasks during neural development, from directing neuronal and axonal positioning to pruning synapses, receptors, and even whole cells. In this chapter, we discuss this literature within the framework that immune activation during discrete windows of neural development can profoundly impact brain function long-term, and thus the risk of neurodevelopmental and neuropsychiatric disorders. In this chapter, we review three sensitive developmental periods – embryonic wiring, early postnatal synaptic pruning, and adolescent circuit refinement – in order to highlight the diversity of functions that microglia perform in building a brain. In reviewing this literature, it becomes obvious that *timing matters*, perhaps more so than the nature of the immune activation itself; largely conserved patterns of microglial response to diverse insults

---

J. E. Dziabis and S. D. Bilbo (✉)
Department of Psychology and Neuroscience, Duke University, Durham, NC, USA

Department of Neurobiology, Duke University, Durham, NC, USA
e-mail: staci.bilbo@duke.edu

Curr Topics Behav Neurosci (2022) 53: 55–78
https://doi.org/10.1007/7854_2021_242
Published Online: 1 September 2021

result in different functional impacts depending on the stage of brain maturation at the time of the challenge.

Keywords Adolescence · Circuit refinement · Embryonic · Inhibitory · Microglia · Neurodevelopment · Neuroimmunology · Neuroinflammation · Sensitive periods · Synaptic pruning

1 Introduction

Our appreciation of the immune system's power to impact brain maturation is not new. Some of the strongest foundational evidence in support of the claim that the immune system can profoundly impact brain development comes from epidemiological studies. Such studies identified maternal infection with a variety of pathogens during pregnancy, including influenza, streptococcus, and toxoplasma, as a strong risk factor for neurodevelopmental disorders such as autism spectrum disorder (ASD) and schizophrenia in their offspring (Atladóttir et al. 2010; Brown 2012; Knuesel et al. 2014). Indeed, initiating maternal immune activation (MIA) by injection of a bacterial or viral compound during gestation in a rodent induces behavioral abnormalities in offspring consistent with human ASD and schizophrenia (Shi et al. 2003; Zuckerman et al. 2003; Malkova et al. 2012; Choi et al. 2016). The diversity of viral, bacterial, and parasitic contributions that influence similar neurological outcomes suggests that perturbation of offspring neural function is not due to the pathogens themselves, but rather the MIA that occurs in response to the pathogen (Estes and McAllister 2016). For example, administration of pro-inflammatory cytokines IL-6 or IL-17a in the absence of any pathogen is sufficient to induce behavioral and neurobiological abnormalities in offspring (Smith et al. 2007; Choi et al. 2016). Further, blocking the activity of these cytokines rescues many MIA-induced deficits (Smith et al. 2007; Choi et al. 2016; Wu et al. 2017; Shin Yim et al. 2017). These findings provide strong evidence that immune molecules must have some avenue through which they can impact the developing brain.

Immune cells that primarily reside in the periphery, such as lymphocytes and myeloid cells, can be found in low densities in the CNS in the absence of illness or disease (Banks and Erickson 2010). The CNS was once considered entirely immune-privileged, although neuroimmune discoveries in recent years have challenged that idea (Louveau et al. 2015). Skin grafting experiments from the 1940s where the rabbit brain parenchyma was found to be immunologically tolerant to allogenic skin grafts served as the foundational evidence for the brain being a unique immunological site (Medawar 1948). The CNS' residence behind the physical protection of the blood-brain barrier supports this claim; in a healthy adult, both potential pathogens and circulating immune cells are generally blocked from entry by a tight system of endothelial cells, pericytes, and astrocytic endfeet (Daneman and Prat 2015). Despite this, there are still a variety of immunocompetent cells, both peripheral and resident, that are present and play active roles in building the brain. In the absence of immune challenge, very sparse numbers of lymphocytes, such as T cells and B cells, and other myeloid cells, such as dendritic cells and mast cells, can be found in the rodent brain during embryonic development (Tanabe and Yamashita 2018; Lenz et al. 2018; Bulloch et al. 2008). They continue their residency postnatally, performing broad functions ranging from B1a cells supporting the proliferation of oligodendrocyte precursor cells to mast cells mediating the masculinization of sex-specific brain regions and behaviors (Tanabe and Yamashita 2018; Lenz et al. 2018).

Like other organs in the body, the brain has its own tissue-resident macrophage: microglia. Comprising approximately 80% of the immune cells in the adult rodent CNS, microglia are the primary immune sentinel of the brain (Mrdjen et al. 2018). Once thought to be "resting" in the absence of a threat to homeostasis, we now recognize microglia as highly active cells, constantly surveilling their environment for changes in homeostasis (Nimmerjahn et al. 2005; Davalos et al. 2005; Wake et al. 2009; Carrier et al. 2020). Microglia rapidly respond to injury or pathogens by changing their morphology, moving to sites of damage, phagocytosing debris, and releasing factors to escalate an immune response (Davalos et al. 2005; Nimmerjahn et al. 2005; Haynes et al. 2006; Wolf et al. 2017; Eyo et al. 2018; Galloway et al. 2019). Although long recognized as critical players in neurodegeneration and trauma, the specific evidence outlining the wide range of microglial contributions across development is relatively underexplored. Here we chose to focus on the roles of microglia across brain maturation, focusing in on three periods when the impact of immune perturbations is particularly strong, and in many cases, long-lasting: embryonic wiring, postnatal experience-dependent pruning, and circuit refinement during adolescence. Microglial activity during these three stages is both highly dynamic and developmentally critical, positioning these immune cells as major facilitators of disruption during key sensitive periods of brain development. The recognition of sex differences in both the specific timing of microglial functions and the mechanisms they employ across brain maturation is of critical importance. While not the focus of this review, we highlight key findings when relevant (see Bordt et al. 2020 and VanRyzin et al. 2020 for detailed reviews). Additionally, we focus primarily on the rodent literature, as this is where most of our mechanistic knowledge comes from, and refer to human evidence where available.

2 Embryonic Wiring

Both their keen surveillance abilities and non-CNS origins make microglia particularly sensitive to perturbations during gestation. Unlike the other glia, oligodendrocytes and astrocytes, which are born from neural stem cells beginning around mouse embryonic day (E)16, most microglia colonize the brain beginning around E9 after traveling there from the yolk sac, either directly through the blood or through the fetal liver and aorta-gonad-mesonephros (AGM) (Ginhoux et al. 2010; Hoeffel and Ginhoux 2015). This ontogeny is almost entirely unique to brain-resident macrophages, both in rodent and human embryos, as other macrophages throughout the body do not share it (Sheng et al. 2015; Bian et al. 2020). The necessary journey of embryonic microglia to the brain, as well as their early residence prior to the closing of the blood-brain barrier at E13.5, leaves microglia vulnerable to react to signals in the periphery (Profaci et al. 2020). In the absence of disease or injury, once the blood-brain barrier closes at E13.5, the microglia that originally seeded the brain will expand and self-renew throughout the lifespan of the organism, suggesting that even small changes to these early microglia could profoundly impact the future brain (Askew and Gomez-Nicola 2018; Thion and Garel 2020). There is also evidence that another, smaller subpopulation of microglia, homeobox (Hoxb8)-lineage microglia, colonize the brain a few days later, traveling from the yolk sac and through the fetal liver and AGM to appear in the brain around E12.5 (Chen et al. 2010; De et al. 2018). Interestingly, loss of Hoxb8+ microglia resulted in an overall reduction of microglia in the brain and a pathological over-grooming phenotype (Chen et al. 2010). While little is known about microglial heterogeneity and its relevance in health and disease, the identification of a developmentally and phenotypically distinct microglial population strongly supports that different subsets of microglia may be programmed to perform specific functions from very early in life and requires further investigation (see Box 1).

Box 1 Microglial Heterogeneity in Development

Microglia are specialized cells that continually sense, interpret, and react to environmental cues. Therefore, the traditional view that microglial functional heterogeneity, or distinct microglial "subtypes," would develop from signals in their surroundings is logical. In the context of disease, microglia display distinct transcriptional patterns that seem to be in response to extrinsic cues (Keren-Shaul et al. 2017; Mrdjen et al. 2018). However, evidence is mounting that microglial heterogeneity may also have some intrinsic predetermination with great relevance for brain development. Hoxb8-lineage microglia display unique early postnatal dynamics relative to non-Hoxb8 cells, and the loss of the Hoxb8+ population results in an overall decrease in microglia in the brain (Chen et al. 2010; De et al. 2018). This is in spite of the fact that both populations have very similar transcriptional profiles (Chen et al. 2010; De et al. 2018). Studies investigating the two ligands for microglial CSF1R reveal a similar requirement for both populations; loss of CSF1 resulted in white-matter microglial density and functional reduction, while IL-34 loss impacted microglial density and function mostly in gray-matter regions (Kondo and Duncan 2009; Wang et al. 2012). Additionally, microglia populate the brain through self-renewal after traveling there during healthy embryonic development, so one study sought to determine how microglia are transcriptionally and epigenetically synchronized to perform all their necessary functions from these original seeding cells. Results from chromatin accessibility and histone modification investigation revealed specific hallmarks across three phases: early (until

(continued)

Box 1 (continued)
E14), pre- (until P0), and adult microglia (Matcovitch-Natan et al. 2016). A later single-cell transcriptomic study that sequenced mouse microglia across development revealed nine distinct subtypes of microglia from E14.5 to adulthood (Hammond et al. 2019). Interestingly, there was the greatest diversity of microglial subtypes in the earliest stages of development, E14.5 and P5 (Hammond et al. 2019). These subtypes were confirmed by localization of distinct microglial populations to particular brain regions, illustrated by single-molecule fluorescent in situ hybridization (smFISH) experiments (Hammond et al. 2019). Together, these studies support that microglial heterogeneity exists in early brain development, although the consequences of disrupting subpopulations of microglia are mostly still unknown. Recent work characterizing the impacts of a combined environmental stress MIA model on the early postnatal anterior cingulate cortex of male offspring found that MIA induced greater microglial heterogeneity, measured by protein expression, compared to controls (Block et al. 2020). Strikingly, the different populations of microglia had varied phagocytic capacity, and therefore their unbalanced presence resulted in a loss of characteristic early postnatal synaptic elimination (Block et al. 2020). This study provides some of the first evidence that microglial heterogeneity changes, such as downregulation of a critical microglial subtype and its functions, may contribute to neurodevelopmental disorder etiology. These and other outstanding questions about microglial heterogeneity highlight the need for more precise tools to modulate microglial function. Promising new work utilizing a Cre-dependent DREADD (designer receptors exclusively activated by designer drugs) system demonstrated that microglial activity can be bi-directionally modulated in a specific brain region through chemogenetic manipulation (Klawonn et al. 2021). This is especially important because we know that microglia in different brain regions vary in a number of ways, including their gene expression patterns, homeostatic signals, phagocytic function, density, physical shape, and, particularly relevant to this discussion, the timing of their colonization, expansion, and maturation (Tan et al. 2020). The ability to change the functional properties of microglia in a region-specific manner opens many doors for future work characterizing microglial heterogeneity.

As immune challenges during this perinatal period fundamentally alter neurobiology for the remainder of the lifespan (for example, in cases of MIA), it could be theorized that dampening microglial function during embryonic development may help prevent damage. However, mice lacking colony-stimulated factor 1 receptor (CSF1R), a receptor necessary for microglial growth and survival, have no microglia and die shortly after birth (Nandi et al. 2012; Erblich et al. 2011; Chitu et al. 2016). $CSF1R^{-/-}$ mice also display anatomical deformities, such as reduced brain size, cortical thickness, and corpus callosum axonal crossing, suggesting that microglia are indispensable for embryonic brain development (Nandi et al. 2012; Erblich et al.

2011; Chitu et al. 2016). Recent examination of the first identified human patients with homozygous CSF1R mutations, individuals almost entirely without microglia, confirms their necessity. One living patient with a homozygous missense mutation in the CSF1R gene did not show severe developmental regression until 12 years old, while an individual with a more severe splice mutation in the CSF1R gene lived to only 10 months old (Oosterhof et al. 2019). Robust white matter irregularities, epilepsy, and incorrect accumulations of neurons in both these patients' brains support that microglia are of particular importance for both the development of oligodendrocytes and myelination, as well as proper interneuron migration in the cortex (Oosterhof et al. 2019). Interestingly, interneuron dysfunction and white matter abnormalities are two of the primary neuropathological hallmarks of schizophrenia, identified in both rodent models and human studies (Lewis et al. 2012; Kochunov and Elliot Hong 2014; Gonzalez-Burgos et al. 2015; Cetin-Karayumak et al. 2020). Together, paired with the strong evidence of MIA risk factors, these studies support immune involvement in the building of the embryonic brain, specifically through some action of microglia.

Once migrated from the yolk sac, microglia localization in the developing embryonic brain is precisely timed. For example, at E12.5, around when microglia are first colonizing the mouse brain, they distribute fairly evenly throughout the forebrain (Squarzoni et al. 2014). However, at embryonic day E14.5, microglia become localized around migrating axonal tract "decision" regions, avoiding the cortical plate until E16.5 (Squarzoni et al. 2014). Around birth, microglia become most associated with progenitor zones and areas of cell death in both rodents and non-human primates (Cunningham et al. 2013). This localization aligns with the critical roles microglia play in embryonic brain development. Using two different microglial depletion methods, Squarzoni and colleagues found that functional microglia are necessary for proper dopaminergic axon extension at E14.5 in mice, and in the absence of microglia, there was an overgrowth of tyrosine-hydroxylase (TH) + axons into the subpallium (Squarzoni et al. 2014). In contrast, an E13.5 lipopolysaccharide (LPS) injection MIA model to "activate" microglia reduced dopaminergic axon outgrowth at E14.5 to below control levels (Squarzoni et al. 2014). The discovery of TH+ fragments of dopaminergic axons inside of microglia in control mice provided further evidence that perturbing microglia in this window impacts their ability to direct, perhaps through phagocytosis, dopaminergic circuit organization during embryonic development (Squarzoni et al. 2014). Beyond this, both depletion and MIA models impacted the organization of neocortical inhibitory neurons, resulting in both altered migration timing and ultimately defective localization in the somatosensory cortex (Squarzoni et al. 2014).

Microglial involvement in the development of inhibitory neurons has long been suspected based on dysfunction consistently identified in both cell types across several neuropsychiatric disorders, in particular those with proposed early developmental origins (Thion and Garel 2020). Both postmortem analysis and live imaging of brains of human patients suffering from schizophrenia point towards aberrant GABAergic interneurons in psychopathology, such as through reduced gene expression of general GABAergic genes in the frontal cortices of schizophrenia patients compared to controls (Nakazawa et al. 2012; Hoftman et al. 2015; de Jonge et al. 2017). Administration of poly I:C, a synthetic double-stranded RNA to mimic viral infection, to mouse dams at E9 to induce MIA reduced parvalbumin (PV) interneuron GABAergic transmission onto pyramidal neurons in brain regions important for schizophrenia pathology in offspring, consistent with human postmortem literature (Canetta et al. 2016; Kaar et al. 2019). In 2019, Thion et al. dissected the embryonic impact of both E13.5 LPS injection and transient macrophage depletion (from E7-birth) on PV interneurons in the somatosensory barrel cortex. Both microglial perturbations revealed a two-phase alteration to PV firing (Thion et al. 2019). In juveniles (P20), there was a mild increase in PV density and an excess of inhibition onto principal excitatory neurons in the barrel cortex in both MIA and depletion offspring. By adulthood, however, the increase in interneuron density was resolved and PV cells were instead hypoactive, exhibiting reduced inhibition onto layer 4 cortical neurons (Thion et al. 2019). These findings are particularly interesting for two reasons. First, although inhibitory dysfunction is a hallmark of neurodevelopmental disorders, this is the first work to directly implicate developmental alterations in microglial activity to a change in inhibitory firing. Second, due to the consistent phenotype across both depletion and MIA, this finding supports the idea that the effect of MIA on microglia may not just be excessive reactivity, but instead a "distraction" from the critical developmental processes performed by microglia during specific periods. These findings also lend support to the hypothesis that the nature of the microglial perturbation may be less important than the specific period during which it occurs, as it is the timing of the challenge that reveals which microglial function and, therefore, which critical aspect of development, is altered.

Microglia are critical for regulating both the precursor cell pool and the functional capabilities of neurons that survive in both rodents and primate species. During the later half of cortical neurogenesis, E80-E100 in a rhesus macaque or E20-P0 in a rat, microglia eliminate excess neural and glial precursor cells through a phagocytosis mechanism in the proliferative zones of the brain (Cunningham et al. 2013). LPS injections administered to pregnant rats on E15 and E16 resulted in a significant reduction in proliferative zone neurons in offspring at E19 and P2 (Cunningham et al. 2013). Interestingly, concurrent exposure of rat offspring to MIA and doxycycline to suppress microglial activation resulted in a persistent increase in neural precursor cells (Cunningham et al. 2013). These findings suggest that conditions that may decrease microglial activation prenatally may result in the long-term survival of excess neurons at a detriment to the developing brain.

Together, these studies highlight the long-term impacts of microglial perturbations during the gestational period, as either dampening or "over-activating" microglial function both result in negative outcomes for developing neurons. Interestingly, these studies do not directly address any sex differences that may emerge in the response to either kind of microglial perturbation. This is despite the strong male sex-bias in disorders that we suspect become programmed during this period, as well as the male-specific androgen surge around P0 and subsequent critical period for sexual differentiation (McCarthy et al. 2018; VanRyzin et al. 2018; Bordt et al. 2020). Additionally, many questions remain about how and why some neuronal populations are differentially impacted by microglial dysfunction. What external cues or intrinsic factors direct microglia to populate certain zones during specific time windows? How do embryonic microglia differentiate between specific subtypes of neurons, such as in their preferential directing of dopaminergic axons or diverse impacts on excitatory- vs inhibitory-fated neuronal populations? These questions highlight just how much there is left to uncover about the heterogeneous roles that microglia play in shaping the embryonic brain.

3 Postnatal Synaptic Pruning and Neuronal Function

The discovery that microglia are constantly mobile and surveying their environment during homeostasis raises questions regarding what kinds of functions they must be performing to justify such a heavy metabolic need (Nimmerjahn et al. 2005; Davalos et al. 2005; Wake et al. 2009). We now know microglia to be attentive gardeners, carefully pruning damage and overgrowth across development in order to shape and encourage healthy and resilient maturation. More than a decade ago, the first evidence that microglial contact with synapses was indeed experience-driven was identified in the primary visual cortex (V1) by modulating light exposure during the visual critical period (Tremblay et al. 2010). Two-photon in vivo imaging revealed that following 6 days of dark adaptation, microglia were less mobile, had more process phagocytic structures, and changed their preference for contacting specific subsets of dendritic elements (Tremblay et al. 2010). Some of these changes could be partially reversed by 2 days of light re-exposure (Tremblay et al. 2010). Soon after, the role of microglia in the engulfment of synaptic material was supported by a study where chemokine fractalkine receptor (CX3CR1), primarily expressed by microglia in the CNS, was knocked out in mice. In the absence of this receptor, microglia still engulfed PSD-95+ puncta in early postnatal synapse refinement, but the overall density of synaptic material and spines was significantly greater in the knockouts at P15 in the hippocampus (Paolicelli et al. 2011). Additional experiments revealed that these extra connections were electrophysiologically immature, strongly pointing to a deficit in synaptic pruning by microglia during this period (Paolicelli et al. 2011). By the time of this discovery, the involvement of the classical complement cascade in developmental pruning of synapses in the dorsal lateral geniculate nucleus (dLGN) had been established (Stevens et al. 2007). Early in development, this region of the thalamus receives overlapping projections from the retinal ganglion cells of both eyes. Eventually the mature pattern of non-overlapping inputs develops following activity-dependent selection of projections from a singular eye, making it a great system to study synapse elimination (Shatz 1990). The complement component 1q (C1q) protein was found to be present on synaptic elements early postnatally, and knockout of C1q resulted in deficient elimination of overlapping inputs from both eyes in the normally developmentally-segregated input areas at P30 (Stevens et al. 2007). Again utilizing the early postnatal developing mouse dLGN system, Schafer and colleagues found that microglia prune presynaptic inputs by engulfment in a complement receptor 3 (CR3)/C3 dependent manner (Schafer et al. 2012). Modulation of neural activity by intraocular injection of tetrodotoxin to weaken inputs from one eye resulted in an increase in microglial phagocytosis of those inputs, while forskolin injection to enhance activity resulted in greater pruning of inputs from the other eye (Schafer et al. 2012). This was one of the first hints that microglia may work to shape circuits in health and disease, and while our understanding of these processes has grown immensely, some of the questions raised almost a decade ago still remain unanswered. For example, do microglia detect the activity levels of nearby neurons in order to "decide" which synaptic elements require removal? Could microglia instead be responding to inputs that have already been identified through a different, unknown mechanism for elimination?

Although the extent of microglial involvement in postnatal developmental synaptic pruning remains unknown, several potential pathways have been identified. We now recognize that the innate immune triggering receptor expressed on myeloid cells 2 (TREM2) is a necessary part of the synaptic pruning process in mice. When expressed on microglia, TREM2 primarily recognizes LPS and other danger molecules and generally performs anti-inflammatory functions (Daws et al. 2003; Wang et al. 2015). TREM2$^{-/-}$ mice showed a CA1-specific decrease in microglia density and activation markers, as well as an increase in excitatory synaptic markers and electrophysiological activity from P18–20 (Filipello et al. 2018). Further, both in vitro and in vivo at P18–20, TREM2$^{-/-}$ microglia were insufficient at phagocytosing synapses, strengthening the finding that these mutants had reduced brain region connectivity, increased repetitive behaviors, and social behavior deficits at P90 (Filipello et al. 2018). Interestingly, a separate study found that TREM2$^{-/-}$ animals had a reduction in synapses in both the cortex and hippocampus at 1 month of age and that KO microglia contained *more* internalized synaptic elements (Jay et al. 2019). Beyond this, loss of TREM2 in microglia seemed to enhance synapse element phagocytosis by astrocytes in brain regions with reduced microglial density (Jay et al. 2019). This finding is not entirely surprising; astrocytes, while only briefly discussed here, play an extremely important role in synaptic construction and maintenance (see Box 2). While the exact contribution of microglial TREM2 in postnatal synaptic refinement is still up for debate, these studies highlight microglia as complicated, interconnected players in the developing brain environment.

Box 2 Astrocytic Involvement in Synaptic Development

Although not from immune cell origins and therefore highly abundant in the CNS, astrocytes are necessary immunocompetent contributors to circuit construction and refinement. Astrocytes are intimately involved in synaptic transmission, where one rodent astrocyte can contact over 100,000 synaptic elements to monitor and modulate signaling, both by releasing their own factors and through regulation of those released into the synaptic cleft (Bushong et al. 2002; Chung et al. 2015). Therefore, it is unsurprising that astrocytes are also

(continued)

Box 2 (continued)
intimately involved in the creation and modulation of synapses across development (Farhy-Tselnicker and Allen 2018). Astrocyte-derived signals, such as thrombospondins, glypicans, and Hevin/Sparcl1, contribute to the proper construction of a synapse, and in some cases, specific signals are necessary for the development of certain kinds of synapses (Risher et al. 2014; Allen and Eroglu 2017). Additionally, evidence that supports the importance of cross-talk between astrocytes and microglia in circuit refinement continues to grow. For example, the presence of astrocytes is necessary for retinal ganglion cells (RGCs) to upregulate C1q, even though it is microglia that respond to the complement call (Stevens et al. 2007; Schafer et al. 2012). More recently, this is highlighted by findings such as the identification of astrocyte-produced cytokine interleukin (IL)-33 as important for modulation of microglial phagocytosis during the early postnatal period (Vainchtein et al. 2018).

Microglia are exposed to and need to interpret many signals from other cells in their environment. For example, neuron-derived CX3CL1 signals through microglial CX3CR1. In order to effectively target the correct synaptic elements for removal, microglia also recognize both "don't eat me" signals, such as CD47, as well as contrasting "eat me" signals (Arcuri et al. 2017; Lehrman et al. 2018). One such recently identified signal is exposed phosphatidylserine (ePS), which is displayed by neurons as an "eat me" signal for microglia to recognize (Scott-Hewitt et al. 2020). Across brain regions, ePS is most present during critical windows of pruning, and in the dLGN this expression is co-localized with C1q labeling (Scott-Hewitt et al. 2020). In vitro, synaptic pruning of ePS by microglia was TREM2-dependent, while in vitro, in C1q-KO mice, there was both an increase in ePS tagged presynaptic elements and a reduction of microglial engulfment without C1q present (Scott-Hewitt et al. 2020). This study highlights one neuronal "tag" that initiates two different microglial pruning mechanisms. Interestingly, astrocytes can also recognize phosphatidylserine through phagocytic receptors MERTK and MEGF10 to eliminate synapses in the dLGN in an activity-dependent manner, both during the first week of life and throughout adulthood (Chung et al. 2013). This alternate phagocytosis pathway is not unique to mammals, suggesting it may have some important function independent of those in microglia (MacDonald et al. 2006; Chung et al. 2013). Further, microglia can also signal through the cannabinoid system. Van Ryzin and colleagues discovered that microglia phagocytose newborn cells in the rat amygdala in the first few days of life, a process that is regulated by androgen and endocannabinoids (VanRyzin et al. 2019). When engulfment of these cells, identified as primarily newborn astrocytes, by microglia is blocked through the administration of an anti-CD11b antibody, dramatic sex differences that are well-characterized in juvenile rat play behaviors are lost (VanRyzin et al. 2019). Even though microglial engulfment of whole cells early in brain development is not new or specific to rodents, this study was one of the first studies to directly implicate microglia in the development of social behavior (Cunningham et al. 2013; VanRyzin et al. 2019).

Recently, a new molecular pathway utilized by microglia to decrease synapses in the absence of phagocytosis was proposed. In 2018, Cheadle et al. discovered that following eye-opening at P7, fibroblast growth factor-inducible 14 (Fn14), a member of the tumor necrosis factor receptor family, was selectively expressed in excitatory thalamocortical neurons (TC) in the dLGN (Cheadle et al. 2018). Its presence was important for the maturation of TC neuron dendritic spines and synaptic refinement during the vision-sensitive period (Cheadle et al. 2020). Tumor necrosis factor-like weak inducer of apoptosis (TWEAK), which is strongly upregulated by microglia only following visual experience, binds postsynaptic Fn14 and contributes to spine elimination. This elimination is not phagocytosis-mediated, as microglia in TWEAK-KO mice do not show phagocytosis deficits at P7 or P27, but instead depends on the proximity of spines to TWEAK-expressing microglia (Cheadle et al. 2020). Interestingly, this novel refinement mechanism is only present after P20, prompting the authors to suggest a 2-phase model. In phase 1, microglia work via complement signaling to engulf presynapses. After eye-opening and sensory experience, there is a transition to phase 2, where postsynaptic refinement happens through the experience-dependent TWEAK-Fn14 mechanism (Cheadle et al. 2020). Perhaps showcasing exactly how much we have left to learn, the mechanisms by which microglia perform this second phase of elimination, as well as whether these are the same microglia that perform phase 1 engulfment or if this is a result of heterogeneity, is still unknown.

What are the long-term consequences for adult behavior when these microglia functions early in postnatal life are perturbed by an immune challenge? Genetic tools have allowed us to look closely at the loss of functional pathways to elucidate their critical functions, but may not elucidate broader disease mechanisms. Despite the clinical relevance of understanding the impacts of early postnatal/neonatal infection on brain development, such as the prevalence of postnatal illness and the potential for increased vulnerability to neuropsychiatric disorders, the animal studies of this nature are limited in comparison to the MIA literature. Those studies that have looked show that early postnatal immune challenges can have long-term consequences on adult behavior (Bilbo and Schwarz 2012). Interestingly, a "second-hit," such as a subsequent immune challenge, seems to be consistently required in order to reveal these long-term changes. For example, administering *E. coli* to rats at P4 primes the system such that if they are exposed to another immune challenge in adulthood, their long-term memory becomes impaired (Bilbo et al. 2006). This effect is mediated by a CD11b + −derived IL-1β response, which becomes sensitized following the initial immune "hit" (Williamson et al. 2011). Indeed, even two LPS injections early in life, administered at P3 and P5 to male pups, are enough to induce spatial memory deficits in adulthood, measured by Morris water maze performance (Peng et al. 2019). Other studies highlight that the second hit in this two-hit model does not need to be a direct immune challenge to the pups. The long-term impact of in utero exposure to maternal high fructose diet was revealed following a P7 LPS injection in rats, where the authors observed aberrant anxiety-like behaviors both during adolescence and adulthood (Bukhari et al. 2018). Interestingly, recent work has shown that a single P4 LPS induced a female-specific decrease in both sociability and social discrimination, which was maintained even in the absence of microglial MyD88-dependent pro-inflammatory signaling (Smith et al. 2020). This finding provides strong support for the investigation of both males and females across brain development, as it seems some microglial mechanisms are not necessarily utilized by both sexes.

Studies such as these also serve as a reminder that changes at single-cell levels in spine density or synaptic refinement may have long-term behavioral consequences despite those perturbations happening early in development. However, immune system perturbations happen throughout the lifespan. From birth to old age, we all suffer illness and stress, which we know impact microglial function (Niraula et al. 2017; Mariani and Kielian 2009). However, not all microglial challenges result in long-term functional changes in our behavior. For example, depletion of microglia in the CNS in adulthood via an antagonist to CSF1R, the loss of which is lethal in development, has no impact on either viability or behavioral function in adult mice (Elmore et al. 2014). This supports the view that there are sensitive periods when microglia are particularly vulnerable to challenges, perhaps due to special windows of rigorous and concentrated actions. Beyond the obvious highly active periods of brain construction detailed above, the literature suggests there is another sensitive window for microglia that extends beyond the initial weeks of life: adolescence.

4 Adolescent Circuit Refinement

The adolescent period, usually defined as around 10–20 years of age in humans, is behaviorally characterized by increased risk-taking behaviors and sensitivity to social cues and stress (Andersen and Teicher 2008; Steinberg 2008; Blakemore and Mills 2014). These changes are neuroanatomically mirrored by increased myelination and synaptic pruning in cortical and limbic areas (Giedd et al. 1999; Paus et al. 2008; Tamnes et al. 2013; Blakemore and Mills 2014). Adolescence is also the time when connectivity between diverse brain regions is refined and strengthened, resulting in changes in synaptic communication (Toga et al. 2006; Whitford et al. 2007). Analogous behavior and brain maturation also occurs in rodents, where adolescence is considered to be from around the time of weaning until sexual maturity, although the precise window is debated and varies by sex (Schneider 2013). The significance of the adolescent period as an important developmental stage is underscored by its identification as the primary age window for the appearance of schizophrenia, some affective disorders, and risk-taking behaviors that increase vulnerability to substance use disorders (Kessler et al. 2007; Paus et al. 2008; Steinberg 2008; Casey and Jones 2010). While the neuroimmune contribution to each of these disorders is not new, the literature investigating the homeostatic functions of microglia during the adolescent period of circuit refinement is still in its infancy.

Despite the relatively young body of adolescent microglia literature, it is clear that the actions of microglia across developmental periods have profound impacts on the brain and behavior during adolescence. Again, the long history of MIA studies reveal that embryonic perturbations of microglial function can culminate in eventual disease onset in adolescence (Brown 2012). As discussed in the previous section, perinatal perturbations of microglia can also profoundly impact the adolescent brain and behavior (VanRyzin et al. 2019). While important to our understanding of microglial roles in circuit refinement and the long-lasting effects of early perturbations of these processes, what these studies involving early life manipulations do not tell us is what may be different about microglia and their functions during the actual adolescent period. In the few transcriptional studies aiming to characterize microglial-specific gene expression changes across development, the adolescent period is entirely excluded (Matcovitch-Natan et al. 2016; Hanamsagar et al. 2017).

Across species, perturbations *during* the sensitive adolescent period have profound impacts on long-term behavior. It is well recognized that the brain at this time is highly neurobiologically vulnerable to addiction and social stress in humans (Chambers et al. 2003; Crews et al. 2007; Blakemore and Mills 2014). In rats, exposure to morphine during adolescence persistently impacted microglial TLR4 signaling and increased susceptibility to robust relapse of drug-seeking behavior later in life (Schwarz et al. 2011). This was not true for rats that were exposed to the same initial morphine paradigm as adults (Schwarz et al. 2011). Similarly, rodents that are exposed to chronic stressors as adolescents have increased anxiety-like behaviors weeks after exposure compared to those that went through the same stressors as adults (Yohn and Blendy 2017; Cotella et al. 2019). This is also true in the case of an acute, intense stressor, though this effect was specific to males (Lovelock and Deak 2019). Stress-based immune-activating stimuli during this period not only change behavior, but also have chronic effects on microglial function and activation (Schwarz and Bilbo 2013; McClain et al. 2011). Interestingly, administration of minocycline, a microglial activation inhibitor, can prevent the development of schizophrenia-like behaviors following an adolescent stressor in mice, providing more support for the critical role of microglia in both the initial wiring and the refinement of circuits up through the adolescent period (S. Giovanoli et al. 2016).

What are microglia actually doing during this time in the brain? Just as synaptic connections are refined early postnatally, pruning continues into rodent and human adolescence in brain structures known to continue to mature later in life, such as the prefrontal cortex (Petanjek et al. 2011; Delevich et al. 2018). While the quantification of this phenomena in the healthy brain is well observed and fairly consistent, only in 2019 were microglia first directly identified as a mechanism through which this reduction in synapses occurs during this period (Premachandran et al. 2020; Markham et al. 2013; Koss et al. 2014; Drzewiecki et al. 2016). Four days following the peak density of spines at P35 in the rat PFC, Mallya and colleagues observed a drastic increase in microglial engulfment of both dendritic spines and presynaptic elements. Interestingly, at P50, while engulfment of spines dropped below levels observed at P24, microglial engulfment of presynaptic elements continued to climb (Mallya et al. 2019). While the precise timing of these observations suggests microglia are at least partially responsible for the pruning of synapses in the PFC in adolescence, whether or not microglia are necessary for this process and the functional implications of their dysfunction remains unanswered.

Microglia appear to perform a different critical function in the adolescent development of reward circuit pathways, which are of particular importance in the context of addiction. In the nucleus accumbens (NAc), dopamine receptors that receive dopaminergic input from the ventral tegmental area drive social play behaviors in adolescent rats (Manduca et al. 2016). In 2018, Kopec and colleagues found that the volume of dopamine D1 receptors (D1rs) in the NAc across adolescence is regulated by complement-dependent microglial phagocytosis in male rats, but not in females (Kopec et al. 2018). Blocking C3 receptors through lateralized NAc injection of a CD11b-subunit competitive antagonist (neutrophil inhibitor factor, NIF) around the peak of D1r expression only impacted male D1r expression (Kopec et al. 2018). Natural male juvenile play behavior peaks at the same time that D1rs peak in the NAc (P30), so a similar paradigm of NIF administration revealed that a decrease in microglial phagocytosis around the D1r peak prolonged male play behaviors beyond their normal timeline (Kopec et al. 2018). Administration of siRNA against D1r alongside NIF confirmed that these changes in social play were D1r-dependent, as the effect of NIF alone was blocked by the siRNA administration (Kopec et al. 2018). Interestingly, the timeline for female rat D1r volume peaks earlier than in males and did not seem to be associated with complement C3, once again showcasing the sex differences in microglial developmental mechanisms (Kopec et al. 2018; Smith et al. 2020).

A number of studies utilizing a pubertal LPS challenge provide evidence that adolescent perturbations have particularly sex-dependent effects. In CD1 mice, administration of an LPS injection at P42 is sufficient to impair learning in a Barnes maze task in both males and females (Kolmogorova et al. 2019). However, the same LPS administration at P42 impacts anxiety-like, depression-like, and sexual behaviors in a female-specific manner long-term (Ismail et al. 2011; Olesen et al. 2011; Ismail et al. 2013). Given that adolescent-emerging affective disorders such as anxiety and depression are female-biased, this female susceptibility to immune challenge hints at an immune-relevant mechanism, although the role of microglia is underexplored (Bekhbat and Neigh 2018). Further, the requirement of a second hit, discussed previously, is not unique to early postnatal challenges. There is a long history to the "two-hit hypothesis" in the context of adolescent-onset schizophrenia, where it was postulated that a genetic or environmental first hit early in brain development primes the system to synergize with a later insult to elicit schizophrenia onset (Bayer et al. 1999; Maynard et al. 2001). Indeed, this can also be modeled in mice, where a study showed a low dose of poly I:C on E9 was not sufficient to induce schizophrenia-relevant behavioral deficits in adult offspring until paired with a chronic variable stress paradigm from P30–40 (Giovanoli et al. 2013). Taken together, it is clear that our understanding of the role of microglia and the immune system during the adolescent period requires more attention.

5 Conclusions

Across each stage of building the brain, microglia perform functions that are critical to the long-term successful functioning of neuronal circuitry and ultimately behavior (Fig. 1). The specific timing of a perturbation, perhaps more so than the nature of it,

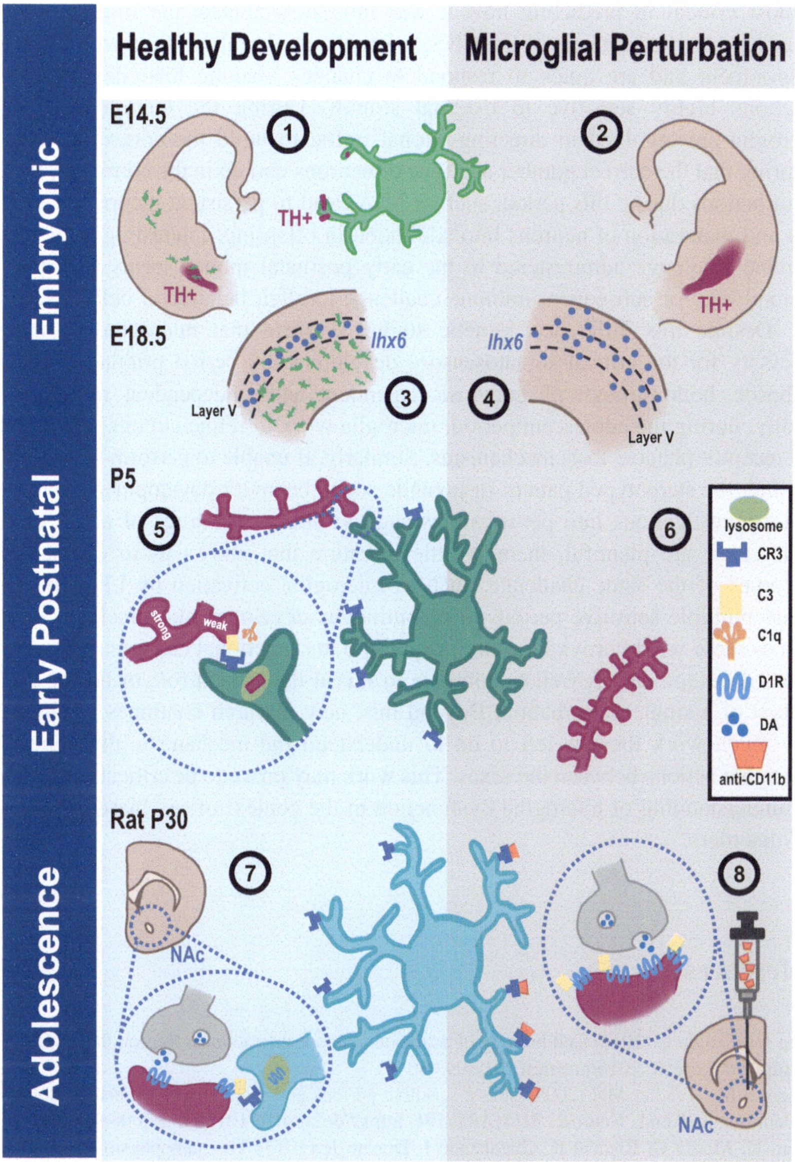

Fig. 1 Embryonic microglia (green) localize to "hotspots" at E14.5 and engulf TH+ dopaminergic fragments (magenta) to modulate and direct axon outgrowth (1). Both anti-CSF1R antibody administration and transgenic myeloid cell to deplete microglia result in an overgrowth of TH+ axons at E14.5 (2). In situ hybridization for *lhx6* mRNA expressed in interneurons (blue) at E18.5 show that in the absence of microglia, specific laminar positioning in cortical layer V is lost (3, 4). In the early postnatal period, microglia (teal) recognize synaptic elements with weak inputs and phagocytose them in the complement-dependent manner (5). Loss of C1q, the complement cascade initiator, C3, or its receptor, CR3, results in deficient synaptic remodeling (6). Microglia engulf dopamine-1 receptors (D1Rs) in the adolescent rat nucleus accumbens (NAc) (7). After administration of a CD11b competitive antagonist to block microglia CR3 activation, there is a greater

is most critical in predicting how it will ultimately impact the function of the organism. As the sentinels of the CNS, microglia are highly in tune with the brain environment and are quick to respond to changes, making their developmental functions highly sensitive to external stimuli. During the embryonic period, microglia are involved in directing axonal traffic to build the cortex, as well as ensuring that the correct number and type of neurons end up in the correct locations. Perturbations during this period, such as MIA, lead to persistent incorrect localization and maturation of neurons into adulthood in offspring. Interestingly, the same immune challenge administered to the early postnatal mouse seems to require a "second hit," or subsequent immune challenge to elicit behavioral deficits later in life. Despite this difference, genetic studies confirm that microglia are indeed necessary for the critical somatosensory developmental period pruning of excess synapses, both through phagocytosis-dependent and -independent mechanisms. Finally, during the adolescent period, microglia work to refine circuits via synaptic and receptor phagocytosis mechanisms. Similarly, if unable to perform this receptor pruning, the stereotyped pattern of juvenile social behavior development is altered. While investigations into perturbations such as stress and drugs of abuse during adolescence are plentiful, there is little literature that allows us to compare the outcomes of the same challenge, such as microglial activation by LPS injection, across multiple sensitive periods in a continuous developmental timeline. Studies such as these would provide further insight into the functional changes of microglia across development, as well as continue to reveal the critical role of timing in the context of a single perturbation. Beyond this, new research continues to highlight how much work there is left to do to understand the mechanistic differences in microglial actions between the sexes. This work may prove to be critical to bettering our understanding of microglial dysfunction in the context of sex-biased neurological disorders.

References

Allen NJ, Eroglu C (2017) Cell biology of astrocyte-synapse interactions. Neuron 96(3):697–708. https://doi.org/10.1016/j.neuron.2017.09.056

Andersen SL, Teicher MH (2008) Stress, sensitive periods and maturational events in adolescent depression. Trends Neurosci 31(4):183–191. https://doi.org/10.1016/j.tins.2008.01.004

Arcuri C, Mecca C, Bianchi R, Giambanco I, Donato R (2017) The pathophysiological role of microglia in dynamic surveillance, phagocytosis and structural remodeling of the developing CNS. Front Mol Neurosci 10. https://doi.org/10.3389/fnmol.2017.00191

Askew K, Gomez-Nicola D (2018) A story of birth and death: insights into the formation and dynamics of the microglial population. Brain Behav Immun 69:9–17. https://doi.org/10.1016/j.bbi.2017.03.009

Fig. 1 (continued) density of D1R (8). Data from Squarzoni et al. (2014), Stevens et al. (2007), Schafer et al. (2012), Kopec et al. (2018)

Atladóttir HO, Thorsen P, Østergaard L, Schendel DE, Lemcke S, Abdallah M, Parner ET (2010) Maternal infection requiring hospitalization during pregnancy and autism spectrum disorders. J Autism Dev Disord 40(12):1423–1430. https://doi.org/10.1007/s10803-010-1006-y

Banks WA, Erickson MA (2010) The blood–brain barrier and immune function and dysfunction. Neurobiol Dis 37(1):26–32. https://doi.org/10.1016/j.nbd.2009.07.031

Bayer TA, Falkai P, Maier W (1999) Genetic and non-genetic vulnerability factors in schizophrenia: the basis of the 'two hit hypothesis'. J Psychiatr Res 33(6):543–548. https://doi.org/10.1016/s0022-3956(99)00039-4

Bekhbat M, Neigh GN (2018) Sex differences in the neuro-immune consequences of stress: focus on depression and anxiety. Brain Behav Immun 67:1–12. https://doi.org/10.1016/j.bbi.2017.02.006

Bian Z, Gong Y, Huang T, Lee CZW, Bian L, Bai Z, Shi H et al (2020) Deciphering human macrophage development at single-cell resolution. Nature 582(7813):571–576. https://doi.org/10.1038/s41586-020-2316-7

Bilbo SD, Schwarz JM (2012) The immune system and developmental programming of brain and behavior. Front Neuroendocrinol 33(3):267–286. https://doi.org/10.1016/j.yfrne.2012.08.006

Bilbo SD, Rudy JW, Watkins LR, Maier SF (2006) A Behavioural characterization of neonatal infection-facilitated memory impairment in adult rats. Behav Brain Res 169(1):39–47. https://doi.org/10.1016/j.bbr.2005.12.002

Blakemore S-J, Mills KL (2014) Is adolescence a sensitive period for sociocultural processing? Annu Rev Psychol 65(1):187–207. https://doi.org/10.1146/annurev-psych-010213-115202

Block CL, Eroglu O, Mague SD, Sriworarat C, Blount C, Malacon KE, Beben KA et al (2020) Prenatal environmental stressors impair postnatal microglia function and adult behavior in males. BioRxiv. https://doi.org/10.1101/2020.10.15.336669

Bordt EA, Ceasrine AM, Bilbo SD (2020) Microglia and sexual differentiation of the developing brain: a focus on ontogeny and intrinsic factors. Glia 68(6):1085–1099. https://doi.org/10.1002/glia.23753

Brown AS (2012) Epidemiologic studies of exposure to prenatal infection and risk of schizophrenia and autism. Dev Neurobiol 72(10):1272–1276. https://doi.org/10.1002/dneu.22024

Bukhari SHF, Clark OE, Williamson LL (2018) Maternal high fructose diet and neonatal immune challenge alter offspring anxiety-like behavior and inflammation across the lifespan. Life Sci 197:114–121. https://doi.org/10.1016/j.lfs.2018.02.010

Bulloch K, Miller MM, Gal-Toth J, Milner TA, Gottfried-Blackmore A, Waters EM, Kaunzner UW et al (2008) CD11c/EYFP transgene illuminates a discrete network of dendritic cells within the embryonic, neonatal, adult, and injured mouse brain. J Comp Neurol 508(5):687–710. https://doi.org/10.1002/cne.21668

Bushong EA, Martone ME, Jones YZ, Ellisman MH (2002) Protoplasmic astrocytes in CA1 stratum radiatum occupy separate anatomical domains. J Neurosci 22(1):183–192. https://doi.org/10.1523/JNEUROSCI.22-01-00183.2002

Canetta S, Bolkan S, Padilla-Coreano N, Song LJ, Sahn R, Harrison NL, Gordon JA, Brown A, Kellendonk C (2016) Maternal immune activation leads to selective functional deficits in offspring parvalbumin interneurons. Mol Psychiatry 21(7):956–968. https://doi.org/10.1038/mp.2015.222

Carrier M, Robert M-È, Ibáñez FG, Desjardins M, Tremblay M-È (2020) Imaging the neuroimmune dynamics across space and time. Front Neurosci 14:903. https://doi.org/10.3389/fnins.2020.00903

Casey BJ, Jones RM (2010) Neurobiology of the adolescent brain and behavior: implications for substance use disorders. J Am Acad Child Adolesc Psychiatry 49(12):1189–1201. https://doi.org/10.1016/j.jaac.2010.08.017

Cetin-Karayumak S, Di Biase MA, Chunga N, Reid B, Somes N, Lyall AE, Kelly S et al (2020) White matter abnormalities across the lifespan of schizophrenia: a harmonized multi-site diffusion MRI study. Mol Psychiatry 25(12):3208–3219. https://doi.org/10.1038/s41380-019-0509-y

Chambers RA, Taylor JR, Potenza MN (2003) Developmental neurocircuitry of motivation in adolescence: a critical period of addiction vulnerability. Am J Psychiatry 160(6):1041–1052. https://doi.org/10.1176/appi.ajp.160.6.1041

Cheadle L, Tzeng CP, Kalish BT, Harmin DA, Rivera S, Ling E, Nagy MA, Hrvatin S, Hu L, Stroud H, Burkly LC, Chen C, Greenberg ME (2018) Visual experience-dependent expression of Fn14 is required for retinogeniculate refinement. Neuron 99(3):525–539.e10. https://doi.org/10.1016/j.neuron.2018.06.036

Cheadle L, Rivera SA, Phelps JS, Ennis KA, Stevens B, Burkly LC, Lee W-CA, Greenberg ME (2020) Sensory experience engages microglia to shape neural connectivity through a non-phagocytic mechanism. Neuron 108(3):451–468.e9. https://doi.org/10.1016/j.neuron.2020.08.002

Chen S-K, Tvrdik P, Peden E, Cho S, Wu S, Spangrude G, Capecchi MR (2010) Hematopoietic origin of pathological grooming in Hoxb8 mutant mice. Cell 141(5):775–785. https://doi.org/10.1016/j.cell.2010.03.055

Chitu V, Gokhan Ş, Nandi S, Mehler MF, Richard Stanley E (2016) Emerging roles for CSF-1 receptor and its ligands in the nervous system. Trends Neurosci 39(6):378–393. https://doi.org/10.1016/j.tins.2016.03.005

Choi GB, Yim YS, Wong H, Kim S, Kim H, Kim SV, Hoeffer CA, Littman DR, Huh JR (2016) The maternal interleukin-17a pathway in mice promotes autism-like phenotypes in offspring. Science 351(6276):933–939. https://doi.org/10.1126/science.aad0314

Chung W-S, Clarke LE, Wang GX, Stafford BK, Sher A, Chakraborty C, Joung J et al (2013) Astrocytes mediate synapse elimination through MEGF10 and MERTK pathways. Nature 504 (7480):394–400. https://doi.org/10.1038/nature12776

Chung W-S, Allen NJ, Eroglu C (2015) Astrocytes control synapse formation, function, and elimination. Cold Spring Harb Perspect Biol 7(9):a020370. https://doi.org/10.1101/cshperspect.a020370

Cotella EM, Gomez AS, Lemen P, Chen C, Fernández G, Hansen C, Herman JP, Paglini MG (2019) Long-term impact of chronic variable stress in adolescence versus adulthood. Prog Neuro-Psychopharmacol Biol Psychiatry 88:303–310. https://doi.org/10.1016/j.pnpbp.2018.08.003

Crews F, He J, Hodge C (2007) Adolescent cortical development: a critical period of vulnerability for addiction. Pharmacol Biochem Behav 86(2):189–199. https://doi.org/10.1016/j.pbb.2006.12.001

Cunningham CL, Martínez-Cerdeño V, Noctor SC (2013) Microglia regulate the number of neural precursor cells in the developing cerebral cortex. J Neurosci Off J Soc Neurosci 33 (10):4216–4233. https://doi.org/10.1523/JNEUROSCI.3441-12.2013

Daneman R, Prat A (2015) The blood–brain barrier. Cold Spring Harb Perspect Biol 7(1):a020412. https://doi.org/10.1101/cshperspect.a020412

Davalos D, Grutzendler J, Yang G, Kim JV, Zuo Y, Jung S, Littman DR, Dustin ML, Gan W-B (2005) ATP mediates rapid microglial response to local brain injury in vivo. Nat Neurosci 8 (6):752–758. https://doi.org/10.1038/nn1472

Daws MR, Sullam PM, Niemi EC, Chen TT, Tchao NK, Seaman WE (2003) Pattern recognition by TREM-2: binding of anionic ligands. J Immunol 171(2):594–599. https://doi.org/10.4049/jimmunol.171.2.594

de Jonge JC, Vinkers CH, Hulshoff Pol HE, Marsman A (2017) GABAergic mechanisms in schizophrenia: linking postmortem and in vivo studies. Front Psych 8. https://doi.org/10.3389/fpsyt.2017.00118

De S, Van Deren D, Peden E, Hockin M, Boulet A, Titen S, Capecchi MR (2018) Two distinct ontogenies confer heterogeneity to mouse brain microglia. Development 145(13). https://doi.org/10.1242/dev.152306

Delevich K, Wren Thomas A, Wilbrecht L (2018) Adolescence and 'late blooming' synapses of the prefrontal cortex. Cold Spring Harb Symp Quant Biol 83:37–43. https://doi.org/10.1101/sqb.2018.83.037507

Drzewiecki CM, Willing J, Juraska JM (2016) Synaptic number changes in the medial prefrontal cortex across adolescence in male and female rats: a role for pubertal onset. Synapse 70 (9):361–368. https://doi.org/10.1002/syn.21909

Elmore MRP, Najafi AR, Koike MA, Dagher NN, Spangenberg EE, Rice RA, Kitazawa M et al (2014) Colony-stimulating factor 1 receptor signaling is necessary for microglia viability, unmasking a microglia progenitor cell in the adult brain. Neuron 82(2):380–397. https://doi.org/10.1016/j.neuron.2014.02.040

Erblich B, Zhu L, Etgen AM, Dobrenis K, Pollard JW (2011) Absence of colony stimulation factor-1 receptor results in loss of microglia, disrupted brain development and olfactory deficits. PLoS One 6(10):e26317. https://doi.org/10.1371/journal.pone.0026317

Estes ML, McAllister AK (2016) Maternal immune activation: implications for neuropsychiatric disorders. Science 353(6301):772–777. https://doi.org/10.1126/science.aag3194

Eyo UB, Mo M, Yi M-H, Murugan M, Liu J, Yarlagadda R, Margolis DJ, Xu P, Long-Jun W (2018) P2Y12R-dependent translocation mechanisms gate the changing microglial landscape. Cell Rep 23(4):959–966. https://doi.org/10.1016/j.celrep.2018.04.001

Farhy-Tselnicker I, Allen NJ (2018) Astrocytes, neurons, synapses: a tripartite view on cortical circuit development. Neural Dev 13(1):7. https://doi.org/10.1186/s13064-018-0104-y

Filipello F, Morini R, Corradini I, Zerbi V, Canzi A, Michalski B, Erreni M et al (2018) The microglial innate immune receptor TREM2 is required for synapse elimination and normal brain connectivity. Immunity 48(5):979–991.e8. https://doi.org/10.1016/j.immuni.2018.04.016

Galloway DA, Phillips AEM, Owen DRJ, Moore CS (2019) Phagocytosis in the brain: homeostasis and disease. Front Immunol 10. https://doi.org/10.3389/fimmu.2019.00790

Giedd JN, Blumenthal J, Jeffries NO, Castellanos FX, Liu H, Zijdenbos A, Paus T, Evans AC, Rapoport JL (1999) Brain development during childhood and adolescence: a longitudinal MRI study. Nat Neurosci 2(10):861–863. https://doi.org/10.1038/13158

Ginhoux F, Greter M, Leboeuf M, Nandi S, See P, Gokhan S, Mehler MF et al (2010) Fate mapping analysis reveals that adult microglia derive from primitive macrophages. Science 330 (6005):841–845. https://doi.org/10.1126/science.1194637

Giovanoli S, Engler H, Engler A, Richetto J, Voget M, Willi R, Winter C et al (2013) Stress in puberty unmasks latent neuropathological consequences of prenatal immune activation in mice. Science 339(6123):1095–1099. https://doi.org/10.1126/science.1228261

Giovanoli S, Engler H, Engler A, Richetto J, Feldon J, Riva MA, Schedlowski M, Meyer U (2016) Preventive effects of minocycline in a neurodevelopmental two-hit model with relevance to schizophrenia. Transl Psychiatry 6:e772. https://doi.org/10.1038/tp.2016.38

Gonzalez-Burgos G, Cho RY, Lewis DA (2015) Alterations in cortical network oscillations and parvalbumin neurons in schizophrenia. Biol Psychiatry 77(12):1031–1040. https://doi.org/10.1016/j.biopsych.2015.03.010

Hammond TR, Dufort C, Dissing-Olesen L, Giera S, Young A, Wysoker A, Walker AJ et al (2019) Single-cell RNA sequencing of microglia throughout the mouse lifespan and in the injured brain reveals complex cell-state changes. Immunity 50(1):253–271.e6. https://doi.org/10.1016/j.immuni.2018.11.004

Hanamsagar R, Alter MD, Block CS, Sullivan H, Bolton JL, Bilbo SD (2017) Generation of a microglial developmental index in mice and in humans reveals a sex difference in maturation and immune reactivity. Glia 65(9):1504–1520. https://doi.org/10.1002/glia.23176

Haynes SE, Hollopeter G, Yang G, Kurpius D, Dailey ME, Gan W-B, Julius D (2006) The P2Y 12 receptor regulates microglial activation by extracellular nucleotides. Nat Neurosci 9 (12):1512–1519. https://doi.org/10.1038/nn1805

Hoeffel G, Ginhoux F (2015) Ontogeny of tissue-resident macrophages. Front Immunol 6:486. https://doi.org/10.3389/fimmu.2015.00486

Hoftman GD, Volk DW, Holly Bazmi H, Li S, Sampson AR, Lewis DA (2015) Altered cortical expression of GABA-related genes in schizophrenia: illness progression vs developmental disturbance. Schizophr Bull 41(1):180–191. https://doi.org/10.1093/schbul/sbt178

Ismail N, Garas P, Blaustein JD (2011) Long-term effects of pubertal stressors on female sexual receptivity and estrogen receptor-α expression in CD-1 female mice. Horm Behav 59 (4):565–571. https://doi.org/10.1016/j.yhbeh.2011.02.010
Ismail N, Kumlin AM, Blaustein JD (2013) A pubertal immune challenge alters the antidepressant-like effects of chronic estradiol treatment in inbred and outbred adult female mice. Neuroscience 249:43–52. https://doi.org/10.1016/j.neuroscience.2012.09.047
Jay TR, von Saucken VE, Muñoz B, Codocedo JF, Atwood BK, Lamb BT, Landreth GE (2019) TREM2 is required for microglial instruction of astrocytic synaptic engulfment in neurodevelopment. Glia 67(10):1873–1892. https://doi.org/10.1002/glia.23664
Kaar SJ, Angelescu I, Marques TR, Howes OD (2019) Pre-frontal Parvalbumin interneurons in schizophrenia: a meta-analysis of post-mortem studies. J Neural Transm (Vienna) 126 (12):1637–1651. https://doi.org/10.1007/s00702-019-02080-2
Keren-Shaul H, Spinrad A, Weiner A, Matcovitch-Natan O, Dvir-Szternfeld R, Ulland TK, David E et al (2017) A unique microglia type associated with restricting development of Alzheimer's disease. Cell 169(7):1276–1290.e17. https://doi.org/10.1016/j.cell.2017.05.018
Kessler RC, Paul Amminger G, Aguilar-Gaxiola S, Alonso J, Lee S, Bedirhan Ustün T (2007) Age of onset of mental disorders: a review of recent literature. Curr Opin Psychiatry 20(4):359–364. https://doi.org/10.1097/YCO.0b013e32816ebc8c
Klawonn AM, Fritz M, Castany S, Pignatelli M, Canal C, Similä F, Tejeda HA et al (2021) Microglial activation elicits a negative affective state through prostaglandin-mediated modulation of striatal neurons. Immunity 54(2):225–234.e6. https://doi.org/10.1016/j.immuni.2020.12.016
Knuesel I, Chicha L, Britschgi M, Schobel SA, Bodmer M, Hellings JA, Toovey S, Prinssen EP (2014) Maternal immune activation and abnormal brain development across CNS disorders. Nat Rev Neurol 10(11):643–660. https://doi.org/10.1038/nrneurol.2014.187
Kochunov P, Elliot Hong L (2014) Neurodevelopmental and neurodegenerative models of schizophrenia: white matter at the center stage. Schizophr Bull 40(4):721–728. https://doi.org/10.1093/schbul/sbu070
Kolmogorova D, Paré C, Kostuck S, Hudson EC, Lebel N, Houlding E, Gregory JG, Ismail N (2019) Pubertal immune stress transiently alters spatial memory processes in adulthood. Psychoneuroendocrinology 102:261–272. https://doi.org/10.1016/j.psyneuen.2018.12.224
Kondo Y, Duncan ID (2009) Selective reduction in microglia density and function in the white matter of Colony-stimulating factor-1-deficient mice. J Neurosci Res 87(12):2686–2695. https://doi.org/10.1002/jnr.22096
Kopec AM, Smith CJ, Ayre NR, Sweat SC, Bilbo SD (2018) Microglial dopamine receptor elimination defines sex-specific nucleus accumbens development and social behavior in adolescent rats. Nat Commun 9(1):3769. https://doi.org/10.1038/s41467-018-06118-z
Koss WA, Belden CE, Hristov AD, Juraska JM (2014) Dendritic remodeling in the adolescent medial prefrontal cortex and the basolateral amygdala of male and female rats. Synapse 68 (2):61–72. https://doi.org/10.1002/syn.21716
Lehrman EK, Wilton DK, Litvina EY, Welsh CA, Chang ST, Frouin A, Walker AJ et al (2018) CD47 protects synapses from excess microglia-mediated pruning during development. Neuron 100(1):120–134.e6. https://doi.org/10.1016/j.neuron.2018.09.017
Lenz KM, Pickett LA, Wright CL, Davis KT, Joshi A, McCarthy MM (2018) Mast cells in the developing brain determine adult sexual behavior. J Neurosci 38(37):8044–8059. https://doi.org/10.1523/JNEUROSCI.1176-18.2018
Lewis DA, Curley AA, Glausier JR, Volk DW (2012) Cortical parvalbumin interneurons and cognitive dysfunction in schizophrenia. Trends Neurosci 35(1):57–67. https://doi.org/10.1016/j.tins.2011.10.004
Louveau A, Harris TH, Kipnis J (2015) Revisiting the mechanisms of CNS immune privilege. Trends Immunol 36(10):569–577. https://doi.org/10.1016/j.it.2015.08.006

Lovelock DF, Deak T (2019) Acute stress imposed during adolescence yields heightened anxiety in Sprague Dawley rats that persists into adulthood: sex differences and potential involvement of the medial amygdala. Brain Res 1723:146392. https://doi.org/10.1016/j.brainres.2019.146392

MacDonald JM, Beach MG, Porpiglia E, Sheehan AE, Watts RJ, Freeman MR (2006) The drosophila cell corpse engulfment receptor draper mediates glial clearance of severed axons. Neuron 50(6):869–881. https://doi.org/10.1016/j.neuron.2006.04.028

Malkova NV, Yu CZ, Hsiao EY, Moore MJ, Patterson PH (2012) Maternal immune activation yields offspring displaying mouse versions of the three core symptoms of autism. Brain Behav Immun 26(4):607–616. https://doi.org/10.1016/j.bbi.2012.01.011

Mallya AP, Wang H-D, Lee HNR, Deutch AY (2019) Microglial pruning of synapses in the prefrontal cortex during adolescence. Cereb Cortex 29(4):1634–1643. https://doi.org/10.1093/cercor/bhy061

Manduca A, Servadio M, Damsteegt R, Campolongo P, Vanderschuren LJ, Trezza V (2016) Dopaminergic neurotransmission in the nucleus accumbens modulates social play behavior in rats. Neuropsychopharmacology 41(9):2215–2223. https://doi.org/10.1038/npp.2016.22

Mariani MM, Kielian T (2009) Microglia in infectious diseases of the central nervous system. J Neuroimmune Pharmacol 4(4):448–461. https://doi.org/10.1007/s11481-009-9170-6

Markham JA, Mullins SE, Koenig JI (2013) Periadolescent maturation of the prefrontal cortex is sex-specific and is disrupted by prenatal stress. J Comp Neurol 521(8):1828–1843. https://doi.org/10.1002/cne.23262

Matcovitch-Natan O, Winter DR, Giladi A, Aguilar SV, Spinrad A, Sarrazin S, Ben-Yehuda H et al (2016) Microglia development follows a stepwise program to regulate brain homeostasis. Science 353(6301):aad8670. https://doi.org/10.1126/science.aad8670

Maynard TM, Sikich L, Lieberman JA, LaMantia AS (2001) Neural development, cell-cell signaling, and the 'two-hit' hypothesis of schizophrenia. Schizophr Bull 27(3):457–476. https://doi.org/10.1093/oxfordjournals.schbul.a006887

McCarthy MM, Herold K, Stockman SL (2018) Fast, furious and enduring: sensitive versus critical periods in sexual differentiation of the brain. Physiol Behav 187:13–19. https://doi.org/10.1016/j.physbeh.2017.10.030

McClain JA, Morris SA, Ayumi Deeny M, Alex Marshall S, Hayes DM, Kiser ZM, Nixon K (2011) Adolescent binge alcohol exposure induces long-lasting partial activation of microglia. Brain Behav Immun 25:S120–S128. https://doi.org/10.1016/j.bbi.2011.01.006

Medawar PB (1948) Immunity to homologous grafted skin. III. The fate of skin homographs transplanted to the brain, to subcutaneous tissue, and to the anterior chamber of the eye. Br J Exp Pathol 29(1):58–69

Mrdjen D, Pavlovic A, Hartmann FJ, Schreiner B, Utz SG, Leung BP, Lelios I et al (2018) High-dimensional single-cell mapping of central nervous system immune cells reveals distinct myeloid subsets in health, aging, and disease. Immunity 48(2):380–395.e6. https://doi.org/10.1016/j.immuni.2018.01.011

Nakazawa K, Zsiros V, Jiang Z, Nakao K, Kolata S, Zhang S, Belforte JE (2012) GABAergic interneuron origin of schizophrenia pathophysiology. Neuropharmacology 62(3):1574–1583. https://doi.org/10.1016/j.neuropharm.2011.01.022

Nandi S, Gokhan S, Dai X-M, Wei S, Enikolopov G, Lin H, Mehler MF, Richard Stanley E (2012) The CSF-1 receptor ligands IL-34 and CSF-1 exhibit distinct developmental brain expression patterns and regulate neural progenitor cell maintenance and maturation. Dev Biol 367 (2):100–113. https://doi.org/10.1016/j.ydbio.2012.03.026

Nimmerjahn A, Kirchhoff F, Helmchen F (2005) Resting microglial cells are highly dynamic surveillants of brain parenchyma in vivo. Science 308(5726):1314–1318. https://doi.org/10.1126/science.1110647

Niraula A, Sheridan JF, Godbout JP (2017) Microglia priming with aging and stress. Neuropsychopharmacology 42(1):318–333. https://doi.org/10.1038/npp.2016.185

Olesen KM, Ismail N, Merchasin ED, Blaustein JD (2011) Long-term alteration of anxiolytic effects of ovarian hormones in female mice by a peripubertal immune challenge. Horm Behav 60(4):318–326. https://doi.org/10.1016/j.yhbeh.2011.06.005
Oosterhof N, Chang IJ, Karimiani EG, Kuil LE, Jensen DM, Daza R, Young E et al (2019) Homozygous mutations in CSF1R cause a pediatric-onset leukoencephalopathy and can result in congenital absence of microglia. Am J Hum Genet 104(5):936–947. https://doi.org/10.1016/j.ajhg.2019.03.010
Paolicelli RC, Bolasco G, Pagani F, Maggi L, Scianni M, Panzanelli P, Giustetto M et al (2011) Synaptic pruning by microglia is necessary for normal brain development. Science 333 (6048):1456–1458. https://doi.org/10.1126/science.1202529
Paus T, Keshavan M, Giedd JN (2008) Why do many psychiatric disorders emerge during adolescence? Nat Rev Neurosci 9(12):947–957. https://doi.org/10.1038/nrn2513
Peng L, Zhu M, Yang Y, Weng Y, Zou W, Zhu X, Guo Q, Zhong T (2019) Neonatal lipopolysaccharide challenge induces long-lasting spatial cognitive impairment and dysregulation of hippocampal histone acetylation in mice. Neuroscience 398:76–87. https://doi.org/10.1016/j.neuroscience.2018.12.001
Petanjek Z, Judaš M, Šimić G, Rašin MR, Uylings HBM, Rakic P, Kostović I (2011) Extraordinary neoteny of synaptic spines in the human prefrontal cortex. Proc Natl Acad Sci 108 (32):13281–13286. https://doi.org/10.1073/pnas.1105108108
Premachandran H, Zhao M, Arruda-Carvalho M (2020) Sex differences in the development of the rodent corticolimbic system. Front Neurosci 14. https://doi.org/10.3389/fnins.2020.583477
Profaci CP, Munji RN, Pulido RS, Daneman R (2020) The blood-brain barrier in health and disease: important unanswered questions. J Exp Med 217(4). https://doi.org/10.1084/jem.20190062
Risher WC, Patel S, Kim IH, Uezu A, Bhagat S, Wilton DK, Pilaz L-J et al (2014) Astrocytes refine cortical connectivity at dendritic spines. eLife 3. https://doi.org/10.7554/eLife.04047
Schafer DP, Lehrman EK, Kautzman AG, Koyama R, Mardinly AR, Yamasaki R, Ransohoff RM, Greenberg ME, Barres BA, Stevens B (2012) Microglia sculpt postnatal neural circuits in an activity and complement-dependent manner. Neuron 74(4):691–705. https://doi.org/10.1016/j.neuron.2012.03.026
Schneider M (2013) Adolescence as a vulnerable period to alter rodent behavior. Cell Tissue Res 354(1):99–106. https://doi.org/10.1007/s00441-013-1581-2
Schwarz JM, Bilbo SD (2013) Adolescent morphine exposure affects long-term microglial function and later-life relapse liability in a model of addiction. J Neurosci Off J Soc Neurosci 33 (3):961–971. https://doi.org/10.1523/JNEUROSCI.2516-12.2013
Schwarz JM, Hutchinson MR, Bilbo SD (2011) Early-life experience decreases drug-induced reinstatement of morphine CPP in adulthood via microglial-specific epigenetic programming of anti-inflammatory IL-10 expression. J Neurosci 31(49):17835–17847. https://doi.org/10.1523/JNEUROSCI.3297-11.2011
Scott-Hewitt N, Perrucci F, Morini R, Erreni M, Mahoney M, Witkowska A, Carey A et al (2020) Local externalization of phosphatidylserine mediates developmental synaptic pruning by microglia. EMBO J 39(16):e105380. https://doi.org/10.15252/embj.2020105380
Shatz CJ (1990) Competitive interactions between retinal ganglion cells during prenatal development. J Neurobiol 21(1):197–211. https://doi.org/10.1002/neu.480210113
Sheng J, Ruedl C, Karjalainen K (2015) Most tissue-resident macrophages except microglia are derived from fetal hematopoietic stem cells. Immunity 43(2):382–393. https://doi.org/10.1016/j.immuni.2015.07.016
Shi L, Hossein Fatemi S, Sidwell RW, Patterson PH (2003) Maternal influenza infection causes marked behavioral and pharmacological changes in the offspring. J Neurosci Off J Soc Neurosci 23(1):297–302
Smith SEP, Li J, Garbett K, Mirnics K, Patterson PH (2007) Maternal immune activation alters fetal brain development through Interleukin-6. J Neurosci Off J Soc Neurosci 27(40):10695–10702. https://doi.org/10.1523/JNEUROSCI.2178-07.2007

Smith CJ, Kingsbury MA, Dziabis JE, Hanamsagar R, Malacon KE, Tran JN, Norris HA, Gulino M, Bordt EA, Bilbo SD (2020) Neonatal immune challenge induces female-specific changes in social behavior and somatostatin cell number. Brain Behav Immun 90:332–345. https://doi.org/10.1016/j.bbi.2020.08.013

Squarzoni P, Oller G, Hoeffel G, Pont-Lezica L, Rostaing P, Low D, Bessis A, Ginhoux F, Garel S (2014) Microglia modulate wiring of the embryonic forebrain. Cell Rep 8(5):1271–1279. https://doi.org/10.1016/j.celrep.2014.07.042

Steinberg L (2008) A social neuroscience perspective on adolescent risk-taking. Dev Rev 28 (1):78–106. https://doi.org/10.1016/j.dr.2007.08.002

Stevens B, Allen NJ, Vazquez LE, Howell GR, Christopherson KS, Nouri N, Micheva KD et al (2007) The classical complement cascade mediates CNS synapse elimination. Cell 131 (6):1164–1178. https://doi.org/10.1016/j.cell.2007.10.036

Tamnes CK, Walhovd KB, Dale AM, Østby Y, Grydeland H, Richardson G, Westlye LT et al (2013) Brain development and aging: overlapping and unique patterns of change. NeuroImage 68:63–74. https://doi.org/10.1016/j.neuroimage.2012.11.039

Tan Y-L, Yuan Y, Tian L (2020) Microglial regional heterogeneity and its role in the brain. Mol Psychiatry 25(2):351–367. https://doi.org/10.1038/s41380-019-0609-8

Tanabe S, Yamashita T (2018) B-1a lymphocytes promote oligodendrogenesis during brain development. Nat Neurosci 21(4):506–516. https://doi.org/10.1038/s41593-018-0106-4

Thion MS, Garel S (2020) Microglial ontogeny, diversity and neurodevelopmental functions. Curr Opin Genet Dev 65:186–194. https://doi.org/10.1016/j.gde.2020.06.013

Thion MS, Mosser C-A, Férézou I, Grisel P, Baptista S, Low D, Ginhoux F, Garel S, Audinat E (2019) Biphasic impact of prenatal inflammation and macrophage depletion on the wiring of neocortical inhibitory circuits. Cell Rep 28(5):1119–1126.e4. https://doi.org/10.1016/j.celrep.2019.06.086

Toga AW, Thompson PM, Sowell ER (2006) Mapping brain maturation. Focus 4(3):378–390. https://doi.org/10.1176/foc.4.3.378

Tremblay M-È, Lowery RL, Majewska AK (2010) Microglial interactions with synapses are modulated by visual experience. PLoS Biol 8(11):e1000527. https://doi.org/10.1371/journal.pbio.1000527

Vainchtein ID, Chin G, Cho FS, Kelley KW, Miller JG, Chien EC, Liddelow SA et al (2018) Astrocyte-derived interleukin-33 promotes microglial synapse engulfment and neural circuit development. Science 359(6381):1269–1273. https://doi.org/10.1126/science.aal3589

VanRyzin JW, Pickett LA, McCarthy MM (2018) Microglia: driving critical periods and sexual differentiation of the brain. Dev Neurobiol 78(6):580–592. https://doi.org/10.1002/dneu.22569

VanRyzin JW, Marquardt AE, Argue KJ, Vecchiarelli HA, Ashton SE, Arambula SE, Hill MN, McCarthy MM (2019) Microglial phagocytosis of newborn cells is induced by endocannabinoids and sculpts sex differences in juvenile rat social play. Neuron 102 (2):435–449.e6. https://doi.org/10.1016/j.neuron.2019.02.006

VanRyzin JW, Marquardt AE, Pickett LA, McCarthy MM (2020) Microglia and sexual differentiation of the developing brain: a focus on extrinsic factors. Glia 68(6):1100–1113. https://doi.org/10.1002/glia.23740

Wake H, Moorhouse AJ, Jinno S, Kohsaka S, Nabekura J (2009) Resting microglia directly monitor the functional state of synapses in vivo and determine the fate of ischemic terminals. J Neurosci Off J Soc Neurosci 29(13):3974–3980. https://doi.org/10.1523/JNEUROSCI.4363-08.2009

Wang Y, Szretter KJ, Vermi W, Gilfillan S, Rossini C, Cella M, Barrow AD, Diamond MS, Colonna M (2012) IL-34 is a tissue-restricted ligand of CSF1R required for the development of Langerhans cells and microglia. Nat Immunol 13(8):753–760. https://doi.org/10.1038/ni.2360

Wang Y, Cella M, Mallinson K, Ulrich JD, Young KL, Robinette ML, Gilfillan S et al (2015) TREM2 lipid sensing sustains the microglial response in an Alzheimer's disease model. Cell 160(6):1061–1071. https://doi.org/10.1016/j.cell.2015.01.049

Whitford TJ, Rennie CJ, Grieve SM, Richard Clark C, Gordon E, Williams LM (2007) Brain maturation in adolescence: concurrent changes in neuroanatomy and neurophysiology. Hum Brain Mapp 28(3):228–237. https://doi.org/10.1002/hbm.20273

Williamson LL, Sholar PW, Mistry RS, Smith SH, Bilbo SD (2011) Microglia and memory: modulation by early-life infection. J Neurosci 31(43):15511–15521. https://doi.org/10.1523/JNEUROSCI.3688-11.2011

Wolf SA, Boddeke HWGM, Kettenmann H (2017) Microglia in physiology and disease. Annu Rev Physiol 79:619–643. https://doi.org/10.1146/annurev-physiol-022516-034406

Wu W-L, Hsiao EY, Yan Z, Mazmanian SK, Patterson PH (2017) The placental interleukin-6 signaling controls fetal brain development and behavior. Brain Behav Immun 62:11–23. https://doi.org/10.1016/j.bbi.2016.11.007

Yim S, Yeong AP, Berrios J, Lafourcade M, Pascual LM, Soares N, Kim JY et al (2017) Reversing behavioural abnormalities in mice exposed to maternal inflammation. Nature 549 (7673):482–487. https://doi.org/10.1038/nature23909

Yohn NL, Blendy JA (2017) Adolescent chronic unpredictable stress exposure is a sensitive window for long-term changes in adult behavior in mice. Neuropsychopharmacology 42 (8):1670–1678. https://doi.org/10.1038/npp.2017.11

Zuckerman L, Rehavi M, Nachman R, Weiner I (2003) Immune activation during pregnancy in rats leads to a PostPubertal emergence of disrupted latent inhibition, dopaminergic hyperfunction, and altered limbic morphology in the offspring: a novel neurodevelopmental model of schizophrenia. Neuropsychopharmacology 28(10):1778–1789. https://doi.org/10.1038/sj.npp.1300248

Is Adolescence a Sensitive Period for the Development of Incentive-Reward Motivation?

Monica Luciana and Paul F. Collins

Contents

Abstract Human adolescence is broadly construed as a time of heightened risk-taking and a vulnerability period for the emergence of psychopathology. These tendencies have been attributed to the age-related development of neural systems that mediate incentive motivation and other aspects of reward processing as well as individual difference factors that interact with ongoing development. Here, we describe the adolescent development of incentive motivation, which we view as an inherently positive developmental progression, and its associated neural mechanisms. We consider challenges in applying the sensitive period concept to these maturational events and discuss future directions that may help to clarify mechanisms of change.

Keywords Agency · Dopamine · Motivation · Neurodevelopment · Prefrontal · Reward

M. Luciana (✉) and P. F. Collins
Department of Psychology, University of Minnesota, Minneapolis, MN, USA
e-mail: lucia003@umn.edu

Curr Topics Behav Neurosci (2022) 53: 79–100
https://doi.org/10.1007/7854_2021_275
Published Online: 17 November 2021

1 Introduction

Within the scientific literature, human adolescence has been broadly construed as a period of heightened risk-taking as well as increased vulnerability for the emergence of psychopathology. These tendencies have been attributed to the development of neural systems that mediate cognitive control versus incentive motivation as well as vulnerability factors that interact with age-related changes, rendering some individuals more likely than others to pursue, engage with, and be compromised by, certain experiences. The field has sought to identify those experiences and to discern the mechanisms through which neural systems are vulnerable to a host of adverse circumstances.

While the emphasis has been on risk and pathology, adolescence is fundamentally a period of increased agency, new learning, flexibility, and long-term goal striving (Luciana et al. 2012), and as such, young people may be inherently prepared to actively pursue and take advantage of opportunities, relationships, and interests during this time. To benefit from these experiences, neural systems that mediate aspects of incentive-reward motivation must be attuned to the environment, flexible and highly resilient.

This resilience achieves a number of species-general goals. From an evolutionary perspective, fundamental imperatives upon reaching sexual maturity include exploring potentially resource-rich new environments and expanding the set of potential sex partners beyond the natal group, which together enhance group fitness and long-term survival. Although this transition to adulthood entails risk-taking, the simultaneous development of high-level cognitive abilities supports adaptive decision making and the acquisition of resources to raise offspring to maturity. Given the slow pace of evolutionary change, this facilitation of reproductive fitness remains the context for adolescent developmental milestones in both behavior and neural systems.

Thus, adolescence represents a distinctive period given the transition to reproductive maturity and the preparedness of incentive motivational systems to engage with the environment to obtain rewards and pursue long-term goals. Given the range of complex socioemotional experiences that accompany this transition, which begins with pubertal onset, adolescence may represent a sensitive period for socioemotional development and incentive-motivated goal-directed behavior. Here, we consider that proposition in relation to how sensitive periods are defined, their temporal contingencies, and long-range outcomes that may be attributable to experiences acquired (or not) during the adolescent period.

2 Defining a Sensitive Period

As described in other chapters of this volume, the sensitive period concept was first advanced in ethology to describe time-limited responses to certain types of experiences that occur usually very early in development, during periods of rapid reorganization, and shape ongoing behavior and neural responsivity (Bornstein 1989; Gabard-Durnam and McLaughlin 2020). A classic example is imprinting, where young birds entrain upon the first moving stimulus that they see (typically the mother) and follow that stimulus. If the stimulus is not encountered within a specific time window, then such following behavior will not occur, and the immature animal may not be adequately nurtured. Studies of animals, and to a significantly lesser extent in humans, have since described similar examples, largely related to sensory, motor, and language system development (though the latter is debated: Werker and Hensch 2015). The term "sensitive period" characterizes developmental periods within which the effects of expected experience on the brain are relatively strong (Knudsen 2004) but not necessarily immutable. In contrast, a critical period is one in which irreversible changes in neural structure or function occur following an expected experience. In specifying the characteristics of a sensitive period, parameters to be evaluated include temporal features such as onset, which is typically rapid, as well as termination, which is more gradual. Neuroscientific investigations at the cellular and molecular levels indicate that alterations in the excitatory-to-inhibitory balance within discrete synapses and neural circuits, as achieved through gabaergic and glutamatergic neurotransmission, enable the opening of sensitive periods, whereas processes such as myelination, the formation of cell adhesion molecules, and elaborations of perineuronal nets inhibit ongoing synaptic alterations and thereby impede synaptic remodeling, terminating sensitive periods (Takesian and Hensch 2013). In addition to these temporal features, sensitive periods require definition of the environmental experience that provokes the time-limited neurobehavioral responses, the intrinsic maturational status that is a prerequisite to express and modify such responses, and finally the specific neural and behavioral systems and processes through which these impacts become embedded (Bornstein 1989). To these sensitive period features, one might add the nature of outcomes (i.e., neurobehavioral effects) linked to variations in the expected environmental experiences. These outcomes appear to be malleable, as are all parameters of a putative sensitive period, and do not necessarily persist after the period ends (Bornstein 1989). Moreover, neurobehavioral effects can be temporally distant from the events that triggered them. Adding even more complexity, there may be individual differences in sensitive period characteristics and the malleability of outcomes.

The types of experiences encountered during a sensitive period are expected based on phylogeny and enable the subsequent development of neurons and connections that may not yet be present or are rudimentarily present. For example, the expected experiences of visual stimulation in infancy provoke the neurodevelopmental organization of ocular dominance columns and the capacity for binocular vision. In contrast, experience-dependent alterations impact neuronal structures that are already present (Kolb et al. 2013). This distinction is relevant, particularly as applied to higher order cognitive and socioemotional behaviors, which are multifaceted, scaffolded by earlier developing perceptual systems but reliant on late-maturing association regions for their full expression. Using reward sensitivity as an example, if it is the case that the neural systems enabling such behaviors are established early in development, then subsequent experience-based alterations are likely to be experience-dependent and variable, leading to a wide range of individual differences in outcomes. Specific experiences are not necessarily essential. Alternatively, if the relevant neural systems are in a state of ongoing dynamic development, then certain types of experiences *must* be acquired to shape or harmonize, across all individuals, basic and necessary aspects of neural structure and function. Relatively late-emerging behaviors and systems allow for both possibilities, which is why it is important to understand their maturational trajectories at a mechanistic level. If our interest is in how reward systems develop their basic structure and function in relatively invariant ways and in how such development might be derailed by injury, adversity, stress or trauma, then the sensitive period concept is relevant. Alternatively, if our focus is on how and when individual differences emerge as a function of unique life experiences and learning, as well as how such differences are biologically determined, then experience-dependent learning and associated mechanisms of lifelong plasticity are more applicable. Thus, the observation of plasticity or even experience-driven change within a given developmental period does not necessarily mean that the period is sensitive. What must be demonstrated is a unique temporally limited association between experience and responsivity that is not similarly evident during other life phases (see Woodard and Pollak 2020 for a similar discussion).

It may seem counterintuitive to apply the idea of a sensitive period, with its inherent inflexibility and specific input–output relations required to establish a given outcome, to adolescent incentive motivation and socioemotional development when the basic life task is to explore the environment, encounter new things, and learn flexibly from that highly varied experience. However, *the motivational push to engage* with the environment in ways that promote exploration, novelty-seeking, sexual strivings, and independence could be construed as an experience-expectant process that triggers a sensitive period, followed thereafter by idiosyncratic experience-dependent learning that builds upon the general foundation established by neurobehavioral responses to the initial incentive-based experiences. In that sense, adolescent reward seeking may involve an interplay of experience-expectant and experience-dependent processes.

In summary, a sensitive period can be understood from an evolutionary perspective as an adaptation that prepares individuals within a species to respond appropriately to an expected type of environmental experience, such as occurs in adolescence during exploration of the physical and social environment with sexually mature peers.

3 Adolescence as a Sensitive Period: General Considerations

There are several ways in which neurobehavioral plasticity might be demonstrated in the adolescent period (Fuhrmann et al. 2015; Woodard and Pollak 2020). In accord with the sensitive period concept, there might be a nonlinear adolescent-specific peak in experiential responsivity that is not evident in childhood or in adulthood. Alternatively, childhood and adolescence might be combined within the same sensitive period with an offset in early adulthood (e.g., following an inverse function of age) (Fuhrmann et al. 2015). There might be wave-like recurring periods of experiential responsivity that punctuate the child and adolescent periods. Or perhaps there is not a sensitive period at all with respect to a given outcome, but plasticity linearly increases or decreases or remains stable, with equiprobable responses to exogenous experience, from infancy to old age. Adolescence is unequivocally conceptualized as a period of heightened plasticity (Lindenberger and Lövdén 2019) and any of these models could apply. A longitudinal lifespan-oriented assessment is needed to determine which developmental model is most appropriate and to achieve a precise analysis of the behavior under investigation (Woodard and Pollak 2020). Complex models likely will be relevant given that the higher order cognitive and affective behaviors of adolescence are multifaceted, network-based, and multiply determined.

With respect to reward sensitivity, several observations have led to the idea that adolescence might represent a sensitive period. First, puberty is a distinctive transitional stage of development during which sexual strivings emerge and susceptibility to peer influence is markedly heightened. Second, studies of early childhood adversity suggest that adolescence may be a period of "second chances" (Gunnar et al. 2019; Masten et al. 2021), during which positive experiences and relationships may be rehabilitative. Third, adolescence represents a risk period for the emergence of psychopathology, including anxiety, affective disorder, psychosis, eating disorders, and substance abuse. Many of these conditions involve deviations in neurocircuitry that supports positive motivation, perhaps representing sensitive period outcomes associated with experiences that departed significantly from normative expectations.

4 Adolescent Trajectory of Reward Sensitivity and Incentive Motivation

Incentive motivation refers to the intrinsic as well as extrinsic energizing of instrumental behavior by anticipation (e.g., wanting: Berridge 2018) of reward acquisition (Depue and Collins 1999; Wise and Robble 2020). Critically, one must be motorically active, engaged and curious to effectively and energetically pursue distal sources of reward and to learn from that experience (Depue and Collins 1999; Gruber and Ranganath 2019). A biologically-based incentive motivational system that mediates approach behaviors, searching/foraging, and environmental engagement has been described, and its activation assures that reward-based learning experiences will be pursued (see Depue and Collins 1999; Luciana 2001; Wahlstrom et al. 2010a for reviews). This system is expressed at multiple levels, incorporating aspects of arousal, motor function, affect, and associated neural structures and functional processes. Within the structure of personality, it is expressed as a dimensional trait encompassing agentic aspects of positive emotionality (PEM: Depue and Collins 1999). In humans, agentic PEM is phenotypically observed as strivings toward social dominance, mastery, and a subjective sense of well-being (Durbin et al. 2016; Patrick et al. 2002). It is conceptually and psychometrically distinct from other facets of personality such as consummatory pleasure following the experience of reward, generalized sensation-seeking (which is arousal-based and non-specific), impulsivity (which reflects deficient levels of inhibitory control rather than reward seeking), and risk-taking. Engagement with, and exploration of, the environment to pursue what one wants or needs but does not currently have is fundamental to survival. Thus, despite genetic influences over aspects of incentive motivation and its expression (de Boer et al. 2019; Frank and Fossella 2011; Solinas et al. 2019), resource limitations should heighten this system's activity.

Incentive-based approach behavior markedly influences both adolescent and toddler behavior, and there is compelling empirical evidence that it is mediated by dopamine (DA) activity in mesocorticolimbic circuits (Depue and Collins 1999; Luciana 2001; Luciana et al. 2012; Wahlstrom et al. 2010b). Key nodes include the ventral tegmental area (VTA) and its ascending projections to the nucleus accumbens (NAcc) region of the ventral striatum, ventral pallidum, amygdala, hippocampus, anterior cingulate cortex, and the medial orbitofrontal cortex (Depue and Collins 1999; Solinas et al. 2019).

In humans, individual differences in traits associated with incentive-reward motivation have been correlated with genetic and neurophysiological variations in DA function (Depue et al. 1994; Dreher et al. 2009; Mueller et al. 2014), though there is ongoing controversy regarding the reliability of these relationships (Wacker and Smillie 2015), some of which may be moderated by resource context (Fischer et al. 2018). Polygenic influences are likely based on recent genome-wide association studies (Ward et al. 2019). A significant source of variation across studies may be that incentive motivation and its associated behavioral traits are operationalized differently across studies.

As predicted by the sensitive period concept, human adolescents, relative to children and adults, show evidence of heightened incentive motivation, including positive affect, reward pursuit, and engagement with rewarding activities. With respect to the latter, researchers have emphasized activities that have a risk-taking component. For instance, epidemiological data indicate accelerating levels of engagement in activities such as sexual behavior and substance use. Data from the US National Survey of Family Growth indicate that from 2015–2017, 42% of never-married female teenagers aged 15–19 and 38% of never-married male teenagers had had sexual intercourse. The probability of engaging in such behavior increases from roughly 20% at age 15–78% by age 20 (Martinez and Abma 2020). Similarly, engagement in alcohol, tobacco, and nicotine use dramatically increases between 8th and 12th grades as indicated by findings from the 2019 Youth Behavior Risk Survey (Center for Disease Control 2020). Though pursuit of rewards does not necessarily imply risk-taking and while absolute levels of engagement in risky behaviors has changed over time and varies across cultural contexts, adolescents across the globe self-report age-related increases in health-related risk propensity (Duell et al. 2018). Peaks in risk propensity appear to be in the early to mid-20s. Self-report questionnaire data from smaller samples indicate a rise in reward sensitivity (Urošević et al. 2012) as well as sensation-seeking (Harden and Tucker-Drob 2011) from early to mid-adolescence; in turn, sensation-seeking is modestly associated (Pearson correlations = 0.13–0.21) with laboratory-based performance on putative measures of reward sensitivity such as the Iowa Gambling Task and Balloon Analog Risk Task (Harden et al. 2018). Durbin et al. (2016) analyzed four waves of data from a longitudinal twin study, finding that aspects of agentic positive emotionality (PEM), as measured by the Multidimensional Personality Questionnaire (MPQ: Patrick et al. 2002) showed a pattern of nonlinear change across adolescence. Specifically, social potency increased across mid to late adolescence followed by a modest decline in the 20s. Achievement motivation increased markedly across adolescence with less substantial increments in early adulthood. These nonlinear changes were paralleled by adolescent-specific decrements in harm avoidance (i.e., risk aversion). In addition to these examples of self-reported reward sensitivity, adolescents experience increases in peer salience and peer influence over decision making (Braams et al. 2019; Ciranka and van den Bos 2019). In parallel with human findings, rodent studies also suggest adolescent-limited increases in novelty preferences, peer salience, and exploratory behaviors (Doremus-Fitzwater and Spear 2016).

Meta-analyses of human task-based functional neuroimaging (fMRI) data indicate increased activation likelihoods in regions such as the ventral and dorsal striatum, amygdala, anterior and posterior cingulate cortices, the orbital frontal cortex, and the lateral occipital cortex in adolescents as compared to adults during the performance of tasks that require reward processing (Silverman et al. 2015). As compared to children, adolescents show increased task-based activation likelihoods in the caudate head, putamen, and posterior cingulate cortex (Yaple et al. 2020). In one particularly elegant study that included three longitudinal waves of assessment, Schreuders et al. (2018) found that NAcc activity in response to rewards received during a non-instrumental gambling task peaks in mid-adolescence, around the age of 16, and this peak level extends well into the 20s. In younger (ages 8–16), but not older adolescents, higher levels of self-reported drive to pursue and achieve personal goals were associated with stronger NAcc activity. Similarly, Murty et al. (2018) observed longitudinal developmental increases in VTA–NAcc functional coupling in children and adolescents who performed a rewarded anti-saccade task, though in that study, the functional peak occurred at ages 9–10. Karlsgodt et al. (2015) used probabilistic tractography to localize the accumbo-to-orbital prefrontal tract, finding an early peak at age 14 for fractional anisotropy (a measure of fiber organization in white matter), followed by a rapid decline and then a leveling off in adulthood.

Males showed a higher and earlier peak relative to females, and these patterns were distinct from what was observed for the superior longitudinal fasciculus, which connects the frontal lobe with posterior cortical regions. Using a computational modeling approach, Willinger et al. (2021) demonstrated that during performance of the Monetary Incentive Delay Task, the ability to increase response vigor for high incentives improves from early adolescence to early adulthood as does information flow within cortico-striatal-thalamic circuitry. The authors conclude that the network that supports incentive-guided action undergoes a fine-tuning of effective connectivity across adolescence. Resting state studies are less consistent (Murty et al. 2018) but with some suggestion of an adolescent peak in ventral striatal connectivity (Porter et al. 2014). Overall, there is ample empirical support for unique patterns of heightened reward-related neural activations in the adolescence, consistent with a sensitive period, though the temporal dynamics (e.g., when the peaks in activation are observed) vary across studies.

Human structural MRI studies that have focused on the maturation of subcortical structures such as the amygdala and ventral striatum suggest heterogeneous maturational patterns across identified regions of interest, and there are pronounced individual differences. Age-related changes are differentially moderated by sex and appear to vary by brain hemisphere (Dennison et al. 2013; Raznahan et al. 2014; Urošević et al. 2014). Associations between gray matter volumes of regions such as the ventral striatum and self-reported incentive-reward motivation have been observed (Urošević et al. 2012) and in children as young as 9–10 years of age (Ide et al. 2020).

To account for these neurodevelopmental patterns that suggest an adolescent acceleration in incentive motivational processes, we have hypothesized that a transient elevation in DA signaling occurs within limbic and striatal circuitry, thereby providing an impetus for the acquisition of reward learning experiences (Luciana 2001; Luciana et al. 2012; Luciana 2013). Specifically, we have suggested that tonic levels of DA, reflective of an individual's genetically-based disposition toward incentive motivation (Goto et al. 2007; Ostlund et al. 2011) increase in adolescence (Luciana et al. 2012; Wahlstrom et al. 2010a, b). Higher levels of incentive motivation accompany higher levels of tonic DA in the ventral striatum and energize exploratory and approach behaviors, encouraging adolescents to engage with novel environments. If rewards are encountered during these explorations, then reward learning, which involves the generation of phasic DA responses (Schultz 2016), will follow. Experiences must be coded as highly salient if the phasic responses are to be detected against background high levels of tonic activity. Thus, due to variations in the specific nature of reward learning experiences that are encountered, there could be inconsistencies and individual differences in the ability to learn from reward prediction error signals (Luciana and Collins 2012; Luciana 2013). Phasic reward-related signals generated in the ventral striatum are conveyed to ventromedial subregions of the prefrontal cortex (vmPFC) where they are integrated with respect to the ongoing environmental structure of reward and punishment magnitudes and probabilities (Schultz 2016). The vmPFC has inhibitory back-projections to the VTA and ventral striatum, as shown by increased release of

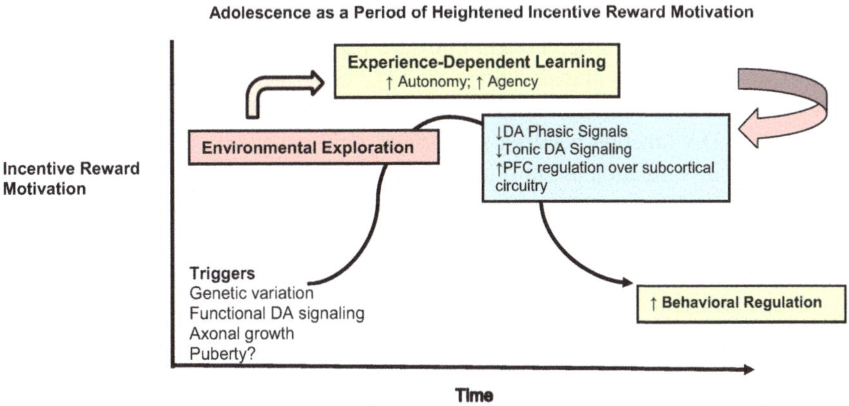

Fig. 1 Framework for the adolescent expression of incentive-reward motivation. Triggered by several individual difference factors, incentive-reward motivation is heightened in the adolescent period, leading to increased environmental exploration and experience-dependent learning that serves to increase personal agency, autonomy, and reward responsivity. This increase is accompanied by an increase in DA signaling in ventral striatal-to-prefrontal circuits. As the neurochemical system adapts to experience-based learning processes, incentive-reward motivation decreases into adulthood, accompanied by increases in capacities for behavioral regulation

DA after vmPFC lesions in rats. As the PFC generally undergoes synaptic maturation during adolescence, the vmPFC inhibitory back-projection to the VTA strengthens, leading to less DA release by early adulthood. Furthermore, with repeated exposures to the same reward stimuli phasic DA signals decline in amplitude as learning reaches an asymptote regarding what can be reliably predicted to occur under probabilistic reward conditions (Schultz 2016). As the initial novelty of the adolescent reward environment wanes and reward learning is consolidated, declines in tonic activity gradually occur and increased top-down prefrontal inhibitory influences are exerted over behavior. This overall dynamic is hypothesized to result in a relative decline in incentive motivational drives in adulthood in the context of more efficient coupling between prefrontal and limbic-striatal systems, which serves to facilitate adaptive motivated behavior (see Fig. 1).

While this formulation has suggested a mechanism for the increased reward sensitivity widely observed in adolescence and assessed neurobehaviorally in human fMRI studies (Silverman et al. 2015), rodent research has continued to produce conflicting findings regarding differences in adolescent versus adult DA system structure and function. For example, presynaptic activity levels of DA have been reported to be lower, not higher, in adolescent dorsal striatum and similar between adolescents and adults in NAcc (Matthews et al. 2013; Simon and Moghaddam 2015); conversely, findings may differ substantially for males vs females and depend on more finely differentiated subregions, e.g., NAcc core vs shell (Pitts et al. 2020). Similarly, differences in adolescent vs adult receptor densities may depend on sex and subregion, and perhaps are more significant for developmental differences in D1:D2 receptor density ratios than for absolute receptor density values (Cullity et al. 2019). More consistent support has been obtained for adolescent differences in PFC DA, e.g., regarding transient D1 receptor over-expression (at least in males; Sonntag et al. 2014), and for higher activity levels of VTA-DA neurons in adolescents (Wong et al. 2013). Thus, the developmental dynamics of DA function between subcortical and, particularly, prefrontal regions merit further investigation as substrates of adolescent reward seeking behavior.

5 Experiential Impacts on Incentive Motivation and its Development

The acceleration in adolescent approach behavior followed by an adult decline suggests the possibility of a time-limited process, similar to what might be predicted to occur during a sensitive period. However, it is not clear that this process has a discrete experiential trigger. It may be intrinsically driven. Accordingly, it seems that the impetus to engage with the environment prepares individuals in an experience-expectant manner (Depue and Collins 1999) for a varying range of social, emotional, and motivational experiences that form the foundation for subsequent learning and facilitate opportunities for reproductive success within a larger social group, and also confer a personal sense of agency in pursuit of adult autonomy.

What happens, then, if an adolescent fails to engage with the reward environment or if the experiential context is impoverished in terms of available resources? A challenge is to differentiate the influence of pre-existing individual difference factors from experience-driven developmental change, i.e., indications of alterations in reward system development associated with aberrant environments. Adaptive developmental plasticity frameworks suggest that cues indicative of harsh or adverse environments have particularly strong effects on the development of dopaminergic excitability within corticostriatal circuits that support incentive-based learning and decision making (Lin et al. 2020). Specifically, animals that experience early stress, such as maternal deprivation or social isolation, experience altered development (c.f., Peña et al. 2019; Sasagawa et al. 2017), though the direction of effects is inconsistent (Lin et al. 2020). Similar patterns are evident in humans as inferred from functional neuroimaging studies (Goff et al. 2013; Guyer et al. 2006). However, most studies focus on severe deprivation or maltreatment experienced early in life and relatively few have adopted dimensional approaches that apply to more normative ranges of experience. Using a population genetics approach, Fischer et al. (2018) demonstrated that a highly functional dopaminergic system at the population level, as indexed by individual differences in genetic indices, is associated with more behavioral approach traits in populations that experience increasing climatic (thermal) stress, indicating possible changes in the phenotypic expression of approach behaviors as a function of environmental demands. Of relevance to adolescence, the experience of adverse environments in childhood may provoke an acceleration of reproductive timing (e.g., earlier pubertal onset), which, in turn, promotes accelerated neurodevelopment of cortical and limbic circuitry involved in incentive salience and executive control (Thijssen et al. 2019; Gee et al. 2013). Accordingly, puberty may introduce opportunities for recalibration following early life stress, allowing high-risk youth to approach more typical levels of functioning as well as resilience (Gunnar et al. 2019; McEwen 2019). There are relatively few studies of neurobiological mechanisms of this type of resilience in adolescence (Feder et al. 2019; Masten et al. 2021). Known protective factors that have emerged from the literature on resilience that could be linked to the provocation of a DA-based incentive system include a sense of agency, mastery, close relationships, and strong levels of executive functioning (Masten et al. 2021).

Overall, these various formulations suggest that there should be reliable associations between aspects of pubertal timing, pubertal development, and the expression of incentive-motivated behavior. However, evidence in favor of such associations is inconsistent at best based on the current human (Dai and Scherf 2019; Goddings et al. 2019) as well as animal (Andersen et al. 2002) literatures, perhaps due to the various methodological challenges in linking pubertal mechanisms to experience-related plasticity in humans (Laube et al. 2020). Moreover, the field's interpretation of neuroimaging-based indices of adolescent neurodevelopment is currently in a state of flux, raising questions about the extent to which observed patterns represent nonlinear maturational changes.

6 Linear Versus Nonlinear Patterns of Adolescent Brain Development

DA-driven functional increases in incentive motivation during adolescence occur against a larger backdrop of changes in brain structure that have been more generally interpreted as indicators of a sensitive period of brain development (Piekarski et al. 2017). In humans and nonhuman primates cortical synapses are produced exuberantly during infancy, reaching approximately 1.5–2 times the adult number, and then are pruned until adult levels are reached by the end of adolescence (Huttenlocher and Dabholkar 1997). Regional differences have been documented whereby primary sensory and motor regions undergo synaptic pruning earlier in development than anterior prefrontal regions, which do not reach adult synaptic levels until mid-adolescence or even later (Petanjek et al. 2011). Synaptic pruning primarily involves asymmetrical junctions on dendritic spines, i.e., excitatory synapses, and outcomes depend on functional competition whereby synapses that are less active are eliminated. Selective elimination of weaker synapses is accompanied by increasing myelination (Monje 2018), which further enhances the functional activity of synaptic networks that were strengthened by experience during adolescent development.

Initial neuroimaging studies of normative adolescent development appeared to conform to this curvilinear "blooming and pruning" pattern of cortical synaptic changes, thereby adding support to the notion that adolescence represents a neurodevelopmental sensitive period. Specifically, the amount of gray matter as measured by gray matter volume (GMV) or cortical thickness (CT) seemed to follow an inverted U-shaped curve, i.e., increasing during childhood and reaching a maximum around the time of puberty onset, then declining toward an asymptote near the beginning of adulthood (Lenroot and Giedd 2006). As with synaptic analyses from human postmortem and animal studies, different cortical regions reached maximum MRI-indicated GMV and CT values at different ages, with the latest peak for PFC. Differences in the maturational timing of regional gray matter peaks were reported to be associated with functional neurocognitive outcomes (Walhovd et al. 2016). Conversely, white matter volume (WMV) appeared to change in a linear increasing pattern from childhood through mid- to late-adolescence, before reaching an asymptote in early adulthood (Paus et al. 2001).

More recent MRI studies have confirmed that cortical white matter development follows a linear increasing trend throughout child and adolescent development, but these studies challenge the existence of an inverted-U pattern for cortical gray matter that parallels synaptic blooming and pruning. When analyses have been based on very large samples (e.g., 700–800 participants or more) and employed sophisticated measurement methods (e.g., combining T_1-weighted and T_2-weighted scans to estimate cortical thickness) along with stringent controls over scan quality, GMV and CT most often have demonstrated monotonic linear declines from early childhood (3–5 years) through late adolescence (18–20 years). More specifically, the approximate time of puberty onset (7–9 years) does not mark an inflection point from increasing to decreasing GMV or CT (Walhovd et al. 2016; Mills et al. 2016).

Methodological factors play a significant role in the discrepancy between initial versus more recent MRI studies of adolescent gray matter development, but a larger issue involves interpretation of the initial findings as reflecting an apparent parallel between in vivo MRI data and synaptic assays in animal research and human postmortem tissue samples. In fact, the T_1-weighted structural MRI scans most commonly used in adolescent brain maturation research cannot detect synaptic changes on the scale reported in postmortem studies, as each voxel contains multiple biological components such as synapses, axons, neuronal cell bodies, glial cells, blood vessels, cerebrospinal fluids, etc., that combine in a complex way to generate a single resonance intensity value. Although reductions in CT values may reflect gray matter changes such as synaptic pruning, they also may be driven by intracortical myelination and/or expansion of the white matter surface boundary, with stretching or compaction of gray matter rather than pruning.

It is possible to obtain information about tissue structure within voxels using multimodal MRI, e.g., diffusion MRI that measures white matter microstructure combined with quantitative MRI that measures macromolecular tissue volume and T_1 relaxation time, which can differentiate apparent CT reductions due to myelination vs elimination of gray matter. For example, recent neuroimaging studies have assessed maturation of areas of the ventral temporal cortex that are specialized for processing faces vs places using multimodal MRI and comparisons with postmortem tissue analyses and found that apparent cortical thinning in face-specific areas was associated with underlying growth in both myelin and gray matter, while changes in cortical surface area rather than thickness were noted in nearby place-specific areas (Natu et al. 2019; Gomez et al. 2017). Similarly, Gennatas et al. (2017) applied novel processing and analysis techniques to T_1-weighted scans in a large sample ($n > 1,000$) of participants ranging from 8–23 years of age and found that age-related decreases in GMV and CT were linked to increases in gray matter density, suggesting that in humans gray matter may be compacted rather than pruned during cortical maturation.

Overall, recent studies suggest that cortical maturation is more complex and varied across functional regions, and involves not only selective gray matter pruning but also growth and microstructural changes within neuropil as well as white matter. In this updated view, cortical maturation appears to be mostly linear from early childhood to late adolescence and does not show dramatic gray matter changes around the time of puberty onset. Research on synaptic development has progressed as well to emphasize synaptic turnover, i.e., a combination of synaptogenesis and pruning, as an ongoing neuroplastic process that can lead to net decreases, increases, or stabilized levels of synaptic density, with varied outcomes particularly in PFC (including projections to subcortical structures such as the basolateral amygdala; Johnson et al. 2016). Thus, the compelling sensitive period model of blooming and pruning at the microstructural (synaptic) and macrostructural (MRI-measured GMV and CT) levels has given way to a more nuanced view that encompasses different forms of microstructural change and reorganization within broadly linear gray and white matter maturation from early childhood to early adulthood. Although we have emphasized cortical development, this more nuanced view may apply to subcortical structures as well, raising questions about the extent to which nonlinear patterns of maturation are evident in the adolescent period.

Without a peripubertal inflection in gray matter development, neuroimaging data do not provide an obvious marker of the opening of a neurodevelopmental sensitive period in adolescence. However, the development of the DA system itself provides a basis for initiating a sensitive period, albeit on a micro- rather than macro-structural level that eludes simple assessment using conventional MRI scans.

As suggested above, animal data indicate numerous changes in functional DA signaling in adolescence (Wahlstrom et al. 2010a, b). While axonal innervations throughout the brain have largely reached adult levels by the onset of adolescence, aspects of DA connectivity and receptor expression in limbic and prefrontal regions continue to be developmentally refined. In particular, DA innervation from the ventral striatum to the medial PFC markedly increases throughout adolescence and into young adulthood, accompanied by increasing numbers of DA synapses onto PFC pyramidal neurons (Hoops and Flores 2017). The DA-related axonal growth from the ventral striatum to the PFC that underpins the increased innervation represents one of the only known cases of long-distance axonal growth during adolescence, and until recently, the mechanisms that facilitate such growth have been elusive. Recent work indicates coordination by the Netrin-1 guidance cue receptor DCC (Vosberg et al. 2020), a finding that has considerable functional significance, particularly for behavioral flexibility. The DCC receptor system is fundamental to the intrinsic development of the PFC, and this development can be altered by experience. For instance, stimulant drug use in adolescence disrupts the pace and tone of PFC maturation through impacts on DCC in DA neurons. Salient experiences in adolescence such as initiation of alcohol and drug use can alter gene expression in a way that impacts DA and DCC receptors and ultimately alters synaptic morphology, e.g., via microRNAs (Hoops and Flores 2017). Variations in DCC receptor expression are also linked to psychiatric disorders, perhaps by altering the balance of resistance vs susceptibility to stressors. Axonal development is evident in adolescence for other subcortical-to-PFC pathways (e.g., from basolateral amygdala to orbitofrontal cortex; Johnson et al. 2016), though these developments are not always linked to DA mechanisms. Nonetheless, DA-related axonal progressions, rather than generalized cortical pruning processes, may represent the morphological substrate for the acceleration in incentive-reward motivated behavior that occurs in adolescence.

7 Conclusions

The period of human adolescence begins with pubertal onset and extends in industrialized societies, based on sociocultural factors and resource availability, well into the third decade of life. From a developmental standpoint, this is a vast period of time in terms of the number of biological, social, cognitive, and affective milestones that are typically achieved. We have argued that a fundamental life task for the adolescent is to distance oneself from parental dependence and develop autonomous, sexually mature relationships with an expanded peer group, ultimately to facilitate reproductive success. This transition is achieved by heightened activation of a biologically-based motivational system that provides a motoric and incentive-based thrust toward novelty and environmental exploration, in search of rewarding experiences of an agentic nature, i.e., associated with striving to act on and master the environment, and to assert social potency in a peer group. This striving for active exploration, mastery, and social status in turn encourages flexible and adaptive reward-based learning as well as pursuit of long-term incentives, processes that are facilitated by the activation of fronto-limbic-striatal DA signaling.

DA responsivity appears to be heightened in adolescence and late-maturing VTA-DA projections innervate the PFC during this time, with effects on neuroplasticity that are driven by environmental experience (and perhaps mediated by recruitment of microRNA). While these elements of a sensitive period are compelling, neither the DA neurobiology nor the heightened incentive-reward behavior is tightly linked to pubertal onset, and thus a validated timing mechanism for the opening of the sensitive period appears to be lacking. While we maintain the centrality of a DA-based mechanism in accounting for heightened incentive-reward behavior in adolescents, it is challenging to assess DA activity reliably and validly in humans, particularly when using low-risk neuroimaging in large-sample research. Although functional MRI during reward-related tasks holds continued promise, there are ongoing concerns about its reliability (Elliott et al. 2020), and extant studies fail to show reliable demarcations between middle childhood and adolescence in brain activation correlates of reward seeking behavior.

The "expected experiences" that are encountered after exploratory tendencies are heightened include social engagement with peers outside of the family group; interest in potential romantic partners; engagement with socially salient educational and athletic goals as well as leisure pursuits; opportunities to earn money; and independent travel/transportation. From this list it is difficult to identify the specific quality of experience that might trigger sensitive period development, or the mechanism by which that experience opens and eventually closes the sensitive period, at least at the level of stringency associated with known critical periods like the development of binocular vision. Moreover, it is not clear that the absence of any of these experiences during adolescence, except perhaps for engaging in autonomous peer social activity away from the natal group, is detrimental to subsequent success in adulthood. It is evident that there are individual differences in outcomes of this experience, but these can be construed as variations within normative limits rather than deprivation-based sensitive period failures. Moreover, "late bloomers" often reach these various milestones later in adulthood, e.g., in association with becoming parents and starting families and achieving vocational successes, apparently without disabling effects given current expectancies regarding extended adolescence. In this regard, human data appear to diverge from animal findings regarding stress-induced developmental disruptions that are more consistent with sensitive period phenomena, although methodological discrepancies remain to be resolved (Watt et al. 2017).

The notion that adolescence represents a period of resilience or "second chances" for those who have experienced early adversity is a compelling alternative to the sensitive period framework. To the extent that early adversity is linked to dysfunctional family dynamics, an adolescent drive to engage deeply with peers away from the family group would seem to introduce a hopeful range of new possibilities to be explored regarding more positive and rewarding interpersonal experiences. We note that such rehabilitation does not require a qualitative shift in neurobehavioral development, but rather a mechanism to promote independent exploration of alternative social environments while ongoing neurobehavioral plasticity remains at a high level. Though not the focus of this review, evidence supports that such plasticity exists not only in adolescence but in childhood and adulthood as well.

With respect to incentive-reward behaviors, while the literature suggests gradual declines from adolescence into adulthood, plasticity of the underlying neural systems remains intact to facilitate lifelong experience-dependent reward learning. Over time and in response to variation in salient risk-reward contingencies, this neuroplasticity manifests in shifting influences of PFC vs limbic-striatal mediation of incentive-reward processing to tailor decision making toward long-term or short-term reward acquisition. Thus, although incentive-reward behavior declines overall toward the end of adolescence, there is little evidence in human research indicating that a functional developmental window closes with the transition to adulthood.

These observations raise questions about the utility of the sensitive period concept in accounting for the maturational milestones accomplished in adolescence regarding incentive-reward behavior, despite the presence of nonlinear increases in behavioral and perhaps dopaminergic responsivity to reward. Part of the issue is definitional: to the extent that the nature of expected experience is variable in tone or intensity, that onset of a sensitive period is temporally variable, that sensitive periods can reoccur, that multiple systems are impacted, that effects are impermanent, and all of the above shaped by inherent individual difference factors, the explanatory "signal" appears reduced against a great deal of background noise and allowable variation, at least as applied to typical human experience. In addition, the higher order cognitive and affective processes that emerge in adolescence are assessed as latent constructs, with wide variation in measured indicator variables that confounds direct description of the specific nature of experiences and behaviors that are central to the sensitive period. In turn, this ambiguity confounds explication of a specific mechanism for the opening and closing of the functional developmental window for incentive-reward behavior, as well as its exact timing. Moreover, the genetic and environmental factors that influence individual differences in the incentive-reward behavioral trait, agentic positive emotionality, are clearly operative well before adolescence and an initial foundation for the behavior already exists. In this context, heightened DA responsivity to reward cues during adolescence may be viewed as a form of juvenile adaptation rather than a sensitive period mechanism, i.e., a transient alteration in a neurobehavioral system that serves to facilitate successful attainment of maturational milestones in incentive-reward behavior by enhancing active engagement with novel environments that must be mastered in the transition to adulthood. We hypothesize that this crucial period of development utilizes the neuroplasticity associated with ongoing brain maturation from early childhood to early adulthood, but without the semi-permanence of brain changes as may occur in a sensitive period. Sensitive periods are striking but rare phenomena in lifespan development (Walhovd and Lövdén 2020), and not always supported by continuing research; as we have described above, there are considerable methodological and interpretive challenges in reconciling, for instance, commonly held "blooming and pruning" sensitive period models of adolescent brain development against more recent structural MRI research.

The field would benefit from a more stringent approach to defining and assessing aspects of reward seeking that are most relevant to our proposed framework. We note again that reward seeking is not synonymous with risk-taking. The limitations of neuroimaging technology should also be explicitly acknowledged, particularly given the desire to enhance rigor and reproducibility of findings. What we are most interested in is a pattern of neurochemical alterations at the synaptic level. Structural and functional neuroimaging at moderate field strengths cannot resolve the dynamics of excitatory and inhibitory neurotransmission. Tools such as positron emission tomography (PET) have significant ethical and practical limitations particularly in developmental applications. A more in-depth exploration of physiological correlates of responses measured by PET and/or using pharmacological challenges might allow this field of inquiry to advance in a more direct manner. Finally, a lifespan perspective from infancy to older age is needed to place adolescent development in its proper position, as one cannot infer the full trajectory of reward seeking behavior from research focused on the adolescent-to-young adult period alone. It is unequivocal that adolescence represents a period of robust development and exciting opportunity. The mechanisms that underlie its unique features remain to be fully determined.

References

Andersen SL, Thompson AP, Krenzel E, Teicher MH (2002) Pubertal changes in gonadal hormones do not underlie adolescent dopamine receptor overproduction. Psychoneuroendocrinology 27 (6):683–691

Berridge KC (2018) Evolving concepts of emotion and motivation. Front Psychol 9:1647. https://doi.org/10.3389/fpsyg.2018.01647

Bornstein MH (1989) Sensitive periods in development: structural characteristics and causal interpretations. Psychol Bull 105(2):179–197

Braams BR, Davidow JY, Somerville LH (2019) Developmental patterns of change in the influence of safe and risky peer choices on risky decision-making. Dev Sci 22:e12717. https://doi.org/10.1111/desc.12717

Center for Disease Control (2020). https://www.cdc.gov/alcohol/fact-sheets/underage-drinking.htm. Accessed 21 Jul 2021

Ciranka S, van den Bos W (2019) Social influence in adolescent decision-making: a formal framework. Front Psychol 10:1915. https://doi.org/10.3389/fpsyg.2019.01915

Cullity ER, Madsen HB, Perry CJ, Kim JH (2019) Postnatal developmental trajectory of dopamine receptor 1 and 2 expression in cortical and striatal brain regions. J Comp Neurol 527:1039–1055

Dai J, Scherf KS (2019) Puberty and functional brain development in humans: convergence in findings? Dev Cogn Neurosci 39:100690

de Boer L, Axelsson J, Chowdhury R et al (2019) Dorsal striatal dopamine D1 receptor availability predicts an instrumental bias in action learning. Proc Natl Acad Sci U S A 116:261–270

Dennison M, Whittle S, Yücel M et al (2013) Mapping subcortical brain maturation during adolescence: evidence of hemisphere- and sex-specific longitudinal changes. Dev Sci 16 (5):772–791

Depue RA, Collins PF (1999) Neurobiology of the structure of personality: dopamine, facilitation of incentive motivation, and extraversion. Behav Brain Sci 22:491–517

Depue RA, Luciana M, Arbisi PA et al (1994) Dopamine and the structure of personality: relation of agonist-induced dopamine activity to positive emotionality. J Pers Soc Psychol 67(3):485–498

Doremus-Fitzwater TL, Spear LP (2016) Reward-centricity and attenuated aversions: an adolescent phenotype emerging from studies in laboratory animals. Neurosci Biobehav Rev 70:121–134

Dreher JC, Kohn P, Kolachana P, Weinberger DR, Berman KF (2009) Variation in dopamine genes influences responsivity of the human reward system. PNAS 106(2):617–622

Duell N, Steinberg L, Icenogle G et al (2018) Age patterns in risk taking across the world. J Youth Adolesc 47(5):1052–1072

Durbin CE, Hicks BM, Blonigen DM et al (2016) Personality trait change across late childhood to young adulthood: evidence for nonlinearity and sex differences in change. Eur J Personal 30 (1):31–44

Elliott ML, Knodt AR, Ireland D et al (2020) What is the test-retest reliability of common task-functional MRI measures? New empirical evidence and a meta-analysis. Psychol Sci 31 (7):792–806

Feder A, Fred-Torres S, Southwick SM, Charney DS (2019) The biology of human resilience: opportunities for enhancing resilience across the life span. Biol Psychiatry 86(6):443–453
Fischer R, Lee A, Verzijden MN (2018) Dopamine genes are linked to extraversion and neuroticism personality traits, but only in demanding climates. Sci Rep 8:1733. https://doi.org/10.1038/s41598-017-18784-y
Frank MJ, Fossella JA (2011) Neurogenetics and pharmacology of learning, motivation, and cognition. Neuropsychopharmacology 36:133–152
Fuhrmann D, Knoll LJ, Blakemore S (2015) Adolescence as a sensitive period of brain development. Trends Cogn Sci 19:558–566
Gabard-Durnam LJ, McLaughlin KA (2020) Sensitive periods in human development: charting a course for the future. Curr Opin Behav Sci 36:120–128
Gee DG, Gabard-Durnam LJ, Flannery J et al (2013) Early developmental emergence of human amygdala-prefrontal connectivity after maternal deprivation. PNAS 110(39):15638–15643
Gennatas ED, Avants BB, Wolf DH et al (2017) Age-related effects and sex differences in gray matter density, volume, mass, and cortical thickness from childhood to young adulthood. J Neurosci 37(20):5065–5073
Goddings AL, Beltz A, Peper JS, Crone EA, Braams BR (2019) Understanding the role of puberty in structural and functional development of the adolescent brain. J Res Adolesc 29(1):32–53
Goff B, Gee DG, Telzer EH et al (2013) Reduced nucleus accumbens reactivity and adolescent depression following early-life stress. Neuroscience 249:129–138
Gomez J, Barnett MA, Natu V et al (2017) Microstructural proliferation in human cortex is coupled with the development of face processing. Science 355(6320):68–71
Goto Y, Otani S, Grace AA (2007) The Yin and Yang of dopamine release: a new perspective. Neuropharmacology 53(5):583–587
Gruber MJ, Ranganath C (2019) How curiosity enhances hippocampus-dependent memory: the prediction, appraisal, curiosity, and exploration (PACE) framework. Trends Cogn Sci 23 (12):1014–1025
Gunnar MR, DePasquale CE, Reid BM, Donzella B (2019) Pubertal stress recalibration reverses the effects of early life stress in post institutionalized children. Proc Natl Acad Sci U S A 116:23984–23988
Guyer AE, Kaufman J, Hodgdon HB et al (2006) Behavioral alterations in reward system function: the role of childhood maltreatment and psychopathology. J Am Acad Child Adolesc Psychiatry 45:1059–1067
Harden KP, Tucker-Drob EM (2011) Individual differences in the development of sensation seeking and impulsivity during adolescence: further evidence for a dual systems model. Dev Psychol 47:739–746
Harden KP, Mann FD, Grotzinger AD et al (2018) Developmental differences in reward sensitivity and sensation seeking in adolescence: testing sex-specific associations with gonadal hormones and pubertal development. J Pers Soc Psychol 115(1):161–178
Hoops D, Flores C (2017) Making dopamine connections in adolescence. Trends Neurosci 40 (12):709–719
Huttenlocher PR, Dabholkar AS (1997) Regional differences in synaptogenesis in human cerebral cortex. J Comp Neurol 387(2):167–178
Ide JS, Li HT, Chen Y et al (2020) Gray matter volumetric correlates of behavioral activation and inhibition system traits in children: an exploratory voxel-based morphometry study of the ABCD project data. NeuroImage 220:117085. https://doi.org/10.1016/j.neuroimage.2020.117085
Johnson CM, Loucks FA, Peckler H et al (2016) Long-range orbitofrontal and amygdala axons show divergent patterns of maturation in the frontal cortex across adolescence. Dev Cogn Neurosci 18:113–120
Karlsgodt KH, John M, Ikuta T et al (2015) The accumbofrontal tract: diffusion tensor imaging characterization and developmental change from childhood to adulthood. Hum Brain Mapp 36 (12):4954–4963

Knudsen EI (2004) Sensitive periods in the development of the brain and behavior. J Cogn Neurosci 16:1412–1425

Kolb B, Mychasiuk R, Muhammad A, Gibb R (2013) Brain plasticity in the developing brain. Prog Brain Res 207:35–64

Laube C, van den Bos W, Fandakova Y (2020) The relationship between pubertal hormones and brain plasticity: implications for cognitive training in adolescence. Dev Cogn Neurosci 42:100753

Lenroot RK, Giedd JN (2006) Brain development in children and adolescents: insights from anatomical magnetic resonance imaging. Neurosci Biobehav Rev 30(6):718–729

Lin WC, Delevich K, Wilbrecht L (2020) A role for adaptive developmental plasticity in learning and decision making. Curr Opin Behav Sci 36:48–54

Lindenberger U, Lövdén M (2019) Brain plasticity in human lifespan development: the exploration–selection–refinement model. Annu Rev Dev Psychol 1(1):197–222

Luciana M (2001) Dopamine-opiate modulations of reward-seeking behavior: implications for the functional assessment of prefrontal development. In: Nelson CA, Luciana M (eds) Handbook of developmental cognitive neuroscience. MIT Press, pp 647–662

Luciana M (2013) Adolescent brain development in normality and psychopathology. Dev Psychopathol 25:1325–1345

Luciana M, Collins PF (2012) Incentive motivation, cognitive control, and the adolescent brain: is it time for a paradigm shift? Child Dev Perspect 6(4):392–399

Luciana M, Wahlstrom D, Collins PF, Porter JN (2012) Dopaminergic modulation of incentive motivation in adolescence: age-related changes in signaling, individual differences, and implications for the development of self-regulation. Dev Psychol 48(3):844–861

Martinez GM, Abma JC (2020) Sexual activity and contraceptive use among teenagers aged 15–19 in the United States, 2015–2017. NCHS data brief no. 366. Updated May 6, 2020. https://www.cdc.gov/nchs/data/databriefs/db366-h.pdf. Accessed 11 May 2020

Masten AS, Lucke CM, Nelson KM, Stallworthy IC (2021) Resilience in development and psychopathology: multisystem perspectives. Annu Rev Clin Psychol 17:521–549

Matthews M, Bondi C, Torres G, Moghaddam B (2013) Reduced presynaptic dopamine activity in adolescent dorsal striatum. Neuropsychopharmacology 38:1344–1351

McEwen B (2019) Resilience of the brain and body. In: Fink G (ed) Stress, vol. 3: physiology, biochemistry, and pathology. Handb. Stress Ser. Academic, Cambridge, pp 19–33

Mills KL, Goddings AL, Herting MM et al (2016) Structural brain development between childhood and adulthood: convergence across four longitudinal samples. NeuroImage 141:273–281

Monje M (2018) Myelin plasticity and nervous system function. Annu Rev Neurosci 41:61–76

Mueller EM, Burgdorf C, Chavanon ML et al (2014) Dopamine modulates frontomedial failure processing of agentic introverts versus extraverts in incentive contexts. Cogn Affect Behav Neurosci 14:756–768

Murty VP, Shah H, Montez D et al (2018) Age-related trajectories of functional coupling between the VTA and nucleus accumbens depend on motivational state. J Neurosci 38(34):7420–7427

Natu VS, Gomez J, Barnett M et al (2019) Apparent thinning of human visual cortex during childhood is associated with myelination. PNAS 116(41):20750–20759

Ostlund SB, Wassum KM, Murphy NP et al (2011) Extracellular dopamine levels in striatal subregions track shifts in motivation and response cost during instrumental conditioning. J Neurosci 31(1):200–207

Patrick CJ, Curtin JJ, Tellegen A (2002) Development and validation of a brief form of the multidimensional personality questionnaire. Psychol Assess 14(2):150–163

Paus T, Collins DL, Evans AC et al (2001) Maturation of white matter in the human brain: a review of magnetic resonance studies. Brain Res Bull 54(3):255–266

Peña CJ, Smith M, Ramakrishnan A et al (2019) Early life stress alters transcriptomic patterning across reward circuitry in male and female mice. Nat Commun 10:5098

Petanjek Z, Judaš M, Šimic G et al (2011) Extraordinary neoteny of synaptic spines in the human prefrontal cortex. PNAS 108(32):13281–13286

Piekarski DJ, Johnson CM, Boivin JR et al (2017) Does puberty mark a transition in sensitive periods for plasticity in the associative neocortex? Brain Res 1654(Pt B):123–144
Pitts EG, Stowe TA, Christensen BA, Ferris MJ (2020) Comparing dopamine release, uptake, and D2 autoreceptor function across the ventromedial to dorsolateral striatum in adolescent and adult male and female rats. Neuropharmacology 175:1081663
Porter JN, Roy AK, Benson B et al (2014) Age-related changes in the intrinsic functional connectivity of the human ventral versus dorsal striatum from childhood to middle age. Dev Cogn Neurosci 11:83–95
Raznahan A, Shaw PW, Lerch JP (2014) Longitudinal four-dimensional mapping of subcortical anatomy in human development. PNAS 111(4):1592–1597
Sasagawa T, Horii-Hayashi N, Okuda A et al (2017) Long-term effects of maternal separation coupled with social isolation on reward seeking and changes in dopamine D1 receptor expression in the nucleus accumbens via DNA methylation in mice. Neurosci Lett 641:33–39
Schreuders E, Braams BR, Blankenstein NE et al (2018) Contributions of reward sensitivity to ventral striatum activity across adolescence and early adulthood. Child Dev 89(3):797–810
Schultz W (2016) Dopamine reward prediction error coding. Dialogues Clin Neurosci 18(1):23–32
Silverman MH, Jedd K, Luciana M (2015) Neural networks involved in adolescent reward processing: an activation likelihood estimation meta-analysis of functional neuroimaging studies. NeuroImage 122:427–439
Simon NW, Moghaddam B (2015) Neural processing of reward in adolescent rodents. Dev Cogn Neurosci 11:145–154
Solinas M, Belujon P, Fernagut PO et al (2019) Dopamine and addiction: what have we learned from 40 years of research. J Neural Transm 126:481–516
Sonntag KC, Brenhouse HC, Freund N, Thompson BS, Puhl M, Andersen SL (2014) Viral over-expression of D1 dopamine receptors in the prefrontal cortex increase high-risk behaviors in adults: comparison with adolescents. Psychopharmacology 231:1615–1626
Takesian AE, Hensch TK (2013) Balancing plasticity/stability across brain development. Prog Brain Res 207:3–34
Thijssen S, Collins PF, Luciana M (2019) Pubertal development mediates the association between family environment and brain development in childhood. Dev Psychopathol. https://doi.org/10.1017/S0954579419000580
Urošević S, Collins P, Muetzel R, Lim K, Luciana M (2012) Longitudinal changes in behavioral approach system sensitivity and brain structures involved in reward processing during adolescence. Dev Psychol 48(5):1488–1500
Urošević S, Collins PF, Muetzel R, Lim KO, Luciana M (2014) Pubertal status associations with reward and threat sensitivities and subcortical brain volumes during adolescence. Brain Cogn 89:15–26
Vosberg DE, Leyton M, Flores C (2020) The Netrin-1/DCC guidance system: dopamine pathway maturation and psychiatric disorders emerging in adolescence. Mol Psychiatry 25:297–307
Wacker J, Smillie LD (2015) Trait extraversion and dopamine function. Soc Personal Psychol Compass 9(6):225–238
Wahlstrom D, Collins PF, White T, Luciana M (2010a) Developmental changes in dopamine neurotransmission in adolescence: behavioral implications and issues in assessment. Brain Cogn 72(1):146–159
Wahlstrom D, White T, Luciana M (2010b) Neurobehavioral evidence for changes in dopamine system activity during adolescence. Neurosci Biobehav Rev 34(5):631–648
Walhovd KB, Lövdén M (2020) A lifespan perspective on human neurocognitive plasticity. In: Poeppel D, Mangun G, Gazzaniga MS (eds) The cognitive neurosciences, 6th edn. MIT Press, Cambridge
Walhovd KB, Fjell AM, Giedd J et al (2016) Through thick and thin: a need to reconcile contradictory results on trajectories in human cortical development. Cereb Cortex 27(2):1472–1481

Ward J, Lyall LM, Bethlehem RAI et al (2019) Novel genome-wide associations for anhedonia, genetic correlation with psychiatric disorders, and polygenic association with brain structure. Transl Psychiatry 9:327. https://doi.org/10.1038/s41398-019-0635-y
Watt MJ, Weber MA, Davies SR, Forster GL (2017) Impact of juvenile chronic stress on adult cortico-accumbal function: implications for cognition and addiction. Prog Neuro-Psychopharmacol Biol Psychiatry 79:136–154
Werker JF, Hensch TK (2015) Critical periods in speech perception: new directions. Annu Rev Psychol 66:173–196
Willinger D, Karipidis II, Dimanova P, Walitza S, Brem S (2021) Neurodevelopment of the incentive network facilitates motivated behaviour from adolescence to adulthood. NeuroImage 237:118186
Wise RA, Robble MA (2020) Dopamine and addiction. Annu Rev Psychol 71:79–106
Wong WC, Ford KA, Pagels NE, McCutcheon JE, Marinelli M (2013) Adolescents are more vulnerable to cocaine addiction: behavioral and electrophysiological evidence. J Neurosci 33 (11):4913–4922
Woodard K, Pollak SD (2020) Is there evidence for sensitive periods in emotional development? Curr Opin Behav Sci 36:1–6
Yaple ZA, Yu R, Arsalidou M (2020) Spatial migration of human reward processing with functional development: evidence from quantitative meta-analyses. Hum Brain Mapp 41 (14):3993–4009

Part II
The Influences of Positive and Negative Factors During Sensitive Periods

Physical Activity, Fitness, and Executive Functions in Youth: Effects, Moderators, and Mechanisms

David R. Lubans, Angus A. Leahy, Myrto F. Mavilidi, and Sarah R. Valkenborghs

Contents

Abstract Over the last decade, a growing body of research has examined the link between physical activity, fitness, and cognitive function in children and adolescents. Physical activity experimental research conducted with children and adolescents has identified selectively greater effects for tasks requiring higher order executive functions. As such, the primary aim of our chapter is to provide an overview of findings from systematic reviews and meta-analyses that have examined the effects of physical activity on measures of executive function in child and

D. R. Lubans (✉), A. A. Leahy, and S. R. Valkenborghs
The University of Newcastle, Callaghan, NSW, Australia
e-mail: David.Lubans@newcastle.edu.au

M. F. Mavilidi
The University of Wollongong, Wollongong, NSW, Australia

Curr Topics Behav Neurosci (2022) 53: 103–130
https://doi.org/10.1007/7854_2021_271
Published Online: 26 October 2021

adolescent populations. We begin our chapter with definitions of key concepts associated with physical activity, fitness and cognitive function. We then provide a synthesis of systematic reviews and meta-analyses that have examined the acute and chronic effect of physical activity on EFs. Following this, we discuss the quantitative (e.g., time, intensity) and qualitative (e.g., type) characteristics of physical activity that may moderate effects. The next section focuses on the neurobiological, psychosocial and behavioral mechanisms responsible for the effect of physical activity on executive functions. We conclude by highlighting the limitations of the existing evidence base and providing recommendations for future research.

Keywords Cognition · Executive function · Exercise · Mechanisms · Physical activity

1 Introduction

Childhood and adolescence represent a critical and sensitive period of brain development (Andersen 2003), characterized by neuronal plasticity, the formulation of identity (Sebastian et al. 2008), and the establishment of health behaviors, such as physical activity (Telama 2009), that may enhance or diminish cognition functioning. Over the last decade there has been an exponential increase in the number of studies examining the links between physical activity, physical fitness, and cognition in young people. Research in children has demonstrated that the effects of physical activity are selectively greater for tasks requiring higher order executive functions (EFs) (Hillman et al. 2014). As such, the primary aim of our chapter is to provide an overview of experimental studies that have examined the effect of physical activity on EFs in child and adolescent populations. We begin our chapter with definitions of key concepts associated with physical activity, fitness, and cognition. Following this, we discuss the acute and chronic effects of physical activity on EFs. In the next section we explore the quantitative (e.g., time and intensity) and qualitative (e.g., type and context of activity) characteristics of physical activity that may moderate effects. We conclude our chapter by providing an overview of the potential neurobiological, psychosocial, and behavioral mechanisms responsible for the effects of physical activity on cognition.

2 Physical Activity and Fitness: Definitions, Guidelines, and Trends

Physical activity can be defined as "any bodily movement produced by skeletal muscles that results in energy expenditure" (Caspersen et al. 1985, p. 126). It is a broad term that includes sport, exercise, recreational pursuits and activities performed as part of daily living (e.g., active transportation). Although there is some variability between national and international physical activity guidelines (Parrish et al. 2020), it is generally accepted that children (5–12 years) and adolescents (13–18 years) should do at least an average of 60 min per day of moderate-to-vigorous physical activity (MVPA), mostly aerobic activity, across the week (Bull et al. 2020). It is also recommended that young people participate in activities of vigorous intensity, including those that strengthen muscle and bone at least three times per week (Bull et al. 2020).

Participation in physical activity is the primary means of improving physical fitness (often referred to as health-related fitness), which consists of cardiorespiratory fitness, muscular fitness, flexibility, and body composition. Cardiorespiratory fitness (CRF), also known as cardiorespiratory endurance, cardiovascular fitness, aerobic capacity, and aerobic fitness, refers to the capacity of the circulatory and respiratory systems to provide oxygen to skeletal muscles for energy production during physical activity (Raghuveer et al. 2020). Participation in aerobic activity involving large muscle groups (e.g., jogging and playing team sports) helps to improve CRF. Muscular fitness refers to an individual's ability to exert maximal force against an external resistance (i.e., muscular strength) or repeatedly under sub-maximal loads (i.e., local muscular endurance). Flexibility refers to an individual's range of motion around a joint or group of joints. Finally, body composition is the relative proportion of total body mass consisting of fat, fat-free tissue, and total body water (Raghuveer et al. 2020; Smith et al. 2020). It is important to note that different types of physical activity will produce different adaptations in physical fitness components.

Physical activity levels typically increase during early childhood, start to decline by age seven (Farooq et al. 2018), and continue to decline during adolescence and early adulthood (Corder et al. 2019; Dumith et al. 2011). For example, Dumith et al. (2011) conducted a meta-analysis of 26 longitudinal studies and found that physical activity levels declined by 7% per year throughout adolescence (defined as 10–19 years). The decline in activity occurs earlier in girls (9–12 years), compared to boys (13–16 years) and continues into adulthood. Corder et al. (2019) examined the decline in activity from adolescence to adulthood and observed a decline of 5.2 min of MVPA/day over a mean study period of 3.4 years.

Physical inactivity has been described as a pandemic and the majority of children and adolescents across the globe are not sufficiently active (Guthold et al. 2020). The Global Matrix of Physical Activity Report Grades was launched in 2014 to provide a better understanding of the global variation in children and adolescent physical activity (Tremblay et al. 2014). In 2018, the Global Matrix 3.0 was released and included physical activity data from 49 countries (Aubert et al. 2018). An average grade of "D" was assigned, indicating that approximately 20 to 40% of children and adolescents globally are meeting physical activity guidelines. In general, countries considered to be very high on the United Nation's human development index (HDI) ranked higher than the low, medium, and high HDI countries.

For the purpose of our chapter, we define cognitive function as the set of mental processes that contribute to perception, memory, intellect, and action (Donnelly et al. 2016). We acknowledge that cognitive tasks exist on a continuum, from basic information processing at one end to tasks requiring high levels of EF at the other. Basic information processing tasks require fewer resources and therefore represents a lower level of cognitive function (e.g., information processing tasks) (Colcombe and Kramer 2003). Alternatively, EFs represent higher order cognitive processes that are critical for planning, problem solving, and learning (Diamond 2013). The core EFs include inhibition (often termed "inhibitory control" or "attentional control"), working memory (also referred to as "updating"), and cognitive flexibility (also known as "set shifting"). A variety of assessments have been used to measure inhibition (e.g., modified flanker task, Stroop-color task, go/no-go task), working memory (e.g., serial n-back task, digit span test, Corsi block task), and cognitive flexibility (e.g., trail making test, Wisconsin card sorting test, switch task) in child and adolescent populations (Wade et al. 2020a).

3 Acute and Chronic Effects of Physical Activity on Executive Functions

Acute physical activity refers to a single bout of physical activity/exercise, while chronic physical activity/exercise can be defined as repeated bouts over a short- or long-term period. A seminal review and meta-analysis conducted by Chang et al. (2012) examined the acute effect of physical activity on cognitive function in children, adolescents, and adults. The authors found transient effects for cognitive performance both during and following a single bout of physical activity. Greater effects were observed for tasks that assessed EFs (as opposed to information processing). More recently, Hillman et al. (2019a) published a narrative review examining the effects of acute bouts of physical activity on EF sub-domains in children. The authors found strong evidence for the benefits of physical activity for inhibition, with selectively larger effects in performance for tasks that require greater inhibitory control (e.g., flanker task performance for incongruent trials). They also identified that few studies have examined the acute effects of physical activity on working memory and cognitive flexibility, with contrasting meta-analytical findings (de Greeff et al. 2018; Ludyga et al. 2016).

The chronic effect of physical activity on EFs is less clear. Numerous systematic reviews and meta-analyses have reported small-to-moderate effects of physical activity on EFs in child and adolescent populations (Álvarez-Bueno et al. 2017; Leahy et al. 2020; Sun et al. 2020). In contrast, an international expert panel recently concluded that there is inconclusive evidence for the beneficial effects of chronic physical activity interventions on cognitive function (not specifically EF) for children and adolescents (Singh et al. 2019).

Considering this lack of clarity, the primary aim of this section is to provide a synthesis of findings from published systematic reviews and meta-analyses. We conducted a search of relevant titles and identified 11 systematic reviews that met the following eligibility criteria: (1) reported the results from meta-analyses examining the acute or chronic effect of physical activity on EFs (overall or sub-domains) and (2) reported results for children (5–12 years), adolescents (13–18 years), or children and adolescents together (5–18 years). Effect size values of 0.2, 0.5, and 0.8 served as thresholds for small, medium, and large effects, respectively. Our secondary aim was to determine if there are certain sub-groups that benefit more or less from physical activity. More specifically, does age (i.e., children vs adolescents) and population type (e.g., healthy vs unhealthy) moderate effects? The quantitative and qualitative aspects of physical activity that may moderate effects are discussed in the next section.

3.1 Acute Effect of Physical Activity on Overall Executive Function

We found six reviews (10 effect sizes) that focused on the acute effect of physical activity on EFs (Fig. 1). Meta-analysis effect sizes ranged from 0.04 (Ludyga et al. 2016) to 0.57 (Verburgh et al. 2014), with an average effect size (ES) of 0.35. One review reported separate performance effects for accuracy and response time (Ludyga et al. 2016). For accuracy, a small effect size was observed for children (ES = 0.29) and a trivial effect observed for adolescents (ES = 0.16). A moderate effect size was found for children's response time (ES = 0.54), while the effect for adolescents was not significant (ES = 0.04). However, only three studies contributed to the effect size for adolescents so these findings should be interpreted with caution.

Effect sizes ranged from 0.13 to 0.57 for children and from 0.04 to 0.52 for adolescents. The average effect sizes for children and adolescents were similar (0.35 vs 0.31, respectively). However, for studies that examined the effects on children and adolescents separately, contrasting findings have been reported. For example, one review found acute physical activity only improved EF in children (Ludyga et al. 2016), while another found that adolescents (and not children) improved EF as a result of acute physical activity (Moreau and Chou 2019). Two reviews found that effects did not significantly differ for children and adolescents, with moderate effect sizes observed (Leahy et al. 2020; Verburgh et al. 2014). Given the current evidence base, age does not appear to be a clear moderator for the acute effects on EFs. No studies examined the acute effects in relation to healthy vs unhealthy populations (i.e., overweight/obese, or cognitive disorders).

3.2 Acute Effect of Physical Activity on Executive Function Sub-domains

Only two reviews (five effect sizes) investigated the acute effect of physical activity on EF sub-domains in children and adolescents (de Greef et al. 2018; Verburgh et al.

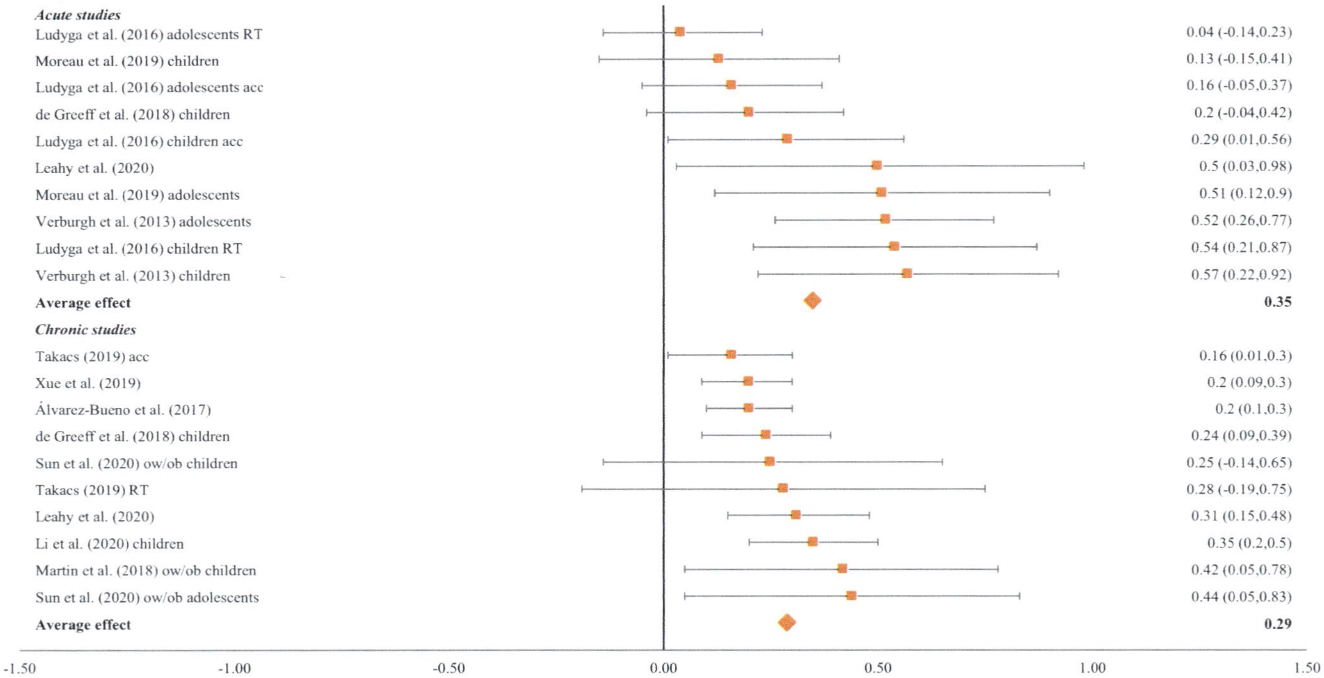

Fig. 1 Summary of meta-analyses examining the acute and chronic effects of physical activity on executive function. *Note: RT* response time, *acc* accuracy, *ow/ob* overweight and obese

2014). Effect sizes for the performance on tasks requiring inhibition ranged from 0.28 to 0.57, with an average effect of 0.46. Verburgh et al. (2014) observed similar effects for children and adolescents (ES = 0.57 and 0.52, respectively), however only a small number of study samples were included (children = 2; adolescents = 3). In a more recent review, de Greeff et al. (2018) observed a smaller effect on inhibition (ES = 0.28). It is worth noting that a larger number of study samples were included (k = 10), and therefore may reflect a more accurate effect. Only one meta-analysis examined the acute effects of physical activity on working memory and cognitive flexibility in children (de Greeff et al. 2018). The authors observed small improvements in working memory (ES = 0.28) and cognitive flexibility (ES = 0.30)

3.3 Chronic Effect of Physical Activity on Overall Executive Functions

We identified a total of eight meta-analyses (10 effect sizes) examining the effect of chronic physical activity (Fig. 1). All reviews were published within the last 5 years. Chronic physical activity interventions appear to produce small-to-moderate positive effects on EF, with effect sizes ranging from 0.16 (Takacs and Kassai 2019) to 0.44 (Sun et al. 2020), with an average effect size of 0.29. Only one review reported specific cognitive performance effects, which revealed a larger effect for response time (ES = 0.28) (rather than accuracy [ES = 0.16]) (Takacs and Kassai 2019).

Chronic physical activity interventions appear to have a small effect for children (ES = 0.29), and a small-to-moderate effect for adolescents (ES = 0.44). This is somewhat consistent with findings from a recent review which found that age was a significant moderator of intervention effects. Specifically, significant effects were only observed for adolescents (ES = 0.44), but not for children (ES = 0.25) (Sun et al. 2020). Other reviews that have examined age as a moderator did not observe any significant differences between children and adolescents. However, these reviews have been limited by the small number of included studies. For example, in the review conducted by Xue et al. (2019) a total of 19 studies (27 comparisons) were included, however only three of the 19 studies (16%) were conducted with adolescents. In a recent HIIT review, only one of the six studies (17%) included in the meta-analysis was conducted with children (Leahy et al. 2020). Given the relatively small number of reviews that have examined age as a moderator (children vs adolescents), further research is needed to confirm these findings.

Li et al. (2020) recently published the first review focused on the effects of physical activity on EFs in younger children (3–7 years). The overall effect on core EFs was slightly larger (ES = 0.35) than other reviews which only included children (de Greeff et al. 2018; Takacs and Kassai 2019). EFs develop rapidly during the preschool years (Anderson and Reidy 2012), therefore younger children may be particularly susceptible to the effects of physical activity interventions.

3.3.1 Special Populations

Two reviews focused on the chronic effects of physical activity on EFs in overweight and obese youth (Martin et al. 2018; Sun et al. 2020). In general, participation in physical activity appears to have a larger effect on EFs in overweight and obese youth, compared to the general population. In their Cochrane systematic review, Martin et al. (2018) observed a small-to-moderate effect on EFs in children (ES = 0.42), however this effect was only from one study. A more recent review by Sun and colleagues examined the effects of physical activity interventions on EFs among overweight and obese youth. The pooled effect from 11 studies ($n = 664$) was ES = 0.30 (Sun et al. 2020), which included both children and adolescents. This was also the first review to exclusively examine the effects of physical activity interventions on EFs among overweight and obese youth.

In their review, Sun and colleagues demonstrated that studies which reported significant intervention effects on adiposity resulted in greater improvements in EFs (ES = 0.44), in comparison with nonsignificant (ES = 0.30) and non-reported (ES = 0.18) intervention effects (Sun et al. 2020). This is somewhat consistent with a previous meta-analysis which demonstrated that higher body mass index at baseline was significantly associated with greater effects on overall EFs (Xue et al. 2019). Contrastingly, another review found a nonsignificant intervention effect when examining only overweight/obese participants from four studies (six samples) (ES = −0.02) (Álvarez-Bueno et al. 2017). One review examined whether intervention effects were different for typically vs non-typically developing youth (Takacs and Kassai 2019). Non-typically developing children were defined as those with a clinical diagnosis, behavioral problems or low EFs. While the summary effect for typically developing children was not significant (ES = 0.03), a small-to-moderate effect was observed for non-typically developing children (ES = 0.40).

3.4 Chronic Effect of Physical Activity on Executive Function Sub-domains

Five reviews examined the effects on specific EF sub-domains, with small-to-moderate effect sizes (Álvarez-Bueno et al. 2017; de Greeff et al. 2018; Li et al. 2020; Takacs and Kassai 2019; Xue et al. 2019). For inhibition, effect sizes ranged from 0.17 to 0.37, with an average effect of 0.25. For working memory, effect sizes ranged from 0.10 to 0.36, with an average effect size of 0.21. For cognitive flexibility, effect sizes ranged from −0.07 to 0.66, with an average effect size of 0.20. These findings suggest that chronic physical activity interventions do not have domain-specific effects on EFs (i.e., chronic physical activity produces small effects for all sub-domains of EF). However, it should be acknowledged that the majority of research has focused on inhibition, with fewer studies examining the impact on working memory and cognitive flexibility (Pontifex et al. 2019; Wade et al. 2020a).

It is not clear why previous studies have focused on inhibition, but future experimental research should aim to examine the effects of chronic physical activity on all three EF domains (i.e., inhibition, working memory, and cognitive flexibility).

4 Moderators of the Effect of Physical Activity on Executive Functions

As noted in the previous section, there is considerable heterogeneity in the magnitude of effects observed in studies examining the acute and chronic effect of physical activity on EFs. In this section we provide a summary of the quantitative (i.e., intensity, session, and study duration) and qualitative (i.e., type, cognitive demand and context) aspects of physical activity that may moderate effects on EFs. To date, the majority of studies examining effect size moderators have been conducted with adults. As such, less is known about the quantitative and qualitative characteristics of physical activity that may enhance or impede effects in children and adolescents.

4.1 Quantitative Moderators of the Effect of Physical Activity on Executive Functions

4.1.1 Intensity of Physical Activity

We use the term "intensity" to describe the physiological demand of the physical activity, typically assessed using percentage of maximal heart rate. A meta-analysis published by Chang et al. (2012) synthesized the findings from 79 studies that examined the effect of acute exercise on cognitive performance. The authors noted that effect sizes were impacted in an interactive fashion by the intensity of the activity and the amount of time between the bout of exercise and cognitive testing. More specifically, when cognitive tests were completed immediately after exercise, activity of very light, light, and moderate intensity was most beneficial. However, when cognitive performance was assessed following a delay of more than 1 min, more intense exercise produced the largest effects. It is important to note that the majority of studies included in the review conducted by Chang et al. (2012) involved adult populations. To the best of our knowledge, no previous systematic review has examined intensity as a moderator of exercise effects on EFs in chronic studies conducted with children and adolescents.

More recently, Moreau and Chou (2019) conducted a review of 28 acute studies of high-intensity exercise in different population groups (participants aged 6 to 60 years). The authors observed a small effect when high-intensity exercise was compared to rest (ES = 0.34), but no effect when it was compared to low-to-moderate intensity exercise (ES = 0.07). Similar findings were observed in a study by Janssen et al. (2014) who tested the effects of three experimental breaks (i.e., passive break, moderate, or vigorous intensity activity) on selective attention in primary school children ($N = 123$). Improvements in attention were observed in all groups, with the strongest effects observed after moderate intensity physical activity. These findings are consistent with the inverted U-hypothesis, which states that physical activity performed at moderate intensity will have the greatest benefit for EFs (Yerkes and Dodson 1908).

4.1.2 Session Duration

Session duration refers to the length of a physical activity or exercise bout. Studies examining the acute effect of physical activity on cognition have involved session durations ranging from 4 to 45 min (with most common duration 10–20 min) (Donnelly et al. 2016; Sibley and Etnier 2003). Session duration did not moderate the acute effects of physical activity on EFs in a systematic review and meta-analysis focused on preadolescent children (de Greeff et al. 2018). Similarly, Xue et al. (2019) concluded that session duration was not a moderator of the effects of chronic physical activity on EFs in children and adolescents.

4.1.3 Study Duration

There is considerable variability in the duration of chronic studies, with time periods ranging from 2 weeks to 1 school year (Donnelly et al. 2016; Vazou et al. 2020). Study duration was not a significant moderator of effects in de Greeff et al. (2018) review which included 14 studies. Similarly, study duration was not a moderator of the effects of HIIT on EFs in the review conducted by Leahy et al. (2020). In contrast, study duration moderated the effect of physical activity on EFs in overweight obese children and adolescents, with the strongest effects observed in studies of less than 12 weeks in duration (Sun et al. 2020). However, only three of the included studies had an intervention period of 12 weeks or less.

4.2 Qualitative Moderators of the Effect of Physical Activity on Executive Functions

4.2.1 Type of Physical Activity

Previous studies have tested the effects of various activity types on young people's EFs. For example, studies have tested the effects of running/walking (Travlos 2010), circuit training (Ludyga et al. 2019), dance (Peruyero et al. 2017), and martial arts (Johnstone and Marí-Beffa 2018). However, few studies have directly compared the effects of activity types on cognitive outcomes in the same population or study. Sibley and Etnier (2003) published one of the first meta-analyses examining the effect of physical activity on young people's cognition (not specifically EFs) and found no significant differences between types of physical activity. Although all effects were significantly greater than zero, resistance/circuit training (ES = 0.64) had larger effects than perceptual-motor training (ES = 0.32), physical education (PE) programs (ES = 0.27), and aerobic exercise (ES = 0.26). More recently, Xue et al. (2019) meta-analyzed the effects from 19 studies and found that activity type was an effect moderator, with the strongest effects observed in studies that involved curricular physical activity (ES = 0.39) and sports/physical activity programs (ES = 0.21). An overall nonsignificant effect (ES = 0.02) on overall EF performance was observed in studies that integrated physical activity into academic classes.

4.2.2 Cognitive Demand of Physical Activity

Cognitively demanding physical activities require individuals to perform cognitive and motor tasks simultaneously (Mavilidi et al. 2018). It has been suggested that this increased demand induces both physiological (e.g., neuroplasticity) and psychological (e.g., self-esteem, stress reduction) changes (Moreau and Tomporowski 2018). A number of recent reviews have examined the moderating effects of physical activity cognitive demand on EFs in young people (Álvarez-Bueno et al. 2017; Leahy et al. 2020; Sun et al. 2020). This research question is clouded by a lack of consensus regarding the definition or threshold of what constitutes "cognitively demanding physical activity" (Leahy et al. 2020), making it difficult to compare findings across individual studies and reviews.

In their systematic review and meta-analysis, Álvarez-Bueno et al. (2017) classified physical activity interventions as: (1) enhanced: focused on increasing activity time, (2) enriched: focused on increasing the motor coordinative and/or cognitive demands of the activity, or (3) both: focusing on increasing activity and the cognitive demands of the activity. The findings were mixed; working memory benefited mostly from enhanced programs (ES = 0.28), whereas selective attention/inhibition benefited mostly from enriched programs (ES = 0.49), and cognitive flexibility did not benefit from any type of intervention. Using a similar classification system, Sun et al. (2020) tested the effects of task characteristics on core EFs in overweight and obese youth. The authors observed no effect for enhanced physical activity, a moderate effect for enriched physical activity (ES = 0.54), and a large effect for both enhanced and enriched physical activity combined (ES = 1.01). It is important to note that a small number of studies examined the effects of enriched ($n = 2$) and both enhanced and enriched ($n = 2$) physical activity interventions in overweight and obese youth, thus limiting the generalizability of findings.

Although early indications suggest that the cognitive demand of physical activity may moderate intervention effects, there have not been enough studies to adequately answer this question. In addition, the effects of cognitively demanding physical activity may differ in acute and chronic studies. For example, Egger et al. (2018) conducted an acute study whereby children were randomly assigned to one of four conditions, consisting of a classroom-based physical activity intervention varying in physical exertion (high physical exertion vs. low physical exertion) and cognitive engagement (high cognitive engagement vs. low cognitive engagement). The researchers observed a significant, negative effect for set shifting in the high cognitive engagement group, while no other significant group-by-time effects were found. Using a similar study design, the same research group examined the long-term effect of activity breaks of varying cognitive engagement over a 20-week period (Egger et al. 2019). At follow-up, students in the high cognitive engagement and high physical exertion group significantly improved their set shifting performance, whereas updating and inhibition remained unaffected in all groups. Thus, it is possible that cognitively engaging activity might have a short term negative effect on EFs due to mental fatigue, but then lead to improvements in EFs over the longer-term.

4.2.3 Context of Physical Activity

The context in which physical activity takes place may moderate the effect on cognitive outcomes, but few studies have tested this hypothesis. The majority of studies have been conducted in schools and have tested the effects of activity delivered during PE, break times and in the classroom settings. Álvarez-Bueno et al. (2017) examined the effect of 36 physical activity interventions on children's cognition and meta-cognition. Curricular PE improved non-executive cognitive functions (ES = 0.42), selective attention/inhibition (ES = 0.41), and higher-level EFs (ES = 0.25). No effects of physical activity interventions were shown for working memory, cognitive flexibility, or language-related skills. Surprisingly, classroom-based physical activity did not impact children's cognition, when compared to extracurricular physical activity and curricular PE.

In one recent study, Wade et al. (2020b) tested the effects of different exercise environments (i.e., degree of nature) on adolescents' rapid visual information processing and spatial working memory. Four groups were randomized to one of the following conditions: (1) non-exercise indoor control, (2) indoor exercise, (3) park exercise, and (4) exercise in a nature reserve. All exercise groups participated in a circuit training lasting ~20 min, which included a mixture of aerobic and body-weight resistance activities. Results showed that indoor exercise improved children's sustained attention more than the park group. There were no differences between groups in terms of working memory. However, due to the sample size and randomization at the class (rather than individual) level, further studies are needed to confirm these findings.

5 Mechanisms Responsible for the Effect of Physical Activity on Executive Functions

As noted by Lubans et al. (2016) there are three broad types of mechanisms that may explain the effect of physical activity on EFs – neurobiological, psychosocial, and behavioral. Neurobiological mechanisms include changes that can be measured at microscopic (e.g., cellular and molecular) and macroscopic (e.g., brain structure and function) levels (Stillman et al. 2016). Psychosocial factors such as motivation, perceptions of novelty, and mood may also act as potential mechanisms. The behavioral mechanism hypothesis proposes that improvements in EFs may result from changes in relevant and associated behaviors, such as quantity and quality of sleep, as well as coping and self-regulation skills.

5.1 Neurobiological Mechanisms

5.1.1 Microscopic Mechanisms

Animal studies have provided much of our initial insights into the molecular and cellular mechanisms by which physical activity may improve cognition. Molecular effects include epigenetic upregulation of expression of genes with potential neuroplastic effects such as brain-derived neurotrophic factor (BDNF), vascular endothelial growth factor (VEGF), and insulin-like growth factor one (IGF-1) (Fernandes et al. 2017; Voss et al. 2013; Wrann et al. 2013). BDNF and VEGF underpin neurogenesis and angiogenesis (i.e., cellular mechanisms), respectively, while IGF-1 influences both neural and angiogenic processes. The downstream cellular effects of these synergistic factors are thought to mediate the relationship between physical activity and cognition (Lista and Sorrentino 2010). While easily investigated in animals, the experimental manipulation and measurement of molecular and cellular outcomes are difficult in humans. Nevertheless, empirical evidence is beginning to emerge in support of some of the same molecular mechanisms mediating the positive effects of physical activity on cognitive health in humans (Stillman et al. 2016, 2020; Voss et al. 2013).

Molecular Mechanisms

BDNF has been the most studied molecular mechanism responsible for the effects of physical activity on cognition (Stillman et al. 2020). Despite this, the evidence for BDNF in humans is not near as unequivocal as it is in animal models. A recent systematic review (Loprinzi 2019) comparing the mediating effects of BDNF on improvements in memory due to physical activity in animals and humans found that only 40% of human studies identified a mediation effect, compared to 100% of animal studies. It is important to note that in animal studies, BDNF concentrations in the brain can be directly measured. In humans, we are limited to inferring brain concentrations indirectly through the measurement of circulating BDNF in blood and saliva samples.

Although studies have observed increases in BDNF (and IGF-1) alongside improvements in memory in adolescents following exercise (Jeon and Ha 2017), a recent meta-analysis found no significant difference in BDNF levels when all adolescent exercise studies were pooled (Azevedo et al. 2020). Thus, equivocal findings in humans may be more reflective of methodological differences than an indication of the role of BDNF. Interestingly, some of the same moderators of the effect of physical activity on cognition mentioned earlier in this chapter have also been identified as potential moderators of the effect of physical activity on BDNF including the type, intensity, duration, and context of physical activity (Walsh et al. 2020). Other potential moderators include participant characteristics such as current fitness levels, habitual physical activity levels, diet, sex, age and genetic profile (Liu et al. 2020; Walsh et al. 2020).

Cellular Mechanisms

Upregulation of the aforementioned molecular processes has been shown to induce complex changes at the cellular level including angiogenesis (i.e., growth of new blood vessels), which acts as a precursor to neurogenesis (i.e., growth of new neurons), and other neuroprotective processes (Stillman et al. 2016). As there are limited techniques available to measure cellular changes in the human brain, almost all we currently know is derived from animal models. However, observations of macroscopic changes in brain structure, function, and blood flow in humans support the cellular adaptations observed in animals.

5.1.2 Macroscopic Mechanisms

Advances in neuroimaging techniques have enabled non-invasive in-vivo investigations into the macroscopic effects of physical activity on brain structure and function. There has been an exponential increase in studies of this nature across the lifespan and while most studies of older adults have examined morphological changes in brain gray and white matter (Firth et al. 2018; Stillman et al. 2016, 2020), studies of changes in functional activation and connectivity have prevailed thus far in the pediatric literature, particularly in children (Valkenborghs et al. 2019). Of note, the majority of the existing evidence pertains to aerobic physical activity in youth, with literature reporting the beneficial effects of resistance training on regions and structures relevant to cognition limited to adults (Herold et al. 2019; Meijer et al. 2020).

Brain Function

The human brain is arranged into many, overlapping networks that are regionally separate, but temporally connected and functionally coordinated (Yeo et al. 2015). By measuring statistical dependencies of neurophysiological events, we can study the functional connectivity between or within different brain regions (Friston 2011). Electroencephalography measures changes in patterns of neuro-electric potentials during response to, or in preparation for, a stimulus or response during a behavioral task (a.k.a. event-related potentials [ERP]) (Fabiani et al. 2007; Hillman et al. 2019a). Instead, functional magnetic resonance imaging (fMRI) detects changes in de-oxyhemoglobin driven by localized changes in blood flow and oxygenation coupled to underlying neuronal activity (Hillman 2014). Interpretation of functional changes is not straightforward and differs depending on region and context (e.g., task or rest). Increased activation of a task-positive region during task performance could be interpreted as greater ability to allocate resources (Chen et al. 2016; Jansma et al. 2001; Voss et al. 2011). However, based upon developmental neuroimaging studies (Casey et al. 1997; Scherf et al. 2006; Squire et al. 1992), decreased activation can also be interpreted as a more efficient use of resources, reflective of a more refined and mature pattern of activation (Chaddock-Heyman et al. 2013; Herting and Nagel 2013; Jansma et al. 2001).

The default mode network, salience network, and frontoparietal network support several domains of cognition (Menon and Uddin 2010; Seeley et al. 2007; Spreng et al. 2013; Uddin 2015). The integrity of these networks is also positively associated with scholastic performance in children (Chaddock-Heyman et al. 2018b). Importantly, regions within these networks in children appear to both respond and adapt to acute (ES = 0.32) and chronic (ES = 0.39) physical activity, respectively, and in patterns that are conducive of more refined and adult-like cognition (Corder et al. 2019; Meijer et al. 2020; Stillman et al. 2020; Valkenborghs et al. 2019). Of note, adaptations to chronic physical activity were accompanied by improved cognitive task performance in all studies included in the meta-analysis by Meijer et al. (2020).

Brain Structure

Morphology is the broad term used to describe several structural characteristics of the brain including the volume of gray and/or white matter, cortical thickness, and white matter integrity. Consistent with, and potentially in response to, the refinement of functional connectivity networks, the development of cognitive processes during childhood and adolescence corresponds with enhancement of the structural organization of the brain by a pruning-like process known as the synaptic fine-tuning hypothesis (Amso and Casey 2006; Chaddock-Heyman et al. 2018b; Changeux and Danchin 1976). Morphologically, this manifests as increased white matter volumes (reflective of ongoing myelination of axons to enhance conduction) (Giedd et al. 1999; Tamnes et al. 2010), and an inverted U-shaped trajectory of cortical volumes with increasing age in typically developing youths (Giedd et al. 1999; Gogtay et al. 2004).

Brain morphology is related to both cognitive function and physical activity across the lifespan (Stillman et al. 2020), and recent evidence elucidates certain morphological characteristics as mediators of the relationship of physical activity and cognition. For example, consistent with animal models, hippocampal volume may mediate the positive association between physical fitness and memory in preadolescent children (Chaddock et al. 2010) and older adults (Erickson et al. 2009). Similarly, a cross-sectional study in adolescents showed that physical fitness was associated with better spatial learning and larger hippocampal volumes (Herting and Nagel 2012).

To the best of our knowledge, only two randomized controlled trials (RCTs) have evaluated the impact of physical activity on white matter in youth and both demonstrated improved white matter integrity of tracts pertinent to cognition (Chaddock-Heyman et al. 2018a; Krafft et al. 2014; Schaeffer et al. 2014). This empirical evidence is supported by observational studies demonstrating a mediation effect of white matter integrity on the relationship between physical fitness and cognition in younger (Opel et al. 2019) and older (Oberlin et al. 2016) adults. Cross-sectional studies in youths have demonstrated that physical fitness is related to white matter microstructure of the corpus callosum (Chaddock-Heyman et al. 2014; Ruotsalainen et al. 2020) and corona radiata (Chaddock-Heyman et al. 2014; Herting et al. 2014; Ruotsalainen et al. 2020), with one study reporting an interesting moderating effect of white matter on the relationship between both physical fitness and physical activity with memory (Ruotsalainen et al. 2020).

5.2 Psychosocial Mechanisms

The microscopic and macroscopic mechanisms outlined above have been the focus of most research in this area. It is plausible that physical activity may also improve cognition through changes in a variety of psychosocial mechanisms (Lubans et al. 2016). For example, short-term changes in mood and longer-term changes in mental health outcomes may act as mediators for the effect of physical activity on cognitive function.

5.2.1 Mood

The acute effect of physical activity on mood may enhance cognitive performance. Participation in physical activity of light-to-moderate intensity is known to increase positive affect, while high-intensity activity typically evokes feelings of displeasure (Ekkekakis 2003). However, studies conducted with adults and adolescents have observed a rebound effect (Jung et al. 2014; Malik et al. 2019), whereby initial decreases in affect during high-intensity activity are followed by improvements after a period of about 20 min. To the best of our knowledge, no previous study has tested if improvements in mood mediate the acute effect of physical activity on cognitive performance.

5.2.2 Mental Health

It is also plausible that improvements in mental health (e.g., depression and anxiety) may mediate the effects of physical activity on cognition. Internalizing problems are associated with poorer cognition across multiple domains (Austin et al. 2001). Interestingly, animal studies have shown that exercise can improve stress-induced depression that coincide with widespread changes in regional homogeneity functional connectivity of brain regions key to cognition, such as the hippocampus and pre-frontal cortex (Dong et al. 2020). It remains to be established if one change invokes the other, if the relationship is bi-directional or if the relationship exists in humans. A conceptual model proposed by Stillman et al. (2016) suggests a bi-directional relationship between physical activity and cognition and mental health. While current literature is not conclusive regarding whether psychosocial mechanisms mediate or moderate the relationship between physical activity and cognition, existing results offer evidence that merits further exploration. In the interim, evidence should be considered holistically across mechanisms as it is likely the result of multi-faceted interactions of neurobiological, psychosocial, and behavioral mechanisms such as sleep.

5.3 Behavioral Mechanisms

In addition to neurobiological and psychosocial mediators, participation in physical activity may enhance cognition via a range of behavioral mechanisms (e.g., sleep, self-regulation and coping skills) (Lubans et al. 2016). Of note, existing evidence suggests that participation in physical activity may improve aspects of sleep, and certain types of physical activity (e.g., martial arts) provide opportunities for young people to develop self-regulation and coping skills that are important for cognitive functioning.

5.3.1 Sleep

The impact of sleep loss on multiple domains of cognition (Goel et al. 2009; Walker 2009) and the importance of sleep quality and quantity for healthy cognitive function (Astill et al. 2012) are well established. Neurobiological studies have also elucidated the negative changes in structural plasticity and synaptic strength (Raven et al. 2018), as well as brain networks that mediate attention and high-order cognitive processes (Thomas et al. 2000). Evidently, sleep, cognition, mental health and brain structure and function are all linked, but directionality is not clear. Participation in physical activity is known to improve sleep duration, sleep efficiency, sleep onset latency and reduce sleepiness in youth (Lang et al. 2013; Master et al. 2019; Mendelson et al. 2016). Therefore, given the relationship between sleep and cognition, and between sleep and physical activity, it is plausible that physical activity can enhance cognition by improving sleep. Sleep efficiency has been shown to mediate the relationship between physical activity and multiple cognitive domains in young and older adults (Cheng et al. 2020). Although evidence of this nature still needs to be replicated, particularly in children and adolescents; in the interim, we can consider sleep as a critical contributor to cognitive performance that can be modified by physical activity.

5.3.2 Self-Regulation Skills

Physical activity that entails physical effort as well as cognitive, social, and emotional demands (e.g., martial arts) may also enhance cognition through the development of self-regulation, defined as "the processes by which the self alters its own responses, including thoughts, emotions, and behaviors" (Baumeister 1997; Diamond and Ling 2019). An early study conducted by Lakes and Hoyt (2004) investigated the impact of school-based martial arts (Tae Kwon Do) training on self-regulatory abilities in kindergarten children. Compared to standard PE classes, martial arts training resulted in greater development in multiple domains of self-regulation, including cognitive self-regulation, as well as performance on a mental arithmetic test (Lakes and Hoyt 2004).

Another empirical study investigated a cognitively demanding physical activity intervention tailored to train self-regulation through inhibition tasks and step-by-step problem solving strategies embedded into child-appropriate physical activity games (Pesce et al. 2020). Compared to standard PE classes of similar intensity and enjoyment, the cognitively demanding physical activity intervention showed greater improvements in self-regulation and "hot" EFs (i.e., those performed under motivationally salient contexts that generate heightened emotion or a tension between prompt and longer-term rewards) (Pesce et al. 2020). Interestingly, the improvements in hot EFs and self-regulation were associated with one another, and there was marginal evidence of causal ordering of improvements over time such that improvements in hot EFs in the first year of the program predicted improvements in self-regulation in the second year. Therefore, until more research has been conducted, it cannot be ruled out that improvements in self-regulation due to physical activity could be mediated by improvements in cognition, or that this too could be a bi-directional relationship.

6 Summary, Limitations, and Emerging Issues

6.1 Summary of the Evidence Base

The primary aim of this chapter was to provide a synthesis of experimental studies that have examined the acute and chronic effect of physical activity on EFs. Evidence supports the view that both acute and chronic physical activity are effective for improving EFs in children and adolescents, with slightly stronger effects observed in acute studies. Although it has been suggested that physical activity may have a larger effect on inhibition compared to working memory and cognitive flexibility (Álvarez-Bueno et al. 2017; Hillman et al. 2019a; Xue et al. 2019), we did not observe this trend. Indeed, the effect sizes were similar for the three EF sub-domains in both acute (ES ~ 0.30) and chronic studies (ES ~ 0.20). However, it is worth noting that the majority of studies examining the effects of physical activity on EF have predominately used tasks that tap aspects of inhibition (Pontifex et al. 2019; Wade et al. 2020a). Therefore, to better understand potential domain-specific effects on EF, future research should include all three EF domains (i.e., inhibition, working memory, and cognitive flexibility).

The available evidence suggests that the effects of chronic physical activity are similar for children and adolescents, but there are specific groups who respond more favorably. For example, a number of reviews included in this chapter suggest effects are stronger among young people who are overweight or obese (Martin et al. 2018; Sun et al. 2020; Xue et al. 2019). However, this finding was not observed in the review conducted by Álvarez-Bueno et al. (2017). Overweight and obese individuals tend to be less active (Hills et al. 2011; Reichert et al. 2009) and have lower levels of CRF (Ortega et al. 2010) than their leaner peers. As such, we would expect to see stronger effects in this sub-group. As noted by Takacs and Kassai (2019), effect sizes appear to be stronger among children who are not typically developing (e.g., those with behavioral problems or low EF skills), compared to their typically developing peers.

A secondary aim of our chapter was to identify potential quantitative and qualitative moderators of effects. Despite the large number of published reviews, there is a lack of consistency in the types of moderators that have been tested in youth populations. In regard to quantitative characteristics, moderate intensity appears to be the optimal level for acute physical activity, with no additional benefits from higher intensity activity. This finding is consistent with the inverted U-hypothesis (Yerkes and Dodson 1908), which states that moderate intensity will have the greatest benefits for EFs. The relationship between physical activity intensity and EFs in chronic studies is less clear. It is likely that both the intensity and duration of activity are important for enhancing EFs in chronic studies. Physical activity of sufficient intensity and duration, designed to improve CRF, is likely to enhance EFs, especially among youth with low levels of fitness or non-typically developing youth.

As stated previously, the assertion that exercise can improve children's EFs is not shared by all. Based on findings from their narrative review of the literature, Diamond and Ling (2016, p. 43) concluded that "physical activity (without cognitive challenge, that brings little joy, and lacks any social component) appears not to improve EFs." In response, Hillman and colleagues (2019b, p. 1) published a rebuttal and described the Diamond and Ling review as "an inaccurate representation of the state of the field." An important aspect of dispute was the categorization of physical activity interventions. For example, Hillman and colleagues' FitKids program (2014) was classified as "mindless" exercise. Although the FitKids afterschool program was focused on developing CRF, it also provided children with opportunities to learn new motor skills and interact socially with each other. This is one example of the challenge faced by researchers when classifying physical activity interventions.

Several reviews included in this chapter compared the effects of enhanced (i.e., designed to increase activity levels), enriched (i.e., designed to increase the motor coordinative and/or cognitive demands of the activity), and combined (i.e., both enhanced and enriched) physical activity interventions (Álvarez-Bueno et al. 2017; de Greeff et al. 2018; Leahy et al. 2020; Li et al. 2020; Sun et al. 2020; Takacs and Kassai 2019). It is perhaps not surprising that physical activity interventions including enhanced and enriched components appear to be most effective (in chronic but not in acute studies). Such interventions may enhance EFs via neurobiological mechanisms (i.e., changes in brain structure and function) and via the skill acquisition process. As noted by Tomporowski and Pesce (2019) the allocation of mental resources during motor skill acquisition, independently or combined with energy expenditure, will reap the largest cognitive benefits. This is a promising line of enquiry, but additional high-quality research is needed to explore the effects of enhanced, enriched, and combined physical activity interventions.

There are a range of neurobiological, psychosocial, and behavioral mechanisms that may explain the effects of physical activity on young people's EFs (Lubans et al. 2016). In regard to microscopic neurobiological mechanisms, animal studies have highlighted the potential role of key molecules, including BDNF, VEGF, and IGF-1 (Fernandes et al. 2017; Voss et al. 2013; Wrann et al. 2013). However, findings from human studies have been more equivocal (Loprinzi 2019). Evidence from nine RCTs indicates that chronic participation in physical activity may enhance white matter tract integrity and activation in regions of the brain associated with EFs (Valkenborghs et al. 2019). Surprisingly, no previous RCTs have reported the impact of physical activity interventions on brain volume in children or adolescents. Little is known regarding the potential psychosocial and behavioral mechanisms, which represent ideal avenues for future research.

6.2 Limitations in the Current Evidence

In general, studies examining the effect of physical activity on EFs in children and adolescents have been described as having moderate-to-high risk of bias. Commonly cited study limitations include small sample sizes, poor implementation fidelity, inadequate assessment of cognitive function (lack of valid and reliable measure), and lack of assessor blinding. In response, a recent expert panel (Singh et al. 2019) identified five recommendations for future research to improve the evidence base: (1) careful selection of the control group to achieve maximal contrast in physical activity (with minimal differences in background variables); (2) monitoring of compliance (e.g., attendance) and physical activity levels during sessions to ensure that participants receive the recommended dose of activity, (3) adequate sample sizes to ensure that studies can accurately detect changes in cognitive outcomes, (4) use of valid and reliable measures of cognitive performance; and (5) accurate reporting of intervention effects, including effect sizes, confidence intervals, and exact *p*-values. In addition, typical studies examining the impact of physical activity on EFs include a large number of outcomes (and sometimes multiple follow-up assessments) and reviews have synthesized findings in different ways. Decisions made post-hoc by authors of reviews may explain some of the heterogeneity in the findings reported in this chapter.

6.3 Recommendations for Future Research

Based on the available evidence, we conclude that both acute and chronic physical activity can improve EFs in children and adolescents, with small-to-moderate effect sizes. We offer the following suggestions for future research that may help improve our understanding of the effects of physical activity on EFs, as well as the potential moderators and mediators of effects:

- The majority of research has been conducted with prepubertal children. As EFs develop rapidly throughout childhood and adolescence, more research in adolescent populations is needed to better understand the effects among different age groups. In addition, non-typically developing children and adolescents, such as those with learning and behavioral difficulties, are under-represented in physical activity and cognition research. More experimental studies with these populations are needed.
- Additional high-quality research is needed to explore the effects of enhanced (i.e., designed to increase activity levels), enriched (i.e., designed to increase the motor coordinative and/or cognitive demands of the activity), and combined (i.e., both enhanced and enriched) physical activity.
- There is a need to elucidate the mediating and moderating roles of BDNF and BDNF genotype, respectively, on the relationship between physical activity and cognition in children and adolescents.

- Further research examining the impact of physical activity intensity, time and type on brain structure and function is warranted. For example, we do not know if enhanced (i.e., designed to increase activity levels), enriched (i.e., designed to increase the motor coordinative and/or cognitive demands of the activity), or combined (i.e., both enhanced and enriched) physical activity is most beneficial.

References

Álvarez-Bueno C, Pesce C, Cavero-Redondo I, Sánchez-López M, Martinez-Hortelano JA, Martínez-Vizcaíno V (2017) The effect of physical activity interventions on children's cognition and metacognition: a systematic review and meta-analysis. J Am Acad Child Adolesc Psychiatry 56 (9):729–738. https://doi.org/10.1016/j.jaac.2017.06.012

Amso D, Casey BJ (2006) Beyond what develops when: neuroimaging may inform how cognition changes with development. Curr Dir Psychol Sci 15(1):24–29. https://doi.org/10.1111/j.0963-7214.2006.00400.x

Andersen SL (2003) Trajectories of brain development: point of vulnerability or window of opportunity? Neurosci Biobehav Rev 27(1–2):3–18

Anderson PJ, Reidy N (2012) Assessing executive function in preschoolers. Neuropsychol Rev 22 (4):345–360. https://doi.org/10.1007/s11065-012-9220-3

Astill RG, Van der Heijden KB, Van Ijzendoorn MH, Van Someren EJW (2012) Sleep, cognition, and behavioral problems in school-age children: a century of research meta-analyzed. Psychol Bull 138(6):1109–1138. https://doi.org/10.1037/a0028204

Aubert S, Barnes JD, Abdeta C, Abi Nader P, Adeniyi AF, Aguilar-Farias N et al (2018) Global matrix 3.0 physical activity report card grades for children and youth: results and analysis from 49 countries. J Phys Act Health 15(S2):S251–s273. https://doi.org/10.1123/jpah.2018-0472

Austin M-P, Mitchell P, Goodwin GM (2001) Cognitive deficits in depression: possible implications for functional neuropathology. Br J Psychiatry 178(3):200–206. https://doi.org/10.1192/bjp.178.3.200

Azevedo KPMD, de Oliveira VH, Medeiros GCBSD, Mata ÁNDS, García DÁ, Martínez DG, Leitão JC, Piuvezam G (2020) The effects of exercise on BDNF levels in adolescents: a systematic review with meta-analysis. Int J Environ Res Public Health 17(17). https://doi.org/10.3390/ijerph17176056

Baumeister RF (1997) Esteem threat, self-regulatory breakdown, and emotional distress as factors in self-defeating behavior. Rev Gen Psychol 1(2):145–174. https://doi.org/10.1037/1089-2680.1.2.145

Bull FC, Al-Ansari SS, Biddle S, Borodulin K, Buman MP, Cardon G et al (2020) World Health Organization 2020 guidelines on physical activity and sedentary behaviour. Br J Sports Med 54 (24):1451–1462. https://doi.org/10.1136/bjsports-2020-102955

Casey BJ, Trainor RJ, Orendi JL, Schubert AB, Nystrom LE, Giedd JN et al (1997) A developmental functional MRI study of prefrontal activation during performance of a go-no-go task. J Cogn Neurosci 9(6):835–847. https://doi.org/10.1162/jocn.1997.9.6.835

Caspersen CJ, Powell KE, Christenson GM (1985) Physical activity, exercise, and physical fitness: definitions and distinctions for health related research. Public Health Rep 100(2):126–131

Chaddock L, Erickson KI, Prakash RS, Kim JS, Voss MW, VanPatter M et al (2010) A neuroimaging investigation of the association between aerobic fitness, hippocampal volume, and memory performance in preadolescent children. Brain Res 1358:172–183. https://doi.org/10.1016/j.brainres.2010.08.049

Chaddock-Heyman L, Erickson KI, Voss M, Knecht A, Pontifex MB, Castelli D et al (2013) The effects of physical activity on functional MRI activation associated with cognitive control in children: a randomized controlled intervention. Front Hum Neurosci 7:72. https://doi.org/10.3389/fnhum.2013.00072

Chaddock-Heyman L, Erickson KI, Holtrop JL, Voss MW, Pontifex MB, Raine LB et al (2014) Aerobic fitness is associated with greater white matter integrity in children. Front Hum Neurosci 8:584. https://doi.org/10.3389/fnhum.2014.00584

Chaddock-Heyman L, Erickson KI, Kienzler C, Drollette ES, Raine LB, Kao S-C et al (2018a) Physical activity increases white matter microstructure in children. Front Neurosci 12:950. https://doi.org/10.3389/fnins.2018.00950

Chaddock-Heyman L, Weng TB, Kienzler C, Erickson KI, Voss MW, Drollette ES et al (2018b) Scholastic performance and functional connectivity of brain networks in children. PLoS One 13 (1):e0190073. https://doi.org/10.1371/journal.pone.0190073

Chang YK, Labban JD, Gapin JI, Etnier JL (2012) The effects of acute exercise on cognitive performance: a meta-analysis. Brain Res 1453:87–101. https://doi.org/10.1016/j.brainres.2012.02.068

Changeux JP, Danchin A (1976) Selective stabilisation of developing synapses as a mechanism for the specification of neuronal networks. Nature 264:705–712. https://doi.org/10.1038/264705a0

Chen A-G, Zhu L-N, Yan J, Yin H-C (2016) Neural basis of working memory enhancement after acute aerobic exercise: fMRI study of preadolescent children. Front Psychol 7. https://doi.org/10.3389/fpsyg.2016.01804

Cheng W, Rolls E, Gong W, Du J, Zhang J, Zhang X-Y et al (2020) Sleep duration, brain structure, and psychiatric and cognitive problems in children. Mol Psychiatry:1–12. https://doi.org/10.1038/s41380-020-0663-2

Colcombe S, Kramer AF (2003) Fitness effects on the cognitive function of older adults: a meta-analytic study. Psychol Sci 14(2):125–130

Corder K, Winpenny E, Love R, Brown HE, White M, Van Sluijs E (2019) Change in physical activity from adolescence to early adulthood: a systematic review and meta-analysis of longitudinal cohort studies. Br J Sports Med 53(8):496–503

de Greeff JW, Bosker RJ, Oosterlaan J, Visscher C, Hartman E (2018) Effects of physical activity on executive functions, attention and academic performance in preadolescent children: a meta-analysis. J Sci Med Sport 21(5):501–507. https://doi.org/10.1016/j.jsams.2017.09.595

Diamond A (2013) Executive functions. Annu Rev Psychol 64:135–168

Diamond A, Ling DS (2016) Conclusions about interventions, programs, and approaches for improving executive functions that appear justified and those that, despite much hype, do not. Dev Cogn Neurosci 18:34–48. https://doi.org/10.1016/j.dcn.2015.11.005

Diamond A, Ling DS (2019) Review of the evidence on, and fundamental questions about, efforts to improve executive functions, including working memory. In: Novick JM, Bunting MF, Dougherty MR, Engle RW (eds) Cognitive and working memory training: perspectives from psychology, neuroscience, and human development. Oxford University Press, Oxford

Dong Z, Liu Z, Liu Y, Zhang R, Mo H, Gao L, Shi Y (2020) Physical exercise rectifies CUMS-induced aberrant regional homogeneity in mice accompanied by the adjustment of skeletal muscle PGC-1a/IDO1 signals and hippocampal function. Behav Brain Res 383:112516. https://doi.org/10.1016/j.bbr.2020.112516

Donnelly J, Hillman C, Castelli D, Etnier J, Lee S, Tomporowski P et al (2016) Physical activity, fitness, cognitive function, and academic achievement in children: a systematic review. Med Sci Sports Exerc 48(6):1223–1224. https://doi.org/10.1249/MSS.0000000000000966

Dumith SC, Gigante DP, Domingues MR, Kohl, H. W. r. (2011) Physical activity change during adolescence: a systematic review and a pooled analysis. Int J Epidemiol 40(3):685–698. https://doi.org/10.1093/ije/dyq272

Egger F, Conzelmann A, Schmidt M (2018) The effect of acute cognitively engaging physical activity breaks on children's executive functions: too much of a good thing? Psychol Sport Exerc 36:178–186. https://doi.org/10.1016/j.psychsport.2018.02.014

Egger F, Benzing V, Conzelmann A, Schmidt M (2019) Boost your brain, while having a break! The effects of long-term cognitively engaging physical activity breaks on children’s executive

functions and academic achievement. PLoS One 14(3):e0212482. https://doi.org/10.1371/journal.pone.0212482
Ekkekakis P (2003) Pleasure and displeasure from the body: perspectives from exercise. Cognit Emot 17(2):213–239
Erickson KI, Prakash RS, Voss MW, Chaddock L, Hu L, Morris KS et al (2009) Aerobic fitness is associated with hippocampal volume in elderly humans. Hippocampus 19(10):1030–1039. https://doi.org/10.1002/hipo.20547
Fabiani M, Gratton G, Federmeier KD (2007) Event-related brain potentials. In: Berntson G, Cacioppo JT, Tassinary LG (eds) Handbook of psychophysiology, 3rd edn. Cambridge University Press, Cambridge
Farooq MA, Parkinson KN, Adamson AJ, Pearce MS, Reilly JK, Hughes AR et al (2018) Timing of the decline in physical activity in childhood and adolescence: Gateshead millennium cohort study. Br J Sports Med 52(15):1002–1006
Fernandes J, Arida RM, Gomez-Pinilla F (2017) Physical exercise as an epigenetic modulator of brain plasticity and cognition. Neurosci Biobehav Rev 80:443–456. https://doi.org/10.1016/j.neubiorev.2017.06.012
Firth J, Stubbs B, Vancampfort D, Schuch F, Lagopoulos J, Rosenbaum S, Ward PB (2018) Effect of aerobic exercise on hippocampal volume in humans: a systematic review and meta-analysis. Neuroimage 166:230–238. https://doi.org/10.1016/j.neuroimage.2017.11.007
Friston KJ (2011) Functional and effective connectivity: a review. Brain Connect 1(1):13–36. https://doi.org/10.1089/brain.2011.0008
Giedd JN, Blumenthal J, Jeffries NO, Castellanos FX, Liu H, Zijdenbos A et al (1999) Brain development during childhood and adolescence: a longitudinal MRI study. Nat Neurosci 2 (10):861–863. https://doi.org/10.1038/13158
Goel N, Rao H, Durmer JS, Dinges DF (2009) Neurocognitive consequences of sleep deprivation. Semin Neurol 25(1):117–129. https://doi.org/10.1055/s-2005-867080
Gogtay N, Giedd JN, Lusk L, Hayashi KM, Greenstein D, Vaituzis AC et al (2004) Dynamic mapping of human cortical development during childhood through early adulthood. Proc Natl Acad Sci U S A 101(21):8174–8179. https://doi.org/10.1073/pnas.0402680101
Guthold R, Stevens GA, Riley LM, Bull FC (2020) Global trends in insufficient physical activity among adolescents: a pooled analysis of 298 population-based surveys with 1·6 million participants. Lancet Child Adolesc Health 4(1):23–35. https://doi.org/10.1016/S2352-4642(19)30323-2
Herold F, Törpel A, Schega L, Müller NG (2019) Functional and/or structural brain changes in response to resistance exercises and resistance training lead to cognitive improvements–a systematic review. Eur Rev Aging Phys Act 16:10. https://doi.org/10.1186/s11556-019-0217-2
Herting MM, Nagel BJ (2012) Aerobic fitness relates to learning on a virtual Morris water task and hippocampal volume in adolescents. Behav Brain Res 233(2):517–525. https://doi.org/10.1016/j.bbr.2012.05.012
Herting MM, Nagel BJ (2013) Differences in brain activity during a verbal associative memory encoding task in high-and low-fit adolescents. J Cogn Neurosci 25(4):595–612. https://doi.org/10.1162/jocn_a_00344
Herting MM, Colby JB, Sowell ER, Nagel BJ (2014) White matter connectivity and aerobic fitness in male adolescents. Dev Cogn Neurosci 7:65–75. https://doi.org/10.1016/j.dcn.2013.11.003
Hillman EM (2014) Coupling mechanism and significance of the BOLD signal: a status report. Annu Rev Neurosci 37:161–181. https://doi.org/10.1146/annurev-neuro-071013-014111
Hillman CH, Pontifex MB, Castelli DM, Khan NA, Raine LB, Scudder MR et al (2014) Effects of the FITKids randomized controlled trial on executive control and brain function. Pediatrics 134 (4):e1063–e1071. https://doi.org/10.1542/peds.2013-3219
Hillman CH, Logan NE, Shigeta TT (2019a) A review of acute physical activity effects on brain and cognition in children. Transl J Am College Sports Med 4(17):132–136. https://doi.org/10.1249/TJX.0000000000000101

Hillman CH, McAuley E, Erickson KI, Liu-Ambrose T, Kramer AF (2019b) On mindful and mindless physical activity and executive function: a response to Diamond and Ling (2016). Dev Cogn Neurosci 37:100529–100529. https://doi.org/10.1016/j.dcn.2018.01.006

Hills AP, Andersen LB, Byrne NM (2011) Physical activity and obesity in children. Br J Sports Med 45(11):866–870. https://doi.org/10.1136/bjsports-2011-090199

Jansma JM, Ramsey NF, Slagter HA, Kahn RS (2001) Functional anatomical correlates of controlled and automatic processing. J Cogn Neurosci 13(6):730–743. https://doi.org/10.1162/08989290152541403

Janssen M, Chinapaw MJM, Rauh SP, Toussaint HM, van Mechelen W, Verhagen EALM (2014) A short physical activity break from cognitive tasks increases selective attention in primary school children aged 10–11. Ment Health Phys Act 7(3):129–134. https://doi.org/10.1016/j.mhpa.2014.07.001

Jeon YK, Ha CH (2017) The effect of exercise intensity on brain derived neurotrophic factor and memory in adolescents. Environ Health Prev Med 22:27. https://doi.org/10.1186/s12199-017-0643-6

Johnstone A, Marí-Beffa P (2018) The effects of martial arts training on attentional networks in typical adults. Front Psychol 9:80. https://doi.org/10.3389/fpsyg.2018.00080

Jung ME, Bourne JE, Little JP (2014) Where does HIT fit? An examination of the affective response to high-intensity intervals in comparison to continuous moderate-and continuous vigorous-intensity exercise in the exercise intensity-affect continuum. PLoS One 9(12):e114541. https://doi.org/10.1371/journal.pone.0114541

Krafft CE, Schaeffer DJ, Schwarz NF, Chi L, Weinberger AL, Pierce JE et al (2014) Improved frontoparietal white matter integrity in overweight children is associated with attendance at an after-school exercise program. Dev Neurosci 36(1):1–9. https://doi.org/10.1159/000356219

Lakes KD, Hoyt WT (2004) Promoting self-regulation through school-based martial arts training. J Appl Dev Psychol 25(3):283–302. https://doi.org/10.1016/j.appdev.2004.04.002

Lang C, Brand S, Feldmeth AK, Holsboer-Trachsler E, Pühse U, Gerber M (2013) Increased self-reported and objectively assessed physical activity predict sleep quality among adolescents. Physiol Behav 120:46–53

Leahy AA, Mavilidi MF, Smith JJ, Hillman CH, Eather N, Barker D, Lubans DR (2020) Review of high-intensity interval training for cognitive and mental health in youth. Med Sci Sports Exerc 52(10):2224–2234

Li L, Zhang J, Cao M, Hu W, Zhou T, Huang T et al (2020) The effects of chronic physical activity interventions on executive functions in children aged 3–7 years: a meta-analysis. J Sci Med Sport 23(10):949–954. https://doi.org/10.1016/j.jsams.2020.03.007

Lista I, Sorrentino G (2010) Biological mechanisms of physical activity in preventing cognitive decline. Cell Mol Neurobiol 30(4):493–503. https://doi.org/10.1007/s10571-009-9488-x

Liu T, Li H, Colton JP, Ge S, Li C (2020) The BDNF Val66Met polymorphism, regular exercise, and cognition: a systematic review. West J Nurs Res 42(8):660–673. https://doi.org/10.1177/0193945920907308

Loprinzi PD (2019) Does brain-derived neurotrophic factor mediate the effects of exercise on memory? Phys Sportsmed 47(4):395–405. https://doi.org/10.1080/00913847.2019.1610255

Lubans D, Richards J, Hillman C, Faulkner G, Beauchamp M, Nilsson M et al (2016) Physical activity for cognitive and mental health in youth: a systematic review of mechanisms. Pediatrics 138(3). https://doi.org/10.1542/peds.2016-1642

Ludyga S, Gerber M, Brand S, Holsboer-Trachsler E, Pühse U (2016) Acute effects of moderate aerobic exercise on specific aspects of executive function in different age and fitness groups: a meta-analysis. Psychophysiology 53(11):1611–1626. https://doi.org/10.1111/psyp.12736

Ludyga S, Puhse U, Lucchi S, Marti J, Gerber M (2019) Immediate and sustained effects of intermittent exercise on inhibitory control and task-related heart rate variability in adolescents. J Sci Med Sport 22(1):96–100. https://doi.org/10.1016/j.jsams.2018.05.027

Malik AA, Williams CA, Weston KL, Barker AR (2019) Perceptual and cardiorespiratory responses to high-intensity interval exercise in adolescents: does work intensity matter? J Sports Sci Med 18(1):1–12
Martin A, Booth JN, Laird Y, Sproule J, Reilly JJ, Saunders DH (2018) Physical activity, diet and other behavioural interventions for improving cognition and school achievement in children and adolescents with obesity or overweight. Cochrane Database Syst Rev 1(1):CD009728. https://doi.org/10.1002/14651858.CD009728.pub3
Master L, Nye RT, Lee S, Nahmod NG, Mariani S, Hale L, Buxton OM (2019) Bidirectional, daily temporal associations between sleep and physical activity in adolescents. Sci Rep 9(1):1–14. https://doi.org/10.1097/AJP.0b013e31829550c6
Mavilidi MF, Ruiter M, Schmidt M, Okely AD, Loyens S, Chandler P, Paas F (2018) A narrative review of school-based physical activity for enhancing cognition and learning: the importance of relevancy and integration. Front Psychol 9:2079. https://doi.org/10.3389/fpsyg.2018.02079
Meijer A, Königs M, Vermeulen GT, Visscher C, Bosker RJ, Hartman E, Oosterlaan J (2020) The effects of physical activity on brain structure and neurophysiological functioning in children: a systematic review and meta-analysis. Dev Cogn Neurosci 45:100828. https://doi.org/10.1016/j.dcn.2020.100828
Mendelson M, Borowik A, Michallet AS, Perrin C, Monneret D, Faure P et al (2016) Sleep quality, sleep duration and physical activity in obese adolescents: effects of exercise training. Pediatr Obes 11(1):26–32. https://doi.org/10.1111/ijpo.12015
Menon V, Uddin LQ (2010) Saliency, switching, attention and control: a network model of insula function. Brain Struct Funct 214(5–6):655–667. https://doi.org/10.1007/s00429-010-0262-0
Moreau D, Chou E (2019) The acute effect of high-intensity exercise on executive function: a meta-analysis. Perspect Psychol Sci 14(5):734–764. https://doi.org/10.1177/1745691619850568
Moreau D, Tomporowski PD (2018) 10 complex motor activities to enhance cognition. In: Cappuccio ML (ed) Handbook of embodied cognition and sport psychology. The MIT Press, London, pp 273–302
Oberlin LE, Verstynen TD, Burzynska AZ, Voss MW, Prakash RS, Chaddock-Heyman L et al (2016) White matter microstructure mediates the relationship between cardiorespiratory fitness and spatial working memory in older adults. Neuroimage 131:91–101. https://doi.org/10.1016/j.neuroimage.2015.09.053
Opel N, Martin S, Meinert S, Redlich R, Enneking V, Richter M et al (2019) White matter microstructure mediates the association between physical fitness and cognition in healthy, young adults. Sci Rep 9(1):1–9. https://doi.org/10.1038/s41598-019-49301-y
Ortega FB, Ruiz JR, Hurtig-Wennlöf A, Vicente-Rodríguez G, Rizzo NS, Castillo MJ, Sjöström M (2010) Cardiovascular fitness modifies the associations between physical activity and abdominal adiposity in children and adolescents: the European Youth Heart Study. Br J Sports Med 44 (4):256–262. https://doi.org/10.1136/bjsm.2008.046391
Parrish A-M, Tremblay MS, Carson S, Veldman SL, Cliff D, Vella S et al (2020) Comparing and assessing physical activity guidelines for children and adolescents: a systematic literature review and analysis. Int J Behav Nutr Phys Act 17(1):16
Peruyero F, Zapata J, Pastor D, Cervelló E (2017) The acute effects of exercise intensity on inhibitory cognitive control in adolescents. Front Psychol 8(921). https://doi.org/10.3389/fpsyg.2017.00921
Pesce C, Lakes KD, Stodden DF, Marchetti R (2020) Fostering self-control development with a designed intervention in physical education: a two-year class-randomized trial. Child Dev:1–22. https://doi.org/10.1111/cdev.13445
Pontifex MB, McGowan AL, Chandler MC, Gwizdala KL, Parks AC, Fenn K, Kamijo K (2019) A primer on investigating the after effects of acute bouts of physical activity on cognition. Psychol Sport Exerc 40:1–22
Raghuveer G, Hartz J, Lubans DR, Takken T, Wiltz J, Mietus-Snyder M et al (2020) Cardiorespiratory fitness in youth: an important marker of health: a scientific statement from the American

Heart Association. Circulation 142(7):e101–e118. https://doi.org/10.1161/CIR.0000000000000866

Raven F, Van der Zee EA, Meerlo P, Havekes R (2018) The role of sleep in regulating structural plasticity and synaptic strength: implications for memory and cognitive function. Sleep Med Rev 39:3–11. https://doi.org/10.1016/j.smrv.2017.05.002

Reichert FF, Menezes AMB, Wells JCK, Dumith SC, Hallal PC (2009) Physical activity as a predictor of adolescent body fatness. Sports Med 39(4):279–294. https://doi.org/10.2165/00007256-200939040-00002

Ruotsalainen I, Gorbach T, Perkola J, Renvall V, Syväoja HJ, Tammelin TH et al (2020) Physical activity, aerobic fitness, and brain white matter: their role for executive functions in adolescence. Dev Cogn Neurosci 42:100765. https://doi.org/10.1016/j.dcn.2020.100765

Schaeffer DJ, Krafft CE, Schwarz NF, Chi L, Rodrigue AL, Pierce JE et al (2014) An 8-month exercise intervention alters frontotemporal white matter integrity in overweight children. Psychophysiology 51(8):728–733. https://doi.org/10.1111/psyp.12227

Scherf KS, Sweeney JA, Luna B (2006) Brain basis of developmental change in visuospatial working memory. J Cogn Neurosci 18(7):1045–1058. https://doi.org/10.1162/jocn.2006.18.7.1045

Sebastian C, Burnett S, Blakemore S-J (2008) Development of the self-concept during adolescence. Trends Cogn Sci 12(11):441–446

Seeley WW, Menon V, Schatzberg AF, Keller J, Glover GH, Kenna H et al (2007) Dissociable intrinsic connectivity networks for salience processing and executive control. J Neurosci 27 (9):2349–2356. https://doi.org/10.1523/jneurosci.5587-06.2007

Sibley BA, Etnier JL (2003) The relationship between physical activity and cognition in children: a meta-analysis. Pediatr Exerc Sci 15(3):243–256. Retrieved from https://journals.humankinetics.com/view/journals/pes/15/3/article-p243.xml

Singh AS, Saliasi E, van den Berg V, Uijtdewilligen L, de Groot RHM, Jolles J et al (2019) Effects of physical activity interventions on cognitive and academic performance in children and adolescents: a novel combination of a systematic review and recommendations from an expert panel. Br J Sports Med 53(10):640–647. https://doi.org/10.1136/bjsports-2017-098136

Smith JJ, Eather N, Harris N, Lubans DR (2020) Exercise for children and adolescents. In: Brusseau T, Fairclough S, Lubans DR (eds) The Routledge handbook of youth physical activity. Routledge, Oxon, pp 738–761

Spreng RN, Sepulcre J, Turner GR, Stevens WD, Schacter DL (2013) Intrinsic architecture underlying the relations among the default, dorsal attention, and frontoparietal control networks of the human brain. J Cogn Neurosci 25(1):74–86. https://doi.org/10.1162/jocn_a_00281

Squire LR, Ojemann JG, Miezin FM, Petersen SE, Videen TO, Raichle ME (1992) Activation of the hippocampus in normal humans: a functional anatomical study of memory. Proc Natl Acad Sci 89(5):1837–1841. https://doi.org/10.1073/pnas.89.5.1837

Stillman CM, Cohen J, Lehman ME, Erickson KI (2016) Mediators of physical activity on neurocognitive function: a review at multiple levels of analysis. Front Hum Neurosci 10:626. https://doi.org/10.3389/fnhum.2016.00626

Stillman CM, Esteban-Cornejo I, Brown B, Bender CM, Erickson KI (2020) Effects of exercise on brain and cognition across age groups and health states. Trends Neurosci 43(7):533–543. https://doi.org/10.1016/j.tins.2020.04.010

Sun X, Li Y, Cai L, Wang Y (2020) Effects of physical activity interventions on cognitive performance of overweight or obese children and adolescents: a systematic review and meta-analysis. Pediatr Res. https://doi.org/10.1038/s41390-020-0941-3

Takacs ZK, Kassai R (2019) The efficacy of different interventions to foster children's executive function skills: a series of meta-analyses. Psychol Bull 145(7):653–697. https://doi.org/10.1037/bul0000195

Tamnes CK, Østby Y, Walhovd KB, Westlye LT, Due-Tønnessen P, Fjell AM (2010) Intellectual abilities and white matter microstructure in development: a diffusion tensor imaging study. Hum Brain Mapp 31(10):1609–1625. https://doi.org/10.1002/hbm.20962

Telama R (2009) Tracking of physical activity from childhood to adulthood: a review. Obes Facts 3:187–195. https://doi.org/10.1159/000222244

Thomas M, Sing H, Belenky G, Holcomb H, Mayberg H, Dannals R et al (2000) Neural basis of alertness and cognitive performance impairments during sleepiness. I. Effects of 24 h of sleep deprivation on waking human regional brain activity. J Sleep Res 9(4):335–352. https://doi.org/10.1046/j.1365-2869.2000.00225.x

Tomporowski PD, Pesce C (2019) Exercise, sports, and performance arts benefit cognition via a common process. Psychol Bull 145(9):929–951. https://doi.org/10.1037/bul0000200

Travlos AK (2010) High intensity physical education classes and cognitive performance in eighth-grade students: an applied study. Int J Sport Exerc Psychol 8(3):302–311. https://doi.org/10.1080/1612197X.2010.9671955

Tremblay MS, Gray CE, Akinroye KK, Harrington DM, Katzmarzyk PT, Lambert EV et al (2014) Physical activity of children: a global matrix of grades comparing 15 countries. J Phys Act Health 11(Supp 1):113–125

Uddin LQ (2015) Salience processing and insular cortical function and dysfunction. Nat Rev Neurosci 16(1):55–61. https://doi.org/10.1038/nrn3857

Valkenborghs SR, Noetel M, Hillman CH, Nilsson M, Smith JJ, Ortega FB, Lubans DR (2019) The impact of physical activity on brain structure and function in youth: a systematic review. Pediatrics 144(4):e20184032. https://doi.org/10.1542/peds.2018-4032

Vazou S, Webster CA, Stewart G, Candal P, Egan CA, Pennell A, Russ LB (2020) A systematic review and qualitative synthesis resulting in a typology of elementary classroom movement integration interventions. Sports Med 6:1. https://doi.org/10.1186/s40798-019-0218-8

Verburgh L, Konigs M, Scherder EJ, Oosterlaan J (2014) Physical exercise and executive functions in preadolescent children, adolescents and young adults: a meta-analysis. Br J Sports Med 48 (12):973–979. https://doi.org/10.1136/bjsports-2012-091441

Voss MW, Chaddock L, Kim JS, VanPatter M, Pontifex MB, Raine LB et al (2011) Aerobic fitness is associated with greater efficiency of the network underlying cognitive control in preadolescent children. Neuroscience 199:166–176. https://doi.org/10.1016/j.neuroscience.2011.10.009

Voss MW, Vivar C, Kramer AF, van Praag H (2013) Bridging animal and human models of exercise-induced brain plasticity. Trends Cogn Sci 17(10):525–544

Wade L, Leahy A, Lubans DR, Smith JJ, Duncan MJ (2020a) A systematic review of cognitive assessment in physical activity research involving children and adolescents. J Sci Med Sport 23 (8):740–745. https://doi.org/10.1016/j.jsams.2019.12.020

Wade L, Lubans DR, Smith JJ, Duncan MJ (2020b) The impact of exercise environments on adolescents' cognitive and psychological outcomes: a randomised controlled trial. Psychol Sport Exerc 49:101707. https://doi.org/10.1016/j.psychsport.2020.101707

Walker MP (2009) The role of sleep in cognition and emotion. Ann N Y Acad Sci 1156 (1):168–197. https://doi.org/10.1111/j.1749-6632.2009.04416.x

Walsh EI, Smith L, Northey J, Rattray B, Cherbuin N (2020) Towards an understanding of the physical activity-BDNF-cognition triumvirate: a review of associations and dosage. Ageing Res Rev 60:101044. https://doi.org/10.1016/j.arr.2020.101044

Wrann CD, White JP, Salogiannnis J, Laznik-Bogoslavski D, Wu J, Ma D et al (2013) Exercise induces hippocampal BDNF through a PGC-1α/FNDC5 pathway. Cell Metab 18(5):649–659. https://doi.org/10.1016/j.cmet.2013.09.008

Xue Y, Yang Y, Huang T (2019) Effects of chronic exercise interventions on executive function among children and adolescents: a systematic review with meta-analysis. Br J Sports Med, bjsports-2018-099825. https://doi.org/10.1136/bjsports-2018-099825

Yeo BT, Krienen FM, Eickhoff SB, Yaakub SN, Fox PT, Buckner RL et al (2015) Functional specialization and flexibility in human association cortex. Cereb Cortex 25(10):3654–3672. https://doi.org/10.1093/cercor/bhu217

Yerkes RM, Dodson JD (1908) The relation of strength of stimulus to rapidity of habit-formation. J Comp Neurol Psychol 18(5):459–482. https://doi.org/10.1002/cne.920180503

Nutrition and Brain Development

Sarah E. Cusick, Amanda Barks, and Michael K. Georgieff

Contents

S. E. Cusick (✉)
Department of Pediatrics, University of Minnesota School of Medicine, Minneapolis, MN, USA
e-mail: scusick@umn.edu

A. Barks and M. K. Georgieff
University of Minnesota Medical School, Minneapolis, MN, USA

Curr Topics Behav Neurosci (2022) 53: 131–166
https://doi.org/10.1007/7854_2021_244
Published Online: 8 October 2021

Abstract All nutrients are essential for brain development, but pre-clinical and clinical studies have revealed sensitive periods of brain development during which key nutrients are critical. An understanding of these nutrient-specific sensitive periods and the accompanying brain regions or processes that are developing can guide effective nutrition interventions as well as the choice of meaningful circuit-specific neurobehavioral tests to best determine outcome. For several nutrients including protein, iron, iodine, and choline, pre-clinical and clinical studies align to identify the same sensitive periods, while for other nutrients, such as long-chain polyunsaturated fatty acids, zinc, and vitamin D, pre-clinical models demonstrate benefit which is not consistently shown in clinical studies. This discordance of pre-clinical and clinical results is potentially due to key differences in the timing, dose, and/or duration of the nutritional intervention as well as the pre-existing nutritional status of the target population. In general, however, the optimal window of success for nutritional intervention to best support brain development is in late fetal and early postnatal life. Lack of essential nutrients during these times can lead to long-lasting dysfunction and significant loss of developmental potential.

Keywords Brain development · Early childhood nutrition · Micronutrients · Nutrition

1 Introduction

Much of brain development is pre-programmed, or experience-independent, but key influences on brain development are external. Stress, environmental toxins, and nutritional deficiencies are preeminent among external influences harmful to the developing brain (Wachs et al. 2014). Each can cause long-term dysfunction if sustained during a sensitive period of brain development. Of these factors, however, nutrition is the one perhaps most amenable to beneficial change with preventative interventions (Wachs et al. 2014; Cusick and Georgieff 2016).

All nutrients are important for brain development, but certain nutrients have substantial pre-clinical and clinical evidence supporting their requirement during specific sensitive periods, when if not present in adequate amount, long-term neurodevelopmental consequences occur (Cusick and Georgieff 2016; Cusick et al. 2018; Georgieff et al. 2018). The brain is not a homogenous organ that develops at the same time. Instead, it is made of different structures, e.g., the hippocampus, and processes, e.g., myelination, which each have unique trajectories that start at different periods of development and continue for different durations (Cusick et al. 2018; Georgieff et al. 2018). The widely used public health message of the importance of the first 1,000 days in laying the foundation of a healthy life is particularly applicable to brain development (Cusick and Georgieff 2016). For the brain, the first 1,000 days extends roughly from conception to the second or third year of postnatal life. The most rapid period of brain development starts at the age of *ex utero* viability, i.e., 23–24 weeks gestation, and continues at a brisk pace during the last trimester of pregnancy and the first year of postnatal life (Cusick and Georgieff 2016; Georgieff et al. 2018). During this rapid period of growth, the brain is vulnerable to damage if critical nutrients aren't present, but it retains a degree of plasticity and amenability to repair (Hensch 2004). There is growing consensus, however, that vulnerability outweighs plasticity, i.e., keeping the brain developing on its set trajectory is preferable to having it fall off of that trajectory and rely on catch up (Hensch 2004).

1.1 Pre-clinical and Clinical Evidence for Nutrient-Specific Sensitive Periods

The demarcation of sensitive periods in brain development has been clearly demonstrated for some nutrients, but not for others (Table 1) (Georgieff et al. 2018). Typically for those nutrients, e.g., iron, which have a well-defined sensitive period, deficits revealed in pre-clinical models are mirrored by corresponding altered behaviors in clinical studies. For example, infants born with iron deficiency anemia have altered recognition memory – a hippocampus-specific task – as measured by auditory-evoked event-related potentials (ERPs) (Geng et al. 2015). This clinically evident deficit is explained by corresponding pre-clinical work demonstrating a lower metabolic rate of iron-deficient hippocampal neurons in rodent models (Bastian et al. 2020). For other nutrients, e.g., vitamin D (Eyles et al. 2011), evidence of specific benefit to neurodevelopment is apparent in pre-clinical models (Eyles et al. 2011) but is largely lacking in clinical studies.

Table 1 Patterns of alignment of pre-clinical and clinical results of studies testing the most commonly studied nutrients and neurobehavioral outcomes[a]

Benefit in both pre-clinical studies and clinical studies	Protein, Iron, Iodine, Choline
Benefit in pre-clinical studies; inconsistent results in clinical supplementation studies	LC-PUFAs, Zinc
Benefit in pre-clinical studies; no clear benefit in clinical supplementation studies	Vitamin D

[a]Pre-clinical and clinical studies investigating nutrient and brain interactions may have discordant results for a number of reasons including: (1) Differences in timing, dose, and duration of nutrient; (2) Differences between supplementation studies vs. studies which correct a deficiency. Pre-clinical studies are almost always the former, while clinical studies can be either or have components of both; and (3) Different metabolic rate of the rodent brain (requires 10% of substrates) as compared to the infant brain (requires 60% of substrates) (Wachs et al. 2014; Kuzawa 1998)

1.1.1 Considerations for Interpretation of Clinical and Pre-clinical Nutrient-Brain Studies

Timing, Dose, and Duration

The explanation behind discordant relationships with regard to incomplete or lacking clinical evidence, despite pre-clinical evidence, underscores the necessity of considering the principal tenets of nutrition–brain interaction, i.e., timing, dose, and duration, with regard to supplemental intervention (Cusick and Georgieff 2016; Georgieff et al. 2018; Kretchmer et al. 1996). Brain structures and circuitry develop at different times and at different rates. Consequently, supplementation or deficit outside of a sensitive period may have a null result, despite having benefit or risk if given during the sensitive period. The dose and duration of the supplement must also be considered. A higher dose of supplement may not result in greater benefit, particularly if the subject is very deficient or completely sufficient at baseline. With regard to duration, a supplementation period that is too short or does not coincide with the sensitive period of the brain structure or process of need may have no effect. Finally, the neurodevelopmental test used to assess the outcome must be carefully chosen. A null result on a global test may mask more granular findings of a circuit-specific test. These rules of engagement that dictate the success or failure of clinical nutrient and brain interactions are difficult to reproduce with complete alignment in pre-clinical models.

Deficiency vs. Supplementation Studies

Related to the rules of nutrient and brain interactions is the difference between supplementation studies, i.e., giving more of a nutrient to an already-sufficient individual, vs. correction of deficiency studies. Deficiency studies typically show greater and more consistent effects in both clinical and pre-clinical studies than supplementation. Most nutrient pre-clinical studies are deficit studies in order to demonstrate an effect or need for the nutrient during a particular time period, often a sensitive period. Pre-clinical supplementation studies are rare because it is assumed that all diets provided to lab animals are sufficient in all nutrients. Thus, pre-clinical supplementation studies of already-sufficient animal models are usually "toxicity" studies. Clinical studies can be in populations that are deficient, sufficient, or a range extending between these two states with regard to a specific nutrient, likely contributing to some misalignment of pre-clinical and clinical results.

Differential Growth and Brain Metabolic Rates of Animal vs. Human Models

Another explanation for discrepant pre-clinical and clinical findings likely lies with the differential growth and brain metabolic rate of animal model vs humans. Many neurobehavioral deficits are due to inadequate cellular and circuit construction (also referred to as neuronal differentiation and synaptogenesis), which requires substrates, i.e., nutrients, that specifically support the metabolic demand of growth and differentiation (e.g., energy, protein, oxygen, iron, copper, iodine). Brain development is an astoundingly highly metabolic process for a human infant, with 60% of a newborn infant's energy going to the brain, as compared to 20% in an adult (Kuzawa 1998). In comparison, only about 10% of a newborn rodent's energy goes to its brain. Thus, using a rodent model where 90% of the energy substrates go to somatic growth instead of brain growth, a deficit may not yield nearly the impact as it would in a human because the rodent brain is "low demand" in comparison, i.e., less sensitive to a nutrient deficit.

1.2 Mechanisms of Long-Term Dysfunction

The mechanisms of how early-life nutritional deficits can have lasting consequences for brain development are still being elucidated, but two primary mechanisms based on substantial pre-clinical evidence have emerged. These mechanisms are not mutually exclusive. The first is an architectural effect in that the brain structures that develop early, e.g., primary sensory systems and the hippocampus, provide the scaffolding for later-developing structures. If the scaffolding is weak, what is built on it will also be weak, making early deficits have long-lasting, rippling consequences even in the face of subsequent correction of the deficiency (Hensch 2004; Fretham et al. 2012). As an example, development of the hippocampus begins in the last trimester of pregnancy and continues at a brisk pace throughout the first 2 years of life (Utsunomiya et al. 1999), laying the foundation for working memory and attention, the buttresses of cognition. The sound formation of this structure is requisite for the subsequent optimal development of structures such as the striatum and frontal lobe, which permit higher-order thinking, set-shifting, and executive functioning, which develop later in infancy and early childhood.

The second mechanism of long-lasting deficits with early-life nutritional deficiency is through alteration of synaptic plasticity genes though epigenetic modification of chromatin. The epigenetic effects of several nutritional deficiencies have been demonstrated in pre-clinical models (Table 2). Iron deficiency modifies activity of the JARID-containing histone demethylases which in turn regulate expression of Brain-Derived Neurotrophic Factor (BDNF) and DNA methylation in the hippocampus (Tran et al. 2015). Epigenetic modifications with deficiencies have also been demonstrated with Intrauterine Growth Restriction (IUGR, via hippocampal H4K20 histone methylation) (Grissom and Reyes 2013), LC-PUFAs (DNA methylation of BDNF) (Tyagi et al. 2015), and choline, which has been shown to reverse the suppression of BDNF repression in animal models (Zeisel 2012; Kennedy et al. 2014; Tran et al. 2016).

Table 2 The most commonly studied nutrients with regard to neurobehavioral development[a]

	Main brain processes affected	Evidence for sensitive periods	Epigenetic mechanism of long-term effects
Protein	Structure Growth factors Neurotransmitters	Yes	Suspected
LCPUFAs	Membrane integrity Signaling	Yes	Yes
Iron	Energetics Myelination Monoamine neurotransmission	Yes	Yes
Iodine	Thyroid-dependent myelination Synaptogenesis Energy metabolism	Yes	No
Choline	Neurotransmitters Myelination	Yes	Yes
Zinc	Growth factors Synaptic efficacy	Yes	No
Vitamin D	Proliferation, apoptosis	No	No

[a]Modified from Source: Georgieff et al. (2018)

In the following sections, the clinical and pre-clinical evidence for the effect of nutrients presented in Tables 1 and 2 will be discussed, starting with macronutrients and continuing with micronutrients. Emphasis is placed on how successful preventative interventions must take into account the sensitive period of the target structure or process, along with the nutrient-specific, evidence-based principles of timing, dose, and duration.

2 Protein

2.1 *Clinical Studies*

Multiple studies have demonstrated that prenatal growth and growth in the first 2 years of life – particularly linear growth reflecting protein intake and accretion and growth of organs including the brain – are associated with cognition and school performance later in life (Perkins et al. 2017; Miller et al. 2016; Alam et al. 2020; Nahar et al. 2020). Recent work combining data from six different low-income countries in the Mal-Ed network demonstrated that the relationship between stunting (impaired linear growth) and later cognition is dose-dependent, in that children with early-onset persistent stunting (first onset of stunting between 0 and 6 months and persisting to 60 months) had significantly lower scores of cognitive development at

5 years of age (Nahar et al. 2020). No such relationship in scores was detected for children with early stunting who recovered or who had late-onset stunting alone.

Growth failure during the fetal period is referred to as intrauterine growth restriction (IUGR), typically defined as fetal weight less than the tenth percentile for gestational age (Cusick and Georgieff 2016). Multiple studies have demonstrated that children with IUGR have long-lasting cognitive deficits, including five times the prevalence of poor neurobehavioral outcomes at 2 years of age, including lower verbal ability and novelty preference scores as compared to children without IUGR (Spinillo et al. 1993; Murray et al. 2015; Wang et al. 2016). A systematic review of 35 studies evaluating IUGR and neurobehavioral outcomes later in childhood found that children with IUGR born at 35 weeks gestation or later scored half a standard deviation unit lower on all neurodevelopmental assessments (Murray et al. 2015). This deficit grew to 0.7 standard deviation units when comparing children with and without IUGR born before 35 weeks' gestation.

A landmark study conducted in Guatemala in the late 1960s and early 1970s clearly demonstrated the importance of early nutrition, in particular protein intake, on later child neurodevelopment (Pollitt et al. 1995). In the study, mothers and children in two villages were assigned twice daily consumption of a beverage containing both protein and energy (Atole 11.5 g protein and 163 kcal/cup). Mothers and children in two other villages were assigned twice daily consumption of a beverage containing energy alone (Fresco, 59 kcal/cup). Adolescents who had been exposed to protein-containing Atole in early life scored better on tests of general intellectual ability as compared to adolescents exposed to Fresco.

The importance of optimal nutrition and growth in early life, specifically in the prenatal period and the first postnatal year, was clearly demonstrated in a study of school-aged children in Thailand (Pongcharoen et al. 2012). Among these children, length at birth and during the first postnatal year, but not after 12 months of age, was positively associated with IQ score at 9 years of age. Growth after 12 months was not associated with any neurodevelopmental outcomes, underscoring the need to intervene in the first 1,000 days to maximize growth.

2.2 *Pre-clinical Studies*

Supporting the findings in clinical studies, amino acids and their source, dietary protein, are essential for brain development because they are the building blocks for structural and functional proteins. They are required for DNA and RNA synthesis and maintenance. Neurotransmitters, re-uptake proteins, neurotransmitter receptors are proteins that confer synaptic plasticity function to the brain. Growth factors (e.g., IGF-1, BDNF, NGF) are small proteins that are essential to drive brain development and to maintain plasticity in adulthood. Myelin, while made up of fatty acids, is placed on a scaffold inhabited by proteins such as myelin basic protein, PLP-1 and PLP-2, thus making neurotransmission speed of processing more efficient. Structural proteins form the basis of neurite extension, e.g., axons, dendrites, and synapse formation and maturation, during development to achieve maximally efficient brain circuit connectivity (Mallard et al. 1999; Rehn et al. 2004; Ruff et al. 2017). IUGR in the rat can induce epigenetic changes to brain chromatin with the potential to alter long-term gene regulation in the brain (Joss-Moore et al. 2011), including altered DNA methylation of brain-derived neurotrophic factor (BDNF), an important growth and differentiation factor during neuronal development and a protein that maintains synaptic plasticity in adulthood (Nishigori et al. 2008). IUGR also disrupts brain histone methylation patterns (Ke et al. 2010, 2014).

In select studies, IUGR is induced through a low protein diet to the rat dam and thus the fetus and offspring (Liu et al. 2011, 2013). This model approximates pure protein undernutrition effects on the brain. As seen in models in which IUGR is simulated with the uterine artery ligation, low protein diet results in hypomyelination and reduced gray matter connectivity. Ultimately, amino acids regulate growth through pathways such as mammalian target of rapamycin (mTOR), an intracellular signaling pathway that senses the availability of other key nutrients (e.g., iron, oxygen, glucose, zinc) to regulate the rate at which the neuron transcribes DNA, translates mRNA to protein, and organizes actin polymerization (Wullschleger et al. 2006; Fretham et al. 2011). As such, restriction of amino acid substrate results in less structural and functional proteins and less complex dendritic arbors. This process underlies the lower synaptic counts, lesser amount of gray and white matter and ultimately the compromised behaviors seen in pre-clinical models, including recognition memory deficits and hyperactivity (Liu et al. 2011, 2013; Naik et al. 2015). Overall, developmentally appropriate pre-clinical models confirm the findings from human studies (Table 1).

Like many nutrients, protein exhibits a U-shaped risk curve with excess protein delivery also constituting a risk. Infusion of amino acids into pregnant sheep at the end of gestation results in fetal hypoxia and acidosis (Tran et al. 2015). Even a direct fetal infusion of amino acids, although better tolerated by the fetus, does not result in increased protein accretion (Maliszewski et al. 2012; Rozance et al. 2009).

3 Long-Chain Polyunsaturated Fatty Acids (LC-PUFAs)

3.1 Clinical Studies

Long-chain fatty acids, specifically the omega-3 fatty acid docosahexaenoic acid (DHA; 6n-3) and the omega-6 fatty acid arachidonic acid (AA; 20:4n-6), play a significant role in brain structure and function and are critical for cortical visual acuity, signal transduction, brain connectivity, hippocampal development, and myelination (Carlson and Colombo 2016; Hadders-Algra 2011; Scholtz et al. 2013). Maternal transfer of LCPUFAs to the fetus occurs primarily during the last trimester of pregnancy, with a full-term infant having a whole-body DHA content of 3,800 mg (Das 2003). Poor LCPUFA status of the mother or factors that shorten pregnancy, such as preterm birth, reduce the amount of LCPUFA transferred to the fetus. After birth, LCPUFAs are transferred from the mother to the infant through breast milk, which is a rich source of DHA amount, a fact that has been posited as an explanation behind the findings of better cognitive outcomes in breastfed vs. formula-fed infants (Scholtz et al. 2013). During the first 6 months of life, DHA increases by 1882 mg in a breastfed infant, but decreases by 993 mg in a formula-fed infant (Das 2003), prompting manufacturers to add DHA and AA to infant formulas in the USA starting in 2002 (Scholtz et al. 2013). The rapid accumulation of LCPUFA in the brain during late pregnancy and early infancy coupled with the importance of maternal LCPUFA status during gestation and breastfeeding highlights these periods as potential times for effective intervention.

A role of LCPUFA's in brain development has been known for three decades, with better cortical visual acuity measured electrophysiologically and behaviorally in term and preterm infants fed formulas supplemented with DHA and AA (Carlson and Colombo 2016; Birch et al. 1992; Carlson et al. 1993; Uauy et al. 2001). Multiple randomized controlled trials of DHA and AA have since confirmed that LCPUFAs improve visual acuity in the first year of life, although whether this benefit persists into early childhood is unclear (Carlson and Colombo 2016; Birch et al. 2010; Colombo et al. 2013).

Studies with neurobehavioral outcomes, both assessed during infancy and later into childhood, have similarly had mixed results. Newborns who received LCPUFA-supplemented formula with DHA at 0.64% of total fatty acids in the first year of life exhibited greater connectivity between prefrontal and parietal regions at 9 years of age compared to infants who received no LCPUFAs or a higher or lower percentage of DHA (Birch et al. 2010; Lepping et al. 2019). However, preterm infants given formula containing LCPUFAs had no cognitive advantage at 9 years of age compared to those infants given control formula, although some advantage in girls vs. boys was noted (Isaacs et al. 2011).

The large number of trials testing LCPUFA supplementation of the mother and/or infant and the inconsistent results have rendered this topic seemingly ideal for meta-analyses. Indeed multiple meta-analyses, including three Cochrane Reviews (Delgado-Noguera et al. 2015; Jasani et al. 2017; Moon et al. 2016), have been done on LCPUFA supplementation in both full-term and preterm infants, with all analyses concluding that there appeared to be no significant association between LCPUFA supplementation and child neurodevelopment. While the conclusions of the analyses may be accurate, meta-analyses are not the ideal approach to assessing nutrient–brain interaction, as the frequent combination of different ages of intervention and outcome assessment, doses of intervention, and neurodevelopmental tests used violates the nutrition-brain rules of timing, dose, and duration (Barnard et al. 2017). Further, most meta-analyses combine studies that have used global tests of cognition, such as the Bayley Scales of Infant Development, since their results can easily be standardized and leave out smaller studies that have used more granular, circuit-specific tests that may best capture nutrient effects (Verfuerden et al. 2020).

The recent KUDOS study exemplifies the disparate results with regard to LCPUFA supplementation and child neurodevelopmental outcomes observed in large number of studies to date (Colombo et al. 2019). This study was a double-blinded, randomized, placebo-controlled trial of 600 mg of daily prenatal DHA supplementation or placebo starting at 14.5 weeks gestation until delivery. Children were followed and had multiple assessment points from birth through 6 years, with multiple global and circuit-specific tests used. Supplementation with DHA significantly reduced the incidence of preterm birth and improved attention during the first year of life but had little effect on neurodevelopmental outcomes assessed between 10 months and 6 years of age (Colombo et al. 2016, 2019). Improved DHA status of mothers and infants was associated with some long-term neurodevelopmental outcomes, but these effects were no longer apparent after adjustment for socioeconomic status. Similarly maternal supplementation with 800 mg DHA or placebo from 20 weeks' gestation until birth had no effect on attention, working memory, or inhibitory control in WIC preschoolers (Gould et al. 2014).

These results of no long-term benefit with prenatal supplementation stand in contrast to studies, including those from the same group, finding long-term benefit with infant supplementation (Colombo et al. 2013). Of note, low-income infants randomly assigned to formula with varying amounts of LCPUFAs for the first year of life demonstrated no difference in neurodevelopmental outcomes at 18 months of age compared to infants who received formula supplemented with no LC-PUFAs, but children who received LCPUFA formula with DHA at 0.32% and 64%, but not 96%, of total fatty acids, scored significantly better from 3 to 5 years of age on tasks of rule-learning and inhibition and on the Wechsler Primary Preschool Scales of Intelligence at 6 years of age (Colombo et al. 2013). The authors posited that these results perhaps collectively suggest that maternal supplementation helps with pregnancy outcomes and neurodevelopmental outcomes early in infancy, but these benefits don't persist, while supplementation during infancy and later has more lasting benefit.

3.2 Pre-clinical Studies

Long-chain polyunsaturated fatty acids (LC-PUFAs), particularly docosahexaenoic acid (DHA) and arachidonic acid (ARA), are critically important for early brain development. Studies in multiple pre-clinical models including mice, rats, pigs and non-human primates of maternal-neonatal LC-PUFA deficiency provide evidence of neurodevelopmental impairments, including deficits in retinal function, retinal acuity development, hippocampal neural transmission, and cognitive function in the offspring (Neuringer et al. 1984, 1986; Reisbick et al. 1997; Benolken et al. 1973; Brenna 2011, 2016). Biochemically, they are integral components of neuronal membrane structures, synapses, and myelin sheaths, especially in the photoreceptor cell membrane of the eye. Severe fatty acid deficiency results in abnormal myelin fatty acid composition, atypical myelination, and loss of neuronal and synaptosomal membrane integrity and signal transduction. Early life LC-PUFA deficiency alters DNA methylation rates in the brain) (Tyagi et al. 2015), thus providing the potential of long-term alterations to brain gene expression. In summary, the evidence in the pre-clinical literature appears to be stronger than the findings from clinical studies in humans (Table 1), but combined both pre-clinical and clinical evidence suggests that these compounds have their strongest effect in the fetal and early postnatal periods. The more consistent pre-clinical results may be due in part to the severity of deficiency induced in the models compared to what is typically found in at-risk human populations.

4 Iron

4.1 Clinical Studies

Iron deficiency is the most common nutritional deficiency worldwide and one of the four leading causes of lost developmental potential among children living in low- and middle-income countries (Walker et al. 2007; WHO 2021). The neurodevelopmental consequences of iron deficiency in children directly parallel the effects identified in pre-clinical models (Table 1), with reduced metabolism of rodent hippocampal neurons reflected by deficits in recognition memory in infants born with iron deficiency (Geng et al. 2015; Bastian et al. 2020; Geng et al. 2020), impaired neuronal myelination apparent as slower speed of processing and impaired motor development in iron-deficient infants (Lozoff et al. 2006; Santos et al. 2018), and alterations in dopaminergic signaling evident as more hesitant, wary, fearful behavior in toddlers (Lozoff et al. 2008). The mirroring of the deleterious effects seen in pre-clinical models with observed clinical consequences is striking with iron and well-documented by the consensus of more than three decades of research. This research not only elucidates the mechanism of damage to the developing brain caused by early-life iron deficiency, but it also provides insight into when to intervene to prevent the diminished neurodevelopment that afflicts more than 200 million children each year (Walker et al. 2007).

As in pre-clinical studies, clinical studies have collectively shown that the optimum time for an iron intervention is before or during a time for peak brain demand for iron so that the brain region, e.g., hippocampus, or process, e.g., myelination, that is developing is fully supported during its critical period (Geng et al. 2015, 2020; Christian et al. 2010; Murray-Kolb et al. 2012). The times of peak risk for iron deficiency are the neonatal period (specifically the last trimester of pregnancy), late infancy through toddlerhood, and adolescence, particularly in menstruating females (Cusick and Georgieff 2016). The first two of these periods overlap with times for peak brain need for iron and thus mark the life periods where the growing brain is most vulnerable to the potentially permanent damage of iron deficiency (Thompson and Nelson 2001).

An important means of ensuring optimal iron for brain development is to optimize the fetal iron endowment, or the amount of body iron an infant has at birth (Georgieff 2020; Fisher and Nemeth 2017). Iron transfer from the mother to the developing fetus occurs primarily in the last trimester of pregnancy, with term fetuses having 75 mg of elemental iron per kilogram body weight (Georgieff 2020; Fisher and Nemeth 2017). If this period is shortened, as with preterm birth, or if iron transfer to or through the placenta is compromised, as with gestational diabetes, IUGR, or maternal inflammation, the infant not only has an insufficient supply of iron to support the rapid growth of the hippocampus and the high speed of myelination occurring during this time, but it is also born without its full iron endowment (Shao et al. 2012; O'Brien et al. 2003; Martínez-Galiano et al. 2019; Georgieff et al. 1990). An insufficient iron endowment sets the stage for continued iron deficiency during the first 2 years of life, notably during late infancy and toddlerhood, the second period of peak brain need for iron (Miller et al. 2003; Scholl 2011).

Interventions that improve fetal iron endowment, such as maternal iron supplementation, delayed cord clamping, and control of gestational diabetes may improve child neurodevelopment (Cusick et al. 2018; Chaparro 2011). Maternal supplementation with 30–60 mg daily is recommended by the CDC and WHO due to its demonstrated benefit in correcting maternal anemia and increasing birth weight (CDC 1998; WHO 2012). Meta-analyses of maternal iron supplementation studies have failed to demonstrate benefit to child neurodevelopmental outcomes, possibly because of differing times in gestation of initiation of therapy or combination of several different global neurobehavioral tests into one outcome measure (Jayasinghe et al. 2018). Non-nutritional interventions such as delayed cord clamping and control of gestational diabetes also increase fetal iron endowment at birth and may improve child neurobehavioral outcomes.

The critical partnership of sufficient iron to support the rapidly brain regions and processing during a sensitive period is exemplified by two sets of studies in Nepal and in China, both designed to separate the effects of prenatal vs. postnatal iron status and supplementation on child neurodevelopment (Geng et al. 2015, 2020; Christian et al. 2010; Murray-Kolb et al. 2012; Angulo-Barroso et al. 2016). In Nepal, the seven-year-old offspring of more than 3,000 women who had been randomized to receive iron/folic acid supplementation from early pregnancy to 12 weeks' postpartum had significantly better scores of working memory, inhibitory control, and fine-motor functioning compared to children whose mothers did not receive prenatal/early postnatal iron, regardless of whether or not the child had received supplemental iron between 12 and 36 months of age (Christian et al. 2010). In fact, a separate analysis found that receiving iron during the second and third years of life conferred no additional benefit in intellectual and motor functioning if the child's mother had received iron (Christian et al. 2011). Finally, iron supplementation of children 12–36 months of age whose mothers did not receive iron during pregnancy did not result in higher neurobehavioral test scores at 7 years of age (Murray-Kolb et al. 2012).

Similarly, Chinese infants who were born with iron deficiency had poorer recognition memory, as assessed by auditory-evoked event-related potentials, as compared to infants born without iron deficiency (Geng et al. 2015). However, children who received iron themselves between 6 and 9 months of age had significantly better gross motor scores at 9 months of age, regardless of whether their mothers had received iron in pregnancy (Angulo-Barroso et al. 2016). This finding is in apparent contrast to that observed in the studies in Nepal, where prenatal iron supplementation was most critical for optimum childhood learning and memory, but motor development and coordination relies on preceding hippocampal development. Motor control shifts to the motor cortex around 3–4 months of age from more primitive reflexes of the brain stem and midbrain. Early infant supplementation would thus likely directly support this process, as would myelination that begins at approximately 36 weeks' gestation and continues through the first 2 years of life.

The corollary to the necessity of intervening early to support iron-dependent brain development is that the neurobehavioral consequences of early-life iron deficiency during a critical period may not be fully corrected with supplemental iron given later in life. In Chile, iron-deficient infants supplemented with 6 months of daily oral iron therapy had slower reaction time and poorer inhibitory control 8–9 years after supplementation, despite having a normal hemoglobin at the end of the supplementation period (Algarín et al. 2013). Similarly, five-year-old Costa Rican children who had had severe iron deficiency anemia as infants had poorer cognitive outcomes than peers who had not been iron deficient in infancy, despite not being iron deficient at the time of neurodevelopmental assessment and having had normal growth (Lozoff et al. 1991).

However, as with protein, a U-shaped curve with iron and neurobehavioral outcomes has been demonstrated with iron. A 10-year follow-up of 835 Chilean infants who had been randomly assigned to iron-fortified formula (mean, 12.7 mg/L) or low-iron formula from 6 to 12 months found that children with a high hemoglobin ($>$12.8 g/dL) who received the iron-fortified formula had significantly poorer outcomes for spatial memory and visual-motor integration, while children with a low hemoglobin ($<$10.5 g/dL) showed better outcomes (Lozoff et al. 2012). The mechanism behind the effects seen in this one study remains unknown and speculative.

4.2 *Pre-clinical Studies*

The neurobiology of iron is well understood because of the more than 40-year history of research in pre-clinical models, predominantly the rat, but also in the sheep, monkey, pig, and mouse. That extensive literature can be broken down into five domains where effects are seen: monoamine neurometabolism, myelination, neuronal structure, and gene regulation. The alterations in those systems relate well to the observed neurobehavioral pathologies in both the models and in humans: mood and affect disorders, slow speed of processing, reduced neural performance, abnormal circuit construction, and reduced expression of synaptic plasticity genes resulting in poorer performance. The timing of ID during brain development is a strong determinant of the ultimate behavioral phenotype. For example, prenatal ID in monkeys results in a hyperactivity phenotype (Golub et al. 2006), while postnatal ID results in a more depressed and anxious phenotype.

The monoamine neurotransmitters are synthesized by iron containing enzymes: tyrosine hydroxylase for dopamine and tryptophan hydroxylase for serotonin and norepinephrine. Iron deficiency alters the concentrations of the neurotransmitters themselves, as well as their receptors and re-uptake mechanisms. Dopamine is found throughout the brain but is particularly rich in the VTA loop. Disruption of striatal dopamine is thought to underlie behavioral deficits ranging from motor to socio-emotional domains. These alterations remain in the adult animal in spite of repletion in the postnatal period (Lozoff et al. 2006; Unger et al. 2012), suggesting a reprogramming of monoamine metabolism. The excessive fear and anxiety of adult rats who were iron deficient as pups relates well to the increased risk of anxiety and depression noted by Lozoff in adults who were iron deficient as young children (Lozoff et al. 2006).

Several of the desaturases that are responsible for the synthesis of fatty acids found in myelin are iron dependent (Connor and Menzies 1996). Rats that are iron-deficient have abnormal myelin fatty acid patterns, reduced myelin basic protein, and reduced myelin content. These abnormalities persist into adulthood (Clardy et al. 2006), consistent with permanent disruption. The persistence of longer latencies on visual evoked responses even after correction of ID in humans indicates slower electrophysiologic processing speed and is consistent with the pre-clinical findings.

Iron is critical for maintaining fundamental neuronal and glial oxygen consumption rate (OCR) through its role in cytochrome-mediated electron transport. Iron-deficient rat pups have abnormal cerebral metabolism (Rao et al. 2003). Mitochondria in iron-deficient neurons have reduced OCR and move less efficiently in developing neurites (Bastian et al. 2016; Bastian 2019). mTOR activity is altered (Fretham et al. 2011). Ultimately, this metabolic compromise results in an abnormally branched and more simplified dendrite arbor (Fretham et al. 2012; Bastian et al. 2016; Carlson et al. 2009); a factor directly related to loss of functional capacity (Jorgenson et al. 2003, 2005). Iron-deficient rat pups have delayed maturation and long-term deficits in long-term potentiation (Jorgenson et al. 2005), the electrophysiologic underpinning of learning activity in the hippocampus. Behaviorally, fetal-neonatal iron-deficient mice and rats have poorer spatial and recognition memory as adults (Fretham et al. 2012; Georgieff et al. 1990; Carlson et al. 2009).

Recent studies demonstrate that iron, like choline and LC-PUFAs, can affect brain gene expression through epigenetic modification of chromatin (Tran et al. 2016). The JARID family of histone demethylases contains a JUMANJI domain with an iron-nickel pocket that is crucial for their activity. Fetal-neonatal iron deficiency alters JARID expression and histone demethylase activity, resulting in suppression of expression of brain-derived neurotrophic factor (BDNF) (Tran et al. 2015). Neonatal iron deficiency causes long-term genome-wide alterations in the hippocampus, activating pathways associated with autism, schizophrenia, mood disorders, and Alzheimer's disease (Tran et al. 2016). The concordance of the pre-clinical data with the human epidemiology data is truly remarkable (Table 1).

5 Iodine

5.1 Clinical Studies

Iodine is critical for brain development due to its role in the thyroid hormones, thyroxine (T_4) and tri-iodothyronine (T_3). In the first trimester, fetal production of T3 is entirely dependent on maternal T4, although maternal thyroid hormone supports fetal development throughout pregnancy. The effect of severe iodine deficiency (median urinary iodine concentration of pregnant population <20 μg/L) is established and leads to cretinism, or congenital hypothyroidism, marked by severe developmental delay, growth retardation, and deficits in speech or hearing (Skeaff 2011). The effectiveness of iodine supplementation during pregnancy in preventing cretinism has been known since the early 1970s when a maternal injection of iodine significantly reduced the incidence of cretinism and improved child neurobehavioral outcomes in an area of severe iodine deficiency in Papua New Guinea (Pharoah et al. 1971, 2012). Subsequent studies in areas of severe iodine deficiency have demonstrated similar results, with the benefits of maternal iodine supplementation still apparent as better child cognitive outcomes after 5 years of age (O'Donnell et al. 2002; Bougma et al. 2013).

Recent programs establishing the universal iodization have been a public health success, with an estimated 88% of the world's population now consuming iodized salt (UNICEF 2019). Consequently, mild-to-moderate iodine deficiency has become more prevalent worldwide than severe iodine deficiency. Although the effect of severe iodine deficiency (urinary iodine <50 μg/L) during pregnancy has been established, the effect of mild-to-moderate iodine deficiency (UI: 50–100 μg/L) on child neurobehavioral outcomes remains unclear (Berbel et al. 2009; Gordon et al. 2009; Verhagen et al. 2020; Zimmermann 2020). Four meta-analyses concluded that iodine supplementation in this population of pregnant women had no detectable benefit on neurobehavioral outcomes in their children (Bougma et al. 2013; Dineva et al. 2020; Harding et al. 2017). The most recent of these meta-analyses included the results of three randomized placebo-controlled clinical trials and again failed to find consistent evidence of benefit (Dineva et al. 2020; Gowachirapant et al. 2017; Zhou et al. 2015; Brucker-Davis et al. 2015). Of the three trials, only one of the trials was adequately powered (Gowachirapant et al. 2017). The trial randomized 832 mildly iodine-deficient women from India and Thailand to start daily supplementation with 200 μg potassium iodine or placebo starting before 12 weeks' gestation. No effect was detected in executive function, behavior, acoustic testing or on the Bayley Scales when the child was 18 months of age. Combination of the data from the two other RCTs, which both compared iodine or placebo started in early pregnancy to placebo and assessed mental development by the Bayley Scales at 1.5–2 years also revealed no differences between the groups (Dineva et al. 2020).

Two observational studies, both in moderately iodine-deficient populations, did reveal that supplementation with iodine earlier rather than later in pregnancy was associated with a greater effect on child neurodevelopment. In one study, iodine supplementation was started in women at either 4–6, 12–14, or 37–40 months gestation (Berbel et al. 2009). The offspring of women who started supplementation at 4–6 months had a higher total development quotient, gross motor quotient, fine-motor quotient, and socialization quotient when they were 18 months of age as compared to children whose mothers started supplementation at 12–14 or 37–-40 months, with no difference in outcomes between the latter two groups. Similarly, 1-year-old children whose mothers started iodine supplementation in the first trimester had higher psychomotor development and a higher behavior rating than children whose mothers received no supplement in pregnancy, but no difference in mental development, as measured by the Bayley Scales was detected between the groups (Velasco et al. 2009).

Observational studies thus indicate that iodine supplementation earlier rather than later in pregnancy produces greater neurodevelopmental benefits, but none of the three randomized placebo-controlled trials demonstrated any benefit of iodine supplementation to pregnant women in areas of low-to-moderate iodine deficiencies. However, only one of the trials was adequately powered, underscoring the need for more adequately powered prospective placebo-controlled trials that account for key factors such as maternal iodine status pre-pregnancy, the timing, i.e., point in gestation, that supplementation began, and use of tests that capture neurobehavioral tasks known to be affected by iodine deficiency, e.g., visual information processing (Bell et al. 2016). Despite this lack of evidence, multiple groups, including the European and American Thyroid Associations recommend iodine supplementation for pregnant women and women planning to become pregnant (Lazarus et al. 2014; Alexander et al. 2017), even those living in areas with access to iodized salt.

5.2 *Pre-clinical Studies*

The neurobiology of iodine is relatively straightforward because the only known function of iodine in the brain is for thyroid hormone synthesis. Thyroid hormone regulates the metabolic rate of the cells in the brain (Giannocco et al. 2020). The human brain's oxygen consumption rate (OCR) is the highest during the late fetal and early postnatal life, accounting for 60% of the total body OCR (Leonard et al. 2003). The high OCR is thought to be due to the metabolic demand of the rapid regional and brain-wide growth and differentiation that occurs at this age. Thus, substrates (nutrients, hormones) that support mitochondrial ATP production are particularly essential during this time frame and deficiencies of those substrates cause more significant deficits than at other times. Thyroid hormone, as the master regulator of cellular OCR, is crucial for normal structural development and is dependent on both iodine and iron for its synthesis.

Iodine deficiency results in the expected brain structure consequences of inadequate neuronal and glial metabolic rate (Bernal 2000). Pre-clinical models confirm the findings from studies in humans in terms of the expected neuropathology and that the timing of iodine deficiency is a critical determinant of the structural and subsequent behavioral phenotype (Table 1). Iodine deficiency from early in pregnancy reduces brain DNA, RNA, and protein content, resulting in a lower brain weight that persists in spite of later iodine repletion. Late fetal and early postnatal iodine deficiency causes reduced dendritic arborization and synaptic counts, factors that are tightly linked to neuronal performance capacity (Redman et al. 2016). Myelination is also reduced because of the requirement of iodine/thyroid hormone for fatty acid synthesis (Lucia et al. 2018). The fatty acid profile of the myelin is abnormal, resulting functionally in slower speed of neuronal processing. Thyroid hormone is also critical in gene regulation because of thyroid sensitive promoter regions. Among genes regulated by thyroid in the brain, many are involved in synaptic plasticity, including sonic hedgehog, cfos, and cjun (Gilbert et al. 2016; Desouza et al. 2011; Dong et al. 2005).

6 Choline

6.1 Clinical Studies

The key role of choline in neurodevelopment has been evident in pre-clinical models for decades, but its importance in human neurodevelopment has only been demonstrated more recently. Several randomized controlled trials have reported positive effects of maternal choline supplementation on infant speed of processing and memory, with benefits still discernible later in childhood (Caudill et al. 2018; Ross et al. 2016; Kable et al. 2015; Derbyshire and Obeid 2020; Boeke et al. 2013). Cross-sectional and observational studies reporting associations between maternal choline status or dietary intake of choline during pregnancy and infant and child cognitive outcomes have had mixed results, with key considerations such as degree of pre-existing deficiency and the accuracy of serum markers of choline status in reflecting brain availability likely affecting outcomes (Caudill et al. 2018; Derbyshire and Obeid 2020; Signore et al. 2008; Wu et al. 2012).

Randomized clinical trials of choline supplementation during pregnancy suggest that a higher supplement dose and testing brain region or process-specific, rather than global, outcomes are critical components of detecting an effect. In a double-blind combined feeding/supplementation study, infants born to mothers who received 930 vs. 480 mg of choline daily starting in the last trimester had faster mean reaction time at 4, 7, 10, and, 13 months of age, likely reflecting choline's role in myelination. However, among children whose mothers received 480 mg of choline daily in late pregnancy, there was still a linear relationship between days of supplementation and reaction speed, suggesting a benefit of choline with longer exposure even at this lower dose. The benefit of this late pregnancy supplementation persisted into childhood. The seven-year-old children of mothers in the 930 mg daily group performed significantly better on a task of color-location memory than children whose mothers had received placebo (Bahnfleth et al. 2019).

Daily maternal supplementation with 900 mg choline beginning in the first trimester followed by infant supplementation with 600 mg of phosphatidylcholine (~100 mg choline) for 52 weeks led to increased suppression of the cerebral evoked response to auditory stimuli and also exhibited fewer parent-reported behavior problems at age 40 months, showing better attention and less social withdrawal (Ross et al. 2013, 2016). Finally, prenatal choline supplementation (750 mg) beginning in the first prenatal visit along with multiple micronutrients compared to multiple micronutrients without choline resulted in significant differences visual habituation at infant age 6–12 months, both in pregnancies affected by alcohol and in those that were not (Kable et al. 2015).

Further work has underscored the efficacy of choline during pregnancy in protecting the developing brain from fetal alcohol exposure. The six-month-old infants of South African mothers who were heavy drinkers but who received 2 g of choline daily starting in mid-pregnancy demonstrated better visual recognition memory as demonstrated by eye blink conditioning and higher novelty preference scores (Jacobson et al. 2018). In all of the randomized controlled trials of maternal choline supplementation, the neurodevelopmental benefit was observed at a dose that was substantially more than or nearly double the current Adequate Intake for choline during pregnancy of 450 mg/day (Zeisel 2006). One randomized controlled trial of maternal supplementation with 750 mg phosphatidylcholine from 18 weeks' gestation to 90 days postpartum, a dose equivalent to approximately 100 mg of choline, had no effect on infant cognitive function, with factors including relatively good pre-existing choline status and lack of follow-up time potentially also affecting the outcomes (Cheatham et al. 2012).

Although fewer in number, randomized controlled trials of choline supplementation in children, in particular, those affected by Fetal Alcohol Spectrum Disorders (FASD) have also demonstrated a neurodevelopmental benefit. Daily supplementation with 500 mg choline to 2- to 5-year-old children with FASD was safe and effective and improved scores on hippocampally mediated memory tasks among children 2–3 years of age (Wozniak et al. 2013; Wozniak et al. 2015). Follow-up of these children after 4 years revealed even more benefits in the group of children who received choline, including better scores in working memory, non-verbal intelligence, and verbal memory and fewer symptoms of ADHD than children who had received placebo 4 years earlier (Wozniak et al. 2020). In contrast, a much smaller daily dose of choline (10.5 mg and other study) for 2 years along with docosahexaenoic acid (DHA) and uridine-5-monophosphate given to infants 1--18 months of age with suspected cerebral palsy had no effect on neurodevelopment, as measured by the Bayley Scales of Infant and Toddler Development (Andrew et al. 2018).

Observational and cross-sectional studies of maternal choline status or dietary intake and infant neurodevelopment have had mixed results (Boeke et al. 2013; Signore et al. 2008; Wu et al. 2012; Villamor et al. 2012; Strain et al. 2013). Many have reported significant positive associations between maternal choline intake in the first and second trimester with offspring visual memory, maternal plasma free choline and early child development (Wu et al. 2012), and serum total choline and reduced risk of neural tube defects (Shaw et al. 2009).

6.2 *Pre-clinical Studies*

Pre-clinical studies over the past 30 years strongly support a role for choline in brain development although the precise mechanisms are not completely understood. Much of the pre-clinical work in rats has concentrated on the role of choline in hippocampal development. The studies at the molecular, cellular, biochemical, electrophysiologic, morphology and behavioral levels provide a cohesive and consistent demonstration of its benefits. Critical periods of choline supplementation are present during mid-gestation during hippocampal neurogenesis and again postnatally during the period of rapid hippocampal neuronal growth and differentiation. Provision of choline during those critical periods to control rat dams and pups results in improved hippocampal morphology, electrical potentiation and, most importantly, recognition learning and memory in adulthood (Meck et al. 2007). The findings illustrate that nutrients provided early in life can have long-lasting and potentially life-span effects. These strong effects in nutritionally replete pre-clinical models are surprising in light of the equivocal studies in young humans (Table 1).

Choline supplementation during either of the two critical periods also improves hippocampal outcomes in pathologic conditions, including rats with fetal alcohol exposure (Ryan et al. 2008), mice with a Rett syndrome mutation (Ward et al. 2009), mice with a Down syndrome mutation characterized by rapid cognitive decline (Moon et al. 2010), and rats with iron deficiency anemia (Kennedy et al. 2014). These findings across multiple species and pathologies (e.g., toxin exposure, genetic mutation, nutritional deficiency) suggest that choline works on a fundamental biology common to all of these conditions.

Zeisel has proposed three potential biological mechanisms related to choline's role in one-carbon metabolism (Zeisel 2017), including the acetylcholine neurotransmitter system, phosphatidylcholine found in myelin, and epigenetic modification with choline acting as a methyl donor. The latter is particularly intriguing in light of the large-scale changes in adult hippocampal gene expression in rats that were iron deficient in early life. Choline supplementation during gestation partially reversed long-term suppression of synaptic plasticity genes induced by early life iron deficiency and, importantly, partially reversed learning and memory deficits (Kennedy et al. 2014; Tran et al. 2016). The particular iron deficiency-induced epigenetic modifications that are reversed are methylation patterns of histones (Tran et al. 2016). Choline therefore appears to be a potential nutritional "workaround" to reverse brain gene suppression caused by adverse early life environmental and genetic factors.

7 Zinc

7.1 Clinical Studies

Zinc is ubiquitous in the body, catalyzing more than 300 metalloenzymes, supporting the structure of more than 2,500 transcription factors, and regulating thousands of genes (King et al. 2015). As such, zinc plays a critical role in the development of the central nervous system, catalyzing numerous enzymes of neuronal metabolism and playing a key role in DNA synthesis. Zinc also plays a role in neuronal pruning during development. High concentrations of zinc are accordingly found in the hippocampus, cerebellum, prefrontal cortex, and in neurons of the limbic system (Gower-Winter and Levenson 2012).

Despite these key role and nearly global presence in the brain, clinical studies of prenatal or early-child zinc supplementation have not demonstrated a clear picture of benefit to brain development. As with iron, several meta-analyses and reviews (Warthon-Medina et al. 2015; Nissensohn et al. 2013), including one Cochrane Review (Gogia and Sachdev 2012), of zinc supplementation and the effect on cognitive and motor function in children have been published, but all found no significant effect of zinc supplementation in infancy or childhood on child cognition or motor development. The authors of these reviews acknowledge a great degree of heterogeneity in the studies of zinc supplementation and child development, with a wide range of effect sizes and study designs.

Individual studies, however, reveal key considerations when interpreting the effect of zinc on child neurobehavioral outcomes, including demonstration of beneficial outcomes when zinc deficiency is prevented in early infancy and also positive impact of zinc when given in combination with iron. Again highlighting the importance of preventing, rather than treating deficiency, Colombo and colleagues demonstrated in a double-blind randomized clinical trial that zinc supplementation to 6-month-old Peruvian infants was associated with the normative decline in both look duration during habituation and in look duration and shifting between objects between free play at 12–18 months of age as compared to infants not supplemented with zinc, supporting a profile of normative information processing and active attention profiles (Colombo et al. 2014). Zinc concentration is high in the hippocampus, the region of the brain that supports memory and attention.

Two studies in Peru demonstrated that zinc supplementation during pregnancy improves markers of fetal activity and heart rate (Merialdi et al. 1999, 2004), but these benefits did not translate into improved neurobehavioral outcomes in later childhood (Caulfield et al. 2010). In the first study, fetuses whose mothers were randomized to 15 mg/kg-day of zinc had an increased fetal heart range, higher number of accelerations, greater time moving, and more large movements at 36 weeks' gestation than fetuses whose mothers received placebo. Moreover, these markers of fetal development improved longitudinally from 32 to 36 weeks' gestation in the zinc-supplemented group, reflecting sustained benefit (Merialdi et al. 1999). In the second study, 25 mg daily prenatal zinc was associated with lower fetal heart rate (reflecting stability and maturation of the autonomic nervous system, greater fetal heart rate variability, and a greater number of accelerations (Merialdi et al. 2004). However, these improvements did not translate to improved cognitive, motor, and behavioral outcomes when the children were assessed at 4.5 years of age, even in this population where zinc deficiency in pregnancy has been estimated to be more than 80% (Caulfield et al. 2010).

Similarly, prenatal zinc supplementation (25 mg-day for second half of gestation) of US women similarly had no effect on tests of cognition, auditory and visual development, and motor development when their children were assessed at 5 years of age (Tamura et al. 2003), while 7- to 9-year-old Bangladeshi children whose mothers received prenatal zinc did not have greater scores in tests of intellectual functioning, working memory, inhibitory control, or fine motor skills compared to children whose mothers did not receive prenatal zinc (Christian et al. 2010).

7.2 *Pre-clinical Studies*

Pre-clinical models demonstrate that zinc is essential for brain development from fetal through postnatal development. As with many nutrients that affect the developing brain, the effects of zinc deficiency are more profound during earlier stages of development and during periods of rapid growth. In general, developmentally appropriately timed pre-clinical models are in agreement with findings in young children (Table 1). Zinc has an important role in enzymes protein and nucleic acid biochemistry, thereby regulating neuronal DNA transcription rates, mRNA translation rates, and synthesis of neurotrophic factors (Kumar et al. 2021). Through these processes, it regulates neuronal differentiation, migration, and structural development (Mackenzie et al. 2007; Adamo and Oteiza 2010). Beyond structure, zinc influences the brain's electrophysiologic function by regulating GABA-stimulated chloride influx into hippocampal neurons, calmodulin-dependent protein kinase II alpha/brain-derived neurotrophic factor (alpha-CaMKII/BDNF) signaling and through its release from presynaptic boutons (Frederickson et al. 2005; Yu et al. 2013).

Zinc deficiency during development of intra-limbic and cortico-limbic structures (e.g., hippocampus, amygdala, prefrontal cortex) results in alterations to circuits underlying neuroemotional and neurocognitive behaviors (Mackenzie et al. 2007; Yu et al. 2013). A social avoidance/autism adult phenotype has been described in a mouse model of fetal zinc deficiency (Grabrucker et al. 2014). Supplementation of zinc to pregnant dams in a genetic mouse model of autism improved synaptic structure and function and mitigated autistic like behaviors in the offspring (Vyas et al. 2020). Postnatal zinc deficiency in the rhesus monkey reduces short-term memory capacity, indicative of hippocampal dysfunction (Golub et al. 1994). Pre-clinical models of early life zinc deficiency indicate that zinc is one of the nutrients that can program brain development into adulthood, although the mechanisms underlying the long-term effects are unknown. Beyond altered structural development, a recent study indicates that alterations to DNA methylation, particularly of brain-derived neurotrophic factor, occur with zinc deficiency (Hu et al. 2017). Thus, zinc joins the ever-expanding list of micronutrients that appear to confer long-term risk to the developing central nervous system through epigenetic modification of synaptic plasticity genes.

8 Vitamin D

8.1 Clinical Studies

Despite abundant pre-clinical evidence demonstrating the presence of vitamin D receptors in the brain along with enzymes required for 25(OH)D activation and inactivation and a demonstrated role for vitamin D in hippocampal neuronal differentiation and apoptosis, clinical evidence for a principal role for vitamin D in brain development is lacking.

To date, two clinical trials have examined a link between child vitamin D supplementation and child neurodevelopmental outcomes (Salas et al. 2018; Wicklow et al. 2016). In the first, extremely preterm infants (23–27 weeks' gestation) were randomized to 200 IU or 400 IU vitamin D or placebo starting from the first day of enteral feeding to postnatal Day 28 (Salas et al. 2018). Neither dose had an effect on the cognitive composite score of the Bayley Scales when the child was 2 years of age. In the second, 55 healthy term, breastfed children were randomized to 400 IU, 800 IU, of 1,200 IU vitamin D (Wicklow et al. 2016). Infants in the 400 IU group scored higher in gross motor achievements as compared to infants in the other two higher-dose supplementation groups.

To date, no randomized clinical trials of maternal vitamin D supplementation have had markers of infant neurobehavioral developments as a primary outcome, but secondary analysis of a randomized clinical trial testing high (400 IU) vs. standard (2,800 IU) maternal vitamin D during the third trimester found no effect on child neurodevelopmental outcomes in the first 6 years of life (Sass et al. 2020). No studies have examined the role of prenatal maternal vitamin D supplementation on child neurobehavioral outcomes, but multiple observational studies have examined associations between maternal vitamin D status and child outcomes after delivery as well as several cross-sectional studies assessing child vitamin D status concurrently with neurodevelopmental outcomes. The results of both types of studies have been equivocal, with multiple studies finding a link between maternal vitamin D status and improved child IQ, motor development, and behavior (Whitehouse et al. 2012; Specht et al. 2020; Melough et al. 2021), but multiple others of similar design finding no such relationships (Mutua et al. 2020; Windham et al. 2019; McCarthy et al. 2018; Chowdhury et al. 2020; Voltas et al. 2020).

Vitamin D could improve child neurobehavioral outcomes indirectly via improved child growth or infection reduction, but supplementation studies testing these hypotheses have also reported mixed results (Pérez-López et al. 2015). Many studies have reported an association between maternal vitamin D supplementation and greater birth weight and head circumference, but a recent prospective randomized trial of high-dose vitamin D in Bangladeshi mothers found no effect on infant length-for-age at 1 year of age.

Vitamin D supplementation of children with autism reduced the core symptoms of autism in 75% of children, and maternal supplementation with vitamin D reduced the incidence of having an autistic child in mothers who already had at least one autistic child (Cannell 2017). The mechanism by which vitamin D may affect autism incidence or severity is not known but may be due in part to increased serotonin synthesis (Patrick and Ames 2014).

While clinical evidence between vitamin D status and brain development is lacking, heterogeneity in study design, length of supplementation, differences in outcome definitions of vitamin D deficiency and insufficiency, and different method of measuring the key biomarker of vitamin D status (25(OH)D) obscure full interpretation of the data that are available.

8.2 Pre-clinical Studies

A strong case can be made that developmental vitamin D deficiency should affect neurodevelopment because vitamin D is critical for fundamental developmental processes such as proliferation and apoptosis (Lisi et al. 2020). Pre-clinical models of early-life vitamin D deficiency demonstrate long-lasting negative effects on cognitive and behavioral development (Martínez-Galiano et al. 2019). Multiple species have been used to try to understand the role of vitamin D sufficiency in normal brain development and to try to relate the effects of vitamin D deficiency during brain development to epidemiologic data in humans suggesting a link to social-cognitive dysfunction, including autism, and anxiety/depression. Zebrafish raised in a vitamin D deficient environment during development showed increased anxiety behavior in adulthood (Oliveri et al. 2020). Neuroinflammation during development has been proposed as a potential mechanism for autistic like behaviors. In a mouse model of maternal immune-activation, the offspring were protected from autistic-like behaviors through vitamin D supplementation (Vuillermot et al. 2017). Vitamin D status modulates hippocampal BDNF levels in adult rodents with consequent effects on learning and memory behavior. Whether it has a profound effect on BDNF and hippocampal function development has only recently been addressed. Recent studies are suggestive of a relationship between early life vitamin D status and risk of adult neurocognitive problems (Lardner 2015), where vitamin D deficiency during gestation alters hippocampal neuronal development and synaptic plasticity in adulthood (Grecksch et al. 2009).

Due to the paucity of convincing and consistent human data and relatively limited exploration in pre-clinical models, there remains a significant disconnect between the findings in pre-clinical models that suggest a major role for vitamin D in brain development and clinical studies that do not show consistent findings (Table 1). As articulated in a recent review, the future research agenda in the pre-clinical venue should include assessments of brain region biochemistry, structure, electrophysiology and behavioral function in models of vitamin D deficiency and supplementation during development (Lardner 2015).

9 Conclusion

The true cost of undernutrition during sensitive periods of brain development is both to the individual and also to society, in terms of lost education and job potential. This effect is particularly marked in populations in which early-life nutritional deficiencies are prevalent and can lead to an entrenched cycle of poverty, undernutrition, and suboptimal brain development. Eradication of iron, iodine, and zinc deficiencies has been estimated to shift the world's IQ 10 points higher (Morris et al. 2008). Iron deficiency in the fetal and early postnatal periods alone can increase the risk of autism, schizophrenia, depression, anxiety, and poorer executive function in adulthood (Schmidt et al. 2014; Insel et al. 2008; Lozoff et al. 2000). Early-life undernutrition thus can cause long-lasting and widespread dysfunction, underscoring the need to introduce preventative interventions before sensitive periods of brain development.

References

Adamo AM, Oteiza PI (2010) Zinc deficiency and neurodevelopment: the case of neurons. BioFactors Oxf Engl 36(2):117–124. https://doi.org/10.1002/biof.91

Alam MA, Richard SA, Fahim SM et al (2020) Impact of early-onset persistent stunting on cognitive development at 5 years of age: results from a multi-country cohort study. PLoS One 15(1):e0227839. https://doi.org/10.1371/journal.pone.0227839

Alexander EK, Pearce EN, Brent GA et al (2017) 2017 guidelines of the American Thyroid Association for the diagnosis and management of thyroid disease during pregnancy and the postpartum. Thyroid Off J Am Thyroid Assoc 27(3):315–389. https://doi.org/10.1089/thy.2016.0457

Algarín C, Nelson CA, Peirano P, Westerlund A, Reyes S, Lozoff B (2013) Iron-deficiency anemia in infancy and poorer cognitive inhibitory control at age 10 years. Dev Med Child Neurol 55 (5):453–458. https://doi.org/10.1111/dmcn.12118

Andrew MJ, Parr JR, Montague-Johnson C et al (2018) Neurodevelopmental outcome of nutritional intervention in newborn infants at risk of neurodevelopmental impairment: the Dolphin neonatal double-blind randomized controlled trial. Dev Med Child Neurol 60(9):897–905. https://doi.org/10.1111/dmcn.13914

Angulo-Barroso RM, Li M, Santos DCC et al (2016) Iron supplementation in pregnancy or infancy and motor development: a randomized controlled trial. Pediatrics 137(4):e20153547. https://doi.org/10.1542/peds.2015-3547

Bahnfleth C, Canfield R, Nevins J, Caudill M, Strupp B (2019) Prenatal choline supplementation improves child color-location memory task performance at 7 Y of age (FS05-01-19). Curr Dev Nutr 3(Suppl 1). https://doi.org/10.1093/cdn/nzz052.FS05-01-19

Barnard ND, Willett WC, Ding EL (2017) The misuse of meta-analysis in nutrition research. JAMA 318(15):1435–1436. https://doi.org/10.1001/jama.2017.12083

Bastian TW (2019) Potential mechanisms driving mitochondrial motility impairments in developing iron-deficient neurons. J Exp Neurosci 13:1179069519858351. https://doi.org/10.1177/1179069519858351

Bastian TW, von Hohenberg WC, Mickelson DJ, Lanier LM, Georgieff MK (2016) Iron deficiency impairs developing hippocampal neuron gene expression, energy metabolism, and dendrite complexity. Dev Neurosci 38(4):264–276. https://doi.org/10.1159/000448514

Bastian TW, Rao R, Tran PV, Georgieff MK (2020) The effects of early-life Iron deficiency on brain energy metabolism. Neurosci Insights 15:2633105520935104. https://doi.org/10.1177/2633105520935104

Bell MA, Ross AP, Goodman G (2016) Assessing infant cognitive development after prenatal iodine supplementation. Am J Clin Nutr 104(Suppl 3):928S–934S. https://doi.org/10.3945/ajcn.115.110411

Benolken RM, Anderson RE, Wheeler TG (1973) Membrane fatty acids associated with the electrical response in visual excitation. Science 182(4118):1253–1254. https://doi.org/10.1126/science.182.4118.1253

Berbel P, Mestre JL, Santamaría A et al (2009) Delayed neurobehavioral development in children born to pregnant women with mild hypothyroxinemia during the first month of gestation: the importance of early iodine supplementation. Thyroid Off J Am Thyroid Assoc 19(5):511–519. https://doi.org/10.1089/thy.2008.0341

Bernal J (2000) Thyroid hormones in brain development and function. In: Feingold KR, Anawalt B, Boyce A et al (eds) Endotext. MDText.com, Inc. http://www.ncbi.nlm.nih.gov/books/NBK285549/

Birch EE, Birch DG, Hoffman DR, Uauy R (1992) Dietary essential fatty acid supply and visual acuity development. Invest Ophthalmol Vis Sci 33(11):3242–3253

Birch EE, Carlson SE, Hoffman DR et al (2010) The DIAMOND (DHA intake and measurement of neural development) study: a double-masked, randomized controlled clinical trial of the maturation of infant visual acuity as a function of the dietary level of docosahexaenoic acid. Am J Clin Nutr 91(4):848–859. https://doi.org/10.3945/ajcn.2009.28557

Boeke CE, Gillman MW, Hughes MD, Rifas-Shiman SL, Villamor E, Oken E (2013) Choline intake during pregnancy and child cognition at age 7 years. Am J Epidemiol 177(12):1338–1347. https://doi.org/10.1093/aje/kws395

Bougma K, Aboud FE, Harding KB, Marquis GS (2013) Iodine and mental development of children 5 years old and under: a systematic review and meta-analysis. Nutrients 5(4):1384–1416. https://doi.org/10.3390/nu5041384

Brenna JT (2011) Animal studies of the functional consequences of suboptimal polyunsaturated fatty acid status during pregnancy, lactation and early post-natal life. Matern Child Nutr 7(Suppl 2):59–79. https://doi.org/10.1111/j.1740-8709.2011.00301.x

Brenna JT (2016) Long-chain polyunsaturated fatty acids and the preterm infant: a case study in developmentally sensitive nutrient needs in the United States. Am J Clin Nutr 103(2):606S–615S. https://doi.org/10.3945/ajcn.114.103994

Brucker-Davis F, Ganier-Chauliac F, Gal J et al (2015) Neurotoxicant exposure during pregnancy is a confounder for assessment of iodine supplementation on neurodevelopment outcome. Neurotoxicol Teratol 51:45–51. https://doi.org/10.1016/j.ntt.2015.07.009

Cannell JJ (2017) Vitamin D and autism, what's new? Rev Endocr Metab Disord 18(2):183–193. https://doi.org/10.1007/s11154-017-9409-0

Carlson SE, Colombo J (2016) Docosahexaenoic acid and arachidonic acid nutrition in early development. Adv Pediatr Infect Dis 63(1):453–471. https://doi.org/10.1016/j.yapd.2016.04.011

Carlson SE, Werkman SH, Rhodes PG, Tolley EA (1993) Visual-acuity development in healthy preterm infants: effect of marine-oil supplementation. Am J Clin Nutr 58(1):35–42. https://doi.org/10.1093/ajcn/58.1.35

Carlson ES, Tkac I, Magid R et al (2009) Iron is essential for neuron development and memory function in mouse hippocampus. J Nutr 139(4):672–679. https://doi.org/10.3945/jn.108.096354

Caudill MA, Strupp BJ, Muscalu L, Nevins JEH, Canfield RL (2018) Maternal choline supplementation during the third trimester of pregnancy improves infant information processing speed: a randomized, double-blind, controlled feeding study. FASEB J Off Publ Fed Am Soc Exp Biol 32(4):2172–2180. https://doi.org/10.1096/fj.201700692RR

Caulfield LE, Putnick DL, Zavaleta N et al (2010) Maternal gestational zinc supplementation does not influence multiple aspects of child development at 54 mo of age in Peru. Am J Clin Nutr 92 (1):130–136. https://doi.org/10.3945/ajcn.2010.29407

CDC (1998) Recommendations to prevent and control iron deficiency in the United States. https://www.cdc.gov/mmwr/preview/mmwrhtml/00051880.htm. Accessed 28 Jan 2021

Chaparro CM (2011) Timing of umbilical cord clamping: effect on iron endowment of the newborn and later iron status. Nutr Rev 69(Suppl 1):S30–S36. https://doi.org/10.1111/j.1753-4887.2011.00430.x

Cheatham CL, Goldman BD, Fischer LM, da Costa K-A, Reznick JS, Zeisel SH (2012) Phosphatidylcholine supplementation in pregnant women consuming moderate-choline diets does not enhance infant cognitive function: a randomized, double-blind, placebo-controlled trial. Am J Clin Nutr 96(6):1465–1472. https://doi.org/10.3945/ajcn.112.037184

Chowdhury R, Taneja S, Kvestad I, Hysing M, Bhandari N, Strand TA (2020) Vitamin D status in early childhood is not associated with cognitive development and linear growth at 6-9 years of age in North Indian children: a cohort study. Nutr J 19(1):14. https://doi.org/10.1186/s12937-020-00530-2

Christian P, Murray-Kolb LE, Khatry SK et al (2010) Prenatal micronutrient supplementation and intellectual and motor function in early school-aged children in Nepal. JAMA 304 (24):2716–2723. https://doi.org/10.1001/jama.2010.1861

Christian P, Morgan ME, Murray-Kolb L et al (2011) Preschool iron-folic acid and zinc supplementation in children exposed to iron-folic acid in utero confers no added cognitive benefit in early school-age. J Nutr 141(11):2042–2048. https://doi.org/10.3945/jn.111.146480

Clardy SL, Wang X, Zhao W et al (2006) Acute and chronic effects of developmental iron deficiency on mRNA expression patterns in the brain. J Neural Transm Suppl 71:173–196. https://doi.org/10.1007/978-3-211-33328-0_19

Colombo J, Carlson SE, Cheatham CL et al (2013) Long-term effects of LCPUFA supplementation on childhood cognitive outcomes. Am J Clin Nutr 98(2):403–412. https://doi.org/10.3945/ajcn.112.040766

Colombo J, Zavaleta N, Kannass KN et al (2014) Zinc supplementation sustained normative neurodevelopment in a randomized, controlled trial of Peruvian infants aged 6-18 months. J Nutr 144(8):1298–1305. https://doi.org/10.3945/jn.113.189365

Colombo J, Gustafson KM, Gajewski BJ et al (2016) Prenatal DHA supplementation and infant attention. Pediatr Res 80(5):656–662. https://doi.org/10.1038/pr.2016.134

Colombo J, Shaddy DJ, Gustafson K et al (2019) The Kansas University DHA Outcomes Study (KUDOS) clinical trial: long-term behavioral follow-up of the effects of prenatal DHA supplementation. Am J Clin Nutr 109(5):1380–1392. https://doi.org/10.1093/ajcn/nqz018

Connor JR, Menzies SL (1996) Relationship of iron to oligodendrocytes and myelination. Glia 17 (2):83–93. https://doi.org/10.1002/(SICI)1098-1136(199606)17:2<83::AID-GLIA1>3.0.CO;2-7

Cusick SE, Georgieff MK (2016) The role of nutrition in brain development: the Golden opportunity of the "first 1000 days". J Pediatr 175:16–21. https://doi.org/10.1016/j.jpeds.2016.05.013

Cusick SE, Georgieff MK, Rao R (2018) Approaches for reducing the risk of early-life iron deficiency-induced brain dysfunction in children. Nutrients 10(2). https://doi.org/10.3390/nu10020227

Das UN (2003) Long-chain polyunsaturated fatty acids in the growth and development of the brain and memory. Nutrition 19(1):62–65. https://doi.org/10.1016/S0899-9007(02)00852-3

Delgado-Noguera MF, Calvache JA, Cosp XB, Kotanidou EP, Galli-Tsinopoulou A (2015) Supplementation with long chain polyunsaturated fatty acids (LCPUFA) to breastfeeding mothers for improving child growth and development. Cochrane Database Syst Rev 7. https://doi.org/10.1002/14651858.CD007901.pub3

Derbyshire E, Obeid R (2020) Choline, neurological development and brain function: a systematic review focusing on the first 1000 days. Nutrients 12(6). https://doi.org/10.3390/nu12061731

Desouza LA, Sathanoori M, Kapoor R et al (2011) Thyroid hormone regulates the expression of the sonic hedgehog signaling pathway in the embryonic and adult mammalian brain. Endocrinology 152(5):1989–2000. https://doi.org/10.1210/en.2010-1396

Dineva M, Fishpool H, Rayman MP, Mendis J, Bath SC (2020) Systematic review and meta-analysis of the effects of iodine supplementation on thyroid function and child neurodevelopment in mildly-to-moderately iodine-deficient pregnant women. Am J Clin Nutr 112(2):389–412. https://doi.org/10.1093/ajcn/nqaa071

Dong J, Yin H, Liu W, Wang P, Jiang Y, Chen J (2005) Congenital iodine deficiency and hypothyroidism impair LTP and decrease C-fos and C-jun expression in rat hippocampus. Neurotoxicology 26(3):417–426. https://doi.org/10.1016/j.neuro.2005.03.003

Eyles D, Burne T, McGrath J (2011) Vitamin D in fetal brain development. Semin Cell Dev Biol 22 (6):629–636. https://doi.org/10.1016/j.semcdb.2011.05.004

Fisher AL, Nemeth E (2017) Iron homeostasis during pregnancy. Am J Clin Nutr 106(Suppl 6):1567S–1574S. https://doi.org/10.3945/ajcn.117.155812

Frederickson CJ, Koh J-Y, Bush AI (2005) The neurobiology of zinc in health and disease. Nat Rev Neurosci 6(6):449–462. https://doi.org/10.1038/nrn1671

Fretham SJB, Carlson ES, Georgieff MK (2011) The role of iron in learning and memory. Adv Nutr Bethesda Md 2(2):112–121. https://doi.org/10.3945/an.110.000190

Fretham SJB, Carlson ES, Wobken J, Tran PV, Petryk A, Georgieff MK (2012) Temporal manipulation of transferrin-receptor-1 dependent iron uptake identifies a sensitive period in mouse hippocampal neurodevelopment. Hippocampus 22(8):1691–1702. https://doi.org/10.1002/hipo.22004

Geng F, Mai X, Zhan J et al (2015) Impact of fetal-neonatal iron deficiency on recognition memory at 2 months of age. J Pediatr 167(6):1226–1232. https://doi.org/10.1016/j.jpeds.2015.08.035

Geng F, Mai X, Zhan J et al (2020) Timing of iron deficiency and recognition memory in infancy. Nutr Neurosci 0(0):1–10. https://doi.org/10.1080/1028415X.2019.1704991

Georgieff MK (2020) Iron deficiency in pregnancy. Am J Obstet Gynecol 223(4):516–524. https://doi.org/10.1016/j.ajog.2020.03.006

Georgieff MK, Landon MB, Mills MM et al (1990) Abnormal iron distribution in infants of diabetic mothers: spectrum and maternal antecedents. J Pediatr 117(3):455–461. https://doi.org/10.1016/s0022-3476(05)81097-2

Georgieff MK, Ramel SE, Cusick SE (2018) Nutritional influences on brain development. Acta Paediatr 107(8):1310–1321. https://doi.org/10.1111/apa.14287

Giannocco G, Kizys MML, Maciel RM, de Souza JS (2020) Thyroid hormone, gene expression, and central nervous system: where we are. Semin Cell Dev Biol. https://doi.org/10.1016/j.semcdb.2020.09.007

Gilbert ME, Sanchez-Huerta K, Wood C (2016) Mild thyroid hormone insufficiency during development compromises activity-dependent neuroplasticity in the hippocampus of adult male rats. Endocrinology 157(2):774–787. https://doi.org/10.1210/en.2015-1643

Gogia S, Sachdev HS (2012) Zinc supplementation for mental and motor development in children. Cochrane Database Syst Rev 12:CD007991. https://doi.org/10.1002/14651858.CD007991.pub2

Golub MS, Takeuchi PT, Keen CL, Gershwin ME, Hendrickx AG, Lonnerdal B (1994) Modulation of behavioral performance of prepubertal monkeys by moderate dietary zinc deprivation. Am J Clin Nutr 60(2):238–243. https://doi.org/10.1093/ajcn/60.2.238

Golub MS, Hogrefe CE, Germann SL, Capitanio JP, Lozoff B (2006) Behavioral consequences of developmental iron deficiency in infant rhesus monkeys. Neurotoxicol Teratol 28(1):3–17. https://doi.org/10.1016/j.ntt.2005.10.005

Gordon RC, Rose MC, Skeaff SA, Gray AR, Morgan KMD, Ruffman T (2009) Iodine supplementation improves cognition in mildly iodine-deficient children. Am J Clin Nutr 90(5):1264–1271. https://doi.org/10.3945/ajcn.2009.28145

Gould JF, Makrides M, Colombo J, Smithers LG (2014) Randomized controlled trial of maternal omega-3 long-chain PUFA supplementation during pregnancy and early childhood development of attention, working memory, and inhibitory control. Am J Clin Nutr 99(4):851–859. https://doi.org/10.3945/ajcn.113.069203

Gowachirapant S, Jaiswal N, Melse-Boonstra A et al (2017) Effect of iodine supplementation in pregnant women on child neurodevelopment: a randomised, double-blind, placebo-controlled trial. Lancet Diabetes Endocrinol 5(11):853–863. https://doi.org/10.1016/S2213-8587(17)30332-7

Gower-Winter SD, Levenson CW (2012) Zinc in the central nervous system: from molecules to behavior. BioFactors Oxf Engl 38(3):186–193. https://doi.org/10.1002/biof.1012

Grabrucker S, Jannetti L, Eckert M et al (2014) Zinc deficiency dysregulates the synaptic ProSAP/Shank scaffold and might contribute to autism spectrum disorders. Brain J Neurol 137(Pt 1):137–152. https://doi.org/10.1093/brain/awt303

Grecksch G, Rüthrich H, Höllt V, Becker A (2009) Transient prenatal vitamin D deficiency is associated with changes of synaptic plasticity in the dentate gyrus in adult rats. Psychoneuroendocrinology 34(Suppl 1):S258–S264. https://doi.org/10.1016/j.psyneuen.2009.07.004

Grissom NM, Reyes TM (2013) Gestational overgrowth and undergrowth affect neurodevelopment: similarities and differences from behavior to epigenetics. Int J Dev Neurosci Off J Int Soc Dev Neurosci 31(6):406–414. https://doi.org/10.1016/j.ijdevneu.2012.11.006

Hadders-Algra M (2011) Prenatal and early postnatal supplementation with long-chain polyunsaturated fatty acids: neurodevelopmental considerations. Am J Clin Nutr 94(6 Suppl):1874S–1879S. https://doi.org/10.3945/ajcn.110.001065

Harding KB, Peña-Rosas JP, Webster AC et al (2017) Iodine supplementation for women during the preconception, pregnancy and postpartum period. Cochrane Database Syst Rev 3: CD011761. https://doi.org/10.1002/14651858.CD011761.pub2

Hensch TK (2004) Critical period regulation. Annu Rev Neurosci 27:549–579. https://doi.org/10.1146/annurev.neuro.27.070203.144327

Hu Y-D, Pang W, He C-C et al (2017) The cognitive impairment induced by zinc deficiency in rats aged 0~2 months related to BDNF DNA methylation changes in the hippocampus. Nutr Neurosci 20(9):519–525. https://doi.org/10.1080/1028415X.2016.1194554

Insel BJ, Schaefer CA, McKeague IW, Susser ES, Brown AS (2008) Maternal iron deficiency and the risk of schizophrenia in offspring. Arch Gen Psychiatry 65(10):1136–1144. https://doi.org/10.1001/archpsyc.65.10.1136

Isaacs EB, Ross S, Kennedy K, Weaver LT, Lucas A, Fewtrell MS (2011) 10-year cognition in preterms after random assignment to fatty acid supplementation in infancy. Pediatrics 128(4): e890–e898. https://doi.org/10.1542/peds.2010-3153

Jacobson SW, Carter RC, Molteno CD et al (2018) Efficacy of maternal choline supplementation during pregnancy in mitigating adverse effects of prenatal alcohol exposure on growth and cognitive function: a randomized, double-blind, placebo-controlled clinical trial. Alcohol Clin Exp Res 42(7):1327–1341. https://doi.org/10.1111/acer.13769

Jasani B, Simmer K, Patole SK, Rao SC (2017) Long chain polyunsaturated fatty acid supplementation in infants born at term. Cochrane Database Syst Rev 3(3):CD000376. https://doi.org/10.1002/14651858.CD000376.pub4

Jayasinghe C, Polson R, van Woerden HC, Wilson P (2018) The effect of universal maternal antenatal iron supplementation on neurodevelopment in offspring: a systematic review and meta-analysis. BMC Pediatr 18. https://doi.org/10.1186/s12887-018-1118-7

Jorgenson LA, Wobken JD, Georgieff MK (2003) Perinatal iron deficiency alters apical dendritic growth in hippocampal CA1 pyramidal neurons. Dev Neurosci 25(6):412–420. https://doi.org/10.1159/000075667

Jorgenson LA, Sun M, O'Connor M, Georgieff MK (2005) Fetal iron deficiency disrupts the maturation of synaptic function and efficacy in area CA1 of the developing rat hippocampus. Hippocampus 15(8):1094–1102. https://doi.org/10.1002/hipo.20128

Joss-Moore LA, Albertine KH, Lane RH (2011) Epigenetics and the developmental origins of lung disease. Mol Genet Metab 104(1–2):61–66. https://doi.org/10.1016/j.ymgme.2011.07.018

Kable JA, Coles CD, Keen CL et al (2015) The impact of micronutrient supplementation in alcohol-exposed pregnancies on information processing skills in Ukrainian infants. Alcohol Fayettev N 49(7):647–656. https://doi.org/10.1016/j.alcohol.2015.08.005

Ke X, Schober ME, McKnight RA et al (2010) Intrauterine growth retardation affects expression and epigenetic characteristics of the rat hippocampal glucocorticoid receptor gene. Physiol Genomics 42(2):177–189. https://doi.org/10.1152/physiolgenomics.00201.2009

Ke X, Xing B, Yu B et al (2014) IUGR disrupts the PPARγ-Setd8-H4K20me(1) and Wnt signaling pathways in the juvenile rat hippocampus. Int J Dev Neurosci Off J Int Soc Dev Neurosci 38:59–67. https://doi.org/10.1016/j.ijdevneu.2014.07.008

Kennedy BC, Dimova JG, Siddappa AJM, Tran PV, Gewirtz JC, Georgieff MK (2014) Prenatal choline supplementation ameliorates the long-term neurobehavioral effects of fetal-neonatal iron deficiency in rats. J Nutr 144(11):1858–1865. https://doi.org/10.3945/jn.114.198739

King JC, Brown KH, Gibson RS et al (2015) Biomarkers of nutrition for development (BOND)-zinc review. J Nutr 146(4):858S–885S. https://doi.org/10.3945/jn.115.220079

Kretchmer N, Beard JL, Carlson S (1996) The role of nutrition in the development of normal cognition. Am J Clin Nutr 63(6):997S–1001S. https://doi.org/10.1093/ajcn/63.6.997

Kumar V, Kumar A, Singh K, Avasthi K, Kim J-J (2021) Neurobiology of zinc and its role in neurogenesis. Eur J Nutr. https://doi.org/10.1007/s00394-020-02454-3

Kuzawa CW (1998) Adipose tissue in human infancy and childhood: an evolutionary perspective. Am J Phys Anthropol Suppl 27:177–209. https://doi.org/10.1002/(sici)1096-8644(1998)107:27+<177::aid-ajpa7>3.0.co;2-b

Lardner AL (2015) Vitamin D and hippocampal development-the story so far. Front Mol Neurosci 8:58. https://doi.org/10.3389/fnmol.2015.00058

Lazarus J, Brown RS, Daumerie C, Hubalewska-Dydejczyk A, Negro R, Vaidya B (2014) 2014 European thyroid association guidelines for the management of subclinical hypothyroidism in pregnancy and in children. Eur Thyroid J 3(2):76–94. https://doi.org/10.1159/000362597

Leonard WR, Robertson ML, Snodgrass JJ, Kuzawa CW (2003) Metabolic correlates of hominid brain evolution. Comp Biochem Physiol A Mol Integr Physiol 136(1):5–15. https://doi.org/10.1016/s1095-6433(03)00132-6

Lepping RJ, Honea RA, Martin LE et al (2019) Long chain polyunsaturated fatty acid supplementation in the first year of life affects brain function, structure, and metabolism at age nine years. Dev Psychobiol 61(1):5–16. https://doi.org/10.1002/dev.21780

Lisi G, Ribolsi M, Siracusano A, Niolu C (2020) Maternal vitamin D and its role in determining fetal origins of mental health. Curr Pharm Des 26(21):2497–2509. https://doi.org/10.2174/1381612826666200506093858

Liu J, Liu L, Chen H (2011) Antenatal taurine supplementation for improving brain ultrastructure in fetal rats with intrauterine growth restriction. Neuroscience 181:265–270. https://doi.org/10.1016/j.neuroscience.2011.02.056

Liu J, Liu Y, Wang X-F, Chen H, Yang N (2013) Antenatal taurine supplementation improves cerebral neurogenesis in fetal rats with intrauterine growth restriction through the PKA-CREB signal pathway. Nutr Neurosci 16(6):282–287. https://doi.org/10.1179/1476830513Y.0000000057

Lozoff B, Jimenez E, Wolf AW (1991) Long-term developmental outcome of infants with iron deficiency. N Engl J Med 325(10):687–694. https://doi.org/10.1056/NEJM199109053251004

Lozoff B, Jimenez E, Hagen J, Mollen E, Wolf AW (2000) Poorer behavioral and developmental outcome more than 10 years after treatment for iron deficiency in infancy. Pediatrics 105(4): e51–e51. https://doi.org/10.1542/peds.105.4.e51

Lozoff B, Beard J, Connor J, Barbara F, Georgieff M, Schallert T (2006) Long-lasting neural and behavioral effects of iron deficiency in infancy. Nutr Rev 64(5 Pt 2):S34–S43; discussion S72–S91. https://doi.org/10.1301/nr.2006.may.s34-s43

Lozoff B, Clark KM, Jing Y, Armony-Sivan R, Angelilli ML, Jacobson SW (2008) Dose-response relationships between iron deficiency with or without anemia and infant social-emotional behavior. J Pediatr 152(5):696–702, 702.31–33. https://doi.org/10.1016/j.jpeds.2007.09.048

Lozoff B, Castillo M, Clark KM, Smith JB (2012) Follow-up of a randomized controlled trial of iron-fortified (12.7 mg/L) vs. low-iron (2.3 mg/L) infant formula: developmental outcome at 10 years. Arch Pediatr Adolesc Med 166(3):208–215. https://doi.org/10.1001/archpediatrics.2011.197

Lucia FS, Pacheco-Torres J, González-Granero S et al (2018) Transient hypothyroidism during lactation arrests myelination in the anterior commissure of rats. A magnetic resonance image and electron microscope study. Front Neuroanat 12:31. https://doi.org/10.3389/fnana.2018.00031

Mackenzie GG, Zago MP, Aimo L, Oteiza PI (2007) Zinc deficiency in neuronal biology. IUBMB Life 59(4–5):299–307. https://doi.org/10.1080/15216540701225966

Maliszewski AM, Gadhia MM, O'Meara MC, Thorn SR, Rozance PJ, Brown LD (2012) Prolonged infusion of amino acids increases leucine oxidation in fetal sheep. Am J Physiol Endocrinol Metab 302(12):E1483–E1492. https://doi.org/10.1152/ajpendo.00026.2012

Mallard EC, Rehn A, Rees S, Tolcos M, Copolov D (1999) Ventriculomegaly and reduced hippocampal volume following intrauterine growth-restriction: implications for the aetiology of schizophrenia. Schizophr Res 40(1):11–21. https://doi.org/10.1016/s0920-9964(99)00041-9

Martínez-Galiano JM, Amezcua-Prieto C, Cano-Ibañez N, Salcedo-Bellido I, Bueno-Cavanillas A, Delgado-Rodriguez M (2019) Maternal iron intake during pregnancy and the risk of small for gestational age. Matern Child Nutr 15(3):e12814. https://doi.org/10.1111/mcn.12814

McCarthy EK, Murray DM, Malvisi L et al (2018) Antenatal vitamin D status is not associated with standard neurodevelopmental assessments at age 5 years in a well-characterized prospective maternal-infant cohort. J Nutr 148(10):1580–1586. https://doi.org/10.1093/jn/nxy150

Meck WH, Williams CL, Cermak JM, Blusztajn JK (2007) Developmental periods of choline sensitivity provide an ontogenetic mechanism for regulating memory capacity and age-related dementia. Front Integr Neurosci 1:7. https://doi.org/10.3389/neuro.07.007.2007

Melough MM, Murphy LE, Graff JC et al (2021) Maternal plasma 25-hydroxyvitamin D during gestation is positively associated with neurocognitive development in offspring at age 4-6 years. J Nutr 151(1):132–139. https://doi.org/10.1093/jn/nxaa309

Merialdi M, Caulfield LE, Zavaleta N, Figueroa A, DiPietro JA (1999) Adding zinc to prenatal iron and folate tablets improves fetal neurobehavioral development. Am J Obstet Gynecol 180(2 Pt 1):483–490. https://doi.org/10.1016/s0002-9378(99)70236-x

Merialdi M, Caulfield LE, Zavaleta N, Figueroa A, Dominici F, Dipietro JA (2004) Randomized controlled trial of prenatal zinc supplementation and the development of fetal heart rate. Am J Obstet Gynecol 190(4):1106–1112. https://doi.org/10.1016/j.ajog.2003.09.072

Miller MF, Stoltzfus RJ, Mbuya NV et al (2003) Total body iron in HIV-positive and HIV-negative Zimbabwean newborns strongly predicts anemia throughout infancy and is predicted by maternal hemoglobin concentration. J Nutr 133(11):3461–3468. https://doi.org/10.1093/jn/133.11.3461

Miller SL, Huppi PS, Mallard C (2016) The consequences of fetal growth restriction on brain structure and neurodevelopmental outcome. J Physiol 594(4):807–823. https://doi.org/10.1113/JP271402

Moon J, Chen M, Gandhy SU et al (2010) Perinatal choline supplementation improves cognitive functioning and emotion regulation in the Ts65Dn mouse model of down syndrome. Behav Neurosci 124(3):346–361. https://doi.org/10.1037/a0019590

Moon K, Rao SC, Schulzke SM, Patole SK, Simmer K (2016) Longchain polyunsaturated fatty acid supplementation in preterm infants. Cochrane Database Syst Rev 2016(12). https://doi.org/10.1002/14651858.CD000375.pub5
Morris SS, Cogill B, Uauy R, Maternal and Child Undernutrition Study Group (2008) Effective international action against undernutrition: why has it proven so difficult and what can be done to accelerate progress? Lancet Lond Engl 371(9612):608–621. https://doi.org/10.1016/S0140-6736(07)61695-X
Murray E, Fernandes M, Fazel M, Kennedy SH, Villar J, Stein A (2015) Differential effect of intrauterine growth restriction on childhood neurodevelopment: a systematic review. BJOG Int J Obstet Gynaecol 122(8):1062–1072. https://doi.org/10.1111/1471-0528.13435
Murray-Kolb LE, Khatry SK, Katz J et al (2012) Preschool micronutrient supplementation effects on intellectual and motor function in school-aged Nepalese children. Arch Pediatr Adolesc Med 166(5):404–410. https://doi.org/10.1001/archpediatrics.2012.37
Mutua AM, Nampijja M, Elliott AM et al (2020) Vitamin D status is not associated with cognitive or motor function in pre-school Ugandan children. Nutrients 12(6). https://doi.org/10.3390/nu12061662
Nahar B, Hossain M, Mahfuz M et al (2020) Early childhood development and stunting: findings from the MAL-ED birth cohort study in Bangladesh. Matern Child Nutr 16(1):e12864. https://doi.org/10.1111/mcn.12864
Naik AA, Patro IK, Patro N (2015) Slow physical growth, delayed reflex ontogeny, and permanent behavioral as well as cognitive impairments in rats following intra-generational protein malnutrition. Front Neurosci 9:446. https://doi.org/10.3389/fnins.2015.00446
Neuringer M, Connor WE, Van Petten C, Barstad L (1984) Dietary omega-3 fatty acid deficiency and visual loss in infant rhesus monkeys. J Clin Invest 73(1):272–276. https://doi.org/10.1172/JCI111202
Neuringer M, Connor WE, Lin DS, Barstad L, Luck S (1986) Biochemical and functional effects of prenatal and postnatal omega 3 fatty acid deficiency on retina and brain in rhesus monkeys. Proc Natl Acad Sci U S A 83(11):4021–4025. https://doi.org/10.1073/pnas.83.11.4021
Nishigori H, Mazzuca DM, Nygard KL, Han VK, Richardson BS (2008) BDNF and TrkB in the preterm and near-term ovine fetal brain and the effect of intermittent umbilical cord occlusion. Reprod Sci Thousand Oaks Calif 15(9):895–905. https://doi.org/10.1177/1933719108324135
Nissensohn M, Sanchez-Villegas A, Lugo D et al (2013) Effect of zinc intake on mental and motor development in infants: a meta-analysis. Int J Vitam Nutr Res 83:203–215. https://doi.org/10.1024/0300-9831/a000161
O'Brien KO, Zavaleta N, Abrams SA, Caulfield LE (2003) Maternal iron status influences iron transfer to the fetus during the third trimester of pregnancy. Am J Clin Nutr 77(4):924–930. https://doi.org/10.1093/ajcn/77.4.924
O'Donnell KJ, Rakeman MA, Zhi-Hong D et al (2002) Effects of iodine supplementation during pregnancy on child growth and development at school age. Dev Med Child Neurol 44(2):76–81. https://doi.org/10.1017/s0012162201001712
Oliveri AN, Knuth M, Glazer L, Bailey J, Kullman SW, Levin ED (2020) Zebrafish show long-term behavioral impairments resulting from developmental vitamin D deficiency. Physiol Behav 224:113016. https://doi.org/10.1016/j.physbeh.2020.113016
Patrick RP, Ames BN (2014) Vitamin D hormone regulates serotonin synthesis. Part 1: relevance for autism. FASEB J 28(6):2398–2413. https://doi.org/10.1096/fj.13-246546
Pérez-López FR, Pasupuleti V, Mezones-Holguin E et al (2015) Effect of vitamin D supplementation during pregnancy on maternal and neonatal outcomes: a systematic review and meta-analysis of randomized controlled trials. Fertil Steril 103(5):1278–1288.e4. https://doi.org/10.1016/j.fertnstert.2015.02.019
Perkins JM, Kim R, Krishna A, McGovern M, Aguayo VM, Subramanian SV (2017) Understanding the association between stunting and child development in low- and middle-income countries: next steps for research and intervention. Soc Sci Med 1982 193:101–109. https://doi.org/10.1016/j.socscimed.2017.09.039

Pharoah PO, Buttfield IH, Hetzel BS (1971) Neurological damage to the fetus resulting from severe iodine deficiency during pregnancy. Lancet Lond Engl 1(7694):308–310. https://doi.org/10.1016/s0140-6736(71)91040-3

Pharoah P, Buttfield IH, Hetzel BS (2012) Neurological damage to the fetus resulting from severe iodine deficiency during pregnancy. Int J Epidemiol 41(3):589–592. https://doi.org/10.1093/ije/dys070

Pollitt E, Gorman KS, Engle PL, Rivera JA, Martorell R (1995) Nutrition in early life and the fulfillment of intellectual potential. J Nutr 125(4 Suppl):1111S–1118S. https://doi.org/10.1093/jn/125.suppl_4.1111S

Pongcharoen T, Ramakrishnan U, DiGirolamo AM et al (2012) Influence of prenatal and postnatal growth on intellectual functioning in school-aged children. Arch Pediatr Adolesc Med 166(5):411–416. https://doi.org/10.1001/archpediatrics.2011.1413

Rao R, Tkac I, Townsend EL, Gruetter R, Georgieff MK (2003) Perinatal iron deficiency alters the neurochemical profile of the developing rat hippocampus. J Nutr 133(10):3215–3221. https://doi.org/10.1093/jn/133.10.3215

Redman K, Ruffman T, Fitzgerald P, Skeaff S (2016) Iodine deficiency and the brain: effects and mechanisms. Crit Rev Food Sci Nutr 56(16):2695–2713. https://doi.org/10.1080/10408398.2014.922042

Rehn AE, Van Den Buuse M, Copolov D, Briscoe T, Lambert G, Rees S (2004) An animal model of chronic placental insufficiency: relevance to neurodevelopmental disorders including schizophrenia. Neuroscience 129(2):381–391. https://doi.org/10.1016/j.neuroscience.2004.07.047

Reisbick S, Neuringer M, Gohl E, Wald R, Anderson GJ (1997) Visual attention in infant monkeys: effects of dietary fatty acids and age. Dev Psychol 33(3):387–395. https://doi.org/10.1037//0012-1649.33.3.387

Ross RG, Hunter SK, McCarthy L et al (2013) Perinatal choline effects on neonatal pathophysiology related to later schizophrenia risk. Am J Psychiatry 170(3):290–298. https://doi.org/10.1176/appi.ajp.2012.12070940

Ross RG, Hunter SK, Hoffman MC et al (2016) Perinatal phosphatidylcholine supplementation and early childhood behavior problems: evidence for CHRNA7 moderation. Am J Psychiatry 173(5):509–516. https://doi.org/10.1176/appi.ajp.2015.15091188

Rozance PJ, Crispo MM, Barry JS et al (2009) Prolonged maternal amino acid infusion in late-gestation pregnant sheep increases fetal amino acid oxidation. Am J Physiol Endocrinol Metab 297(3):E638–E646. https://doi.org/10.1152/ajpendo.00192.2009

Ruff CA, Faulkner SD, Rumajogee P et al (2017) The extent of intrauterine growth restriction determines the severity of cerebral injury and neurobehavioural deficits in rodents. PLoS One 12(9):e0184653. https://doi.org/10.1371/journal.pone.0184653

Ryan SH, Williams JK, Thomas JD (2008) Choline supplementation attenuates learning deficits associated with neonatal alcohol exposure in the rat: effects of varying the timing of choline administration. Brain Res 1237:91–100. https://doi.org/10.1016/j.brainres.2008.08.048

Salas AA, Woodfin T, Phillips V, Peralta-Carcelen M, Carlo WA, Ambalavanan N (2018) Dose-response effects of early vitamin D supplementation on neurodevelopmental and respiratory outcomes of extremely preterm infants at 2 years of age: a randomized trial. Neonatology 113(3):256–262. https://doi.org/10.1159/000484399

Santos DC, Angulo-Barroso RM, Li M et al (2018) Timing, duration, and severity of iron deficiency in early development and motor outcomes at 9 months. Eur J Clin Nutr 72(3):332–341. https://doi.org/10.1038/s41430-017-0015-8

Sass L, Vinding RK, Stokholm J et al (2020) High-dose vitamin D supplementation in pregnancy and neurodevelopment in childhood: a prespecified secondary analysis of a randomized clinical trial. JAMA Netw Open 3(12):e2026018. https://doi.org/10.1001/jamanetworkopen.2020.26018

Schmidt RJ, Tancredi DJ, Krakowiak P, Hansen RL, Ozonoff S (2014) Maternal intake of supplemental iron and risk of autism spectrum disorder. Am J Epidemiol 180(9):890–900. https://doi.org/10.1093/aje/kwu208

Scholl TO (2011) Maternal iron status: relation to fetal growth, length of gestation, and iron endowment of the neonate. Nutr Rev 69(s1):S23–S29. https://doi.org/10.1111/j.1753-4887.2011.00429.x

Scholtz SA, Colombo J, Carlson SE (2013) Clinical overview of effects of dietary long-chain polyunsaturated fatty acids during the perinatal period. Nestle Nutr Inst Workshop Ser 77:145–154. https://doi.org/10.1159/000351397

Shao J, Lou J, Rao R et al (2012) Maternal serum ferritin concentration is positively associated with newborn iron stores in women with low ferritin status in late pregnancy. J Nutr 142 (11):2004–2009. https://doi.org/10.3945/jn.112.162362

Shaw GM, Finnell RH, Blom HJ et al (2009) Choline and risk of neural tube defects in a folate-fortified population. Epidemiology 20(5):714–719. https://doi.org/10.1097/EDE.0b013e3181ac9fe7

Signore C, Ueland PM, Troendle J, Mills JL (2008) Choline concentrations in human maternal and cord blood and intelligence at 5 y of age. Am J Clin Nutr 87(4):896–902. https://doi.org/10.1093/ajcn/87.4.896

Skeaff SA (2011) Iodine deficiency in pregnancy: the effect on neurodevelopment in the child. Nutrients 3(2):265–273. https://doi.org/10.3390/nu3020265

Specht IO, Janbek J, Thorsteinsdottir F, Frederiksen P, Heitmann BL (2020) Neonatal vitamin D levels and cognitive ability in young adulthood. Eur J Nutr 59(5):1919–1928. https://doi.org/10.1007/s00394-019-02042-0

Spinillo A, Stronati M, Ometto A, Fazzi E, Lanzi G, Guaschino S (1993) Infant neurodevelopmental outcome in pregnancies complicated by gestational hypertension and intra-uterine growth retardation. J Perinat Med 21(3):195–203. https://doi.org/10.1515/jpme.1993.21.3.195

Strain JJ, McSorley EM, van Wijngaarden E et al (2013) Choline status and neurodevelopmental outcomes at 5 years of age in the Seychelles Child Development Nutrition Study. Br J Nutr 110 (2):330–336. https://doi.org/10.1017/S0007114512005077

Tamura T, Goldenberg RL, Ramey SL, Nelson KG, Chapman VR (2003) Effect of zinc supplementation of pregnant women on the mental and psychomotor development of their children at 5 y of age. Am J Clin Nutr 77(6):1512–1516. https://doi.org/10.1093/ajcn/77.6.1512

Thompson RA, Nelson CA (2001) Developmental science and the media. Early brain development. Am Psychol 56(1):5–15. https://doi.org/10.1037/0003-066x.56.1.5

Tran PV, Kennedy BC, Lien Y-C, Simmons RA, Georgieff MK (2015) Fetal iron deficiency induces chromatin remodeling at the Bdnf locus in adult rat hippocampus. Am J Physiol Regul Integr Comp Physiol 308(4):R276–R282. https://doi.org/10.1152/ajpregu.00429.2014

Tran PV, Kennedy BC, Pisansky MT et al (2016) Prenatal choline supplementation diminishes early-life Iron deficiency-induced reprogramming of molecular networks associated with behavioral abnormalities in the adult rat hippocampus. J Nutr 146(3):484–493. https://doi.org/10.3945/jn.115.227561

Tyagi E, Zhuang Y, Agrawal R, Ying Z, Gomez-Pinilla F (2015) Interactive actions of Bdnf methylation and cell metabolism for building neural resilience under the influence of diet. Neurobiol Dis 73:307–318. https://doi.org/10.1016/j.nbd.2014.09.014

Uauy R, Hoffman DR, Peirano P, Birch DG, Birch EE (2001) Essential fatty acids in visual and brain development. Lipids 36(9):885–895. https://doi.org/10.1007/s11745-001-0798-1

Unger EL, Hurst AR, Georgieff MK et al (2012) Behavior and monoamine deficits in prenatal and perinatal iron deficiency are not corrected by early postnatal moderate-Iron or high-Iron diets in rats. J Nutr 142(11):2040–2049. https://doi.org/10.3945/jn.112.162198

UNICEF (2019) Iodine. UNICEF DATA. https://data.unicef.org/topic/nutrition/iodine/. Accessed 29 Jan 2021

Utsunomiya H, Takano K, Okazaki M, Mitsudome A (1999) Development of the temporal lobe in infants and children: analysis by MR-based volumetry. AJNR Am J Neuroradiol 20(4):717–723

Velasco I, Carreira M, Santiago P et al (2009) Effect of iodine prophylaxis during pregnancy on neurocognitive development of children during the first two years of life. J Clin Endocrinol Metab 94(9):3234–3241. https://doi.org/10.1210/jc.2008-2652

Verfuerden ML, Dib S, Jerrim J, Fewtrell M, Gilbert RE (2020) Effect of long-chain polyunsaturated fatty acids in infant formula on long-term cognitive function in childhood: a systematic review and meta-analysis of randomised controlled trials. PLoS One 15(11):e0241800. https://doi.org/10.1371/journal.pone.0241800

Verhagen NJE, Gowachirapant S, Winichagoon P, Andersson M, Melse-Boonstra A, Zimmermann MB (2020) Iodine supplementation in mildly iodine-deficient pregnant women does not improve maternal thyroid function or child development: a secondary analysis of a randomized controlled trial. Front Endocrinol 11. https://doi.org/10.3389/fendo.2020.572984

Villamor E, Rifas-Shiman SL, Gillman MW, Oken E (2012) Maternal intake of methyl-donor nutrients and child cognition at 3 years of age. Paediatr Perinat Epidemiol 26(4):328–335. https://doi.org/10.1111/j.1365-3016.2012.01264.x

Voltas N, Canals J, Hernández-Martínez C, Serrat N, Basora J, Arija V (2020) Effect of vitamin D status during pregnancy on infant neurodevelopment: the ECLIPSES study. Nutrients 12 (10):3196. https://doi.org/10.3390/nu12103196

Vuillermot S, Luan W, Meyer U, Eyles D (2017) Vitamin D treatment during pregnancy prevents autism-related phenotypes in a mouse model of maternal immune activation. Mol Autism 8:9. https://doi.org/10.1186/s13229-017-0125-0

Vyas Y, Lee K, Jung Y, Montgomery JM (2020) Influence of maternal zinc supplementation on the development of autism-associated behavioural and synaptic deficits in offspring Shank3-knockout mice. Mol Brain 13(1):110. https://doi.org/10.1186/s13041-020-00650-0

Wachs TD, Georgieff M, Cusick S, McEwen B (2014) Issues in the timing of integrated early interventions: contributions from nutrition, neuroscience and psychological research. Ann N Y Acad Sci 1308:89–106. https://doi.org/10.1111/nyas.12314

Walker SP, Wachs TD, Meeks Gardner J et al (2007) Child development: risk factors for adverse outcomes in developing countries. Lancet 369(9556):145–157. https://doi.org/10.1016/S0140-6736(07)60076-2

Wang Y, Fu W, Liu J (2016) Neurodevelopment in children with intrauterine growth restriction: adverse effects and interventions. J Matern-Fetal Neonatal Med 29(4):660–668. https://doi.org/10.3109/14767058.2015.1015417

Ward BC, Kolodny NH, Nag N, Berger-Sweeney JE (2009) Neurochemical changes in a mouse model of Rett syndrome: changes over time and in response to perinatal choline nutritional supplementation. J Neurochem 108(2):361–371. https://doi.org/10.1111/j.1471-4159.2008.05768.x

Warthon-Medina M, Moran VH, Stammers A-L et al (2015) Zinc intake, status and indices of cognitive function in adults and children: a systematic review and meta-analysis. Eur J Clin Nutr 69(6):649–661. https://doi.org/10.1038/ejcn.2015.60

Whitehouse AJO, Holt BJ, Serralha M, Holt PG, Kusel MMH, Hart PH (2012) Maternal serum vitamin D levels during pregnancy and offspring neurocognitive development. Pediatrics 129 (3):485–493. https://doi.org/10.1542/peds.2011-2644

WHO (2021) Anaemia. https://www.who.int/data/gho/data/indicators/indicator-details/GHO/prevalence-of-anaemia-in-children-under-5-years-(-). Accessed 28 Jan 2021

WHO (2012) Daily iron and folic acid supplementation during pregnancy. http://www.who.int/elena/titles/guidance_summaries/daily_iron_pregnancy/en/. Accessed 28 Jan 2021

Wicklow B, Gallo S, Majnemer A et al (2016) Impact of vitamin D supplementation on gross motor development of healthy term infants: a randomized dose-response trial. Phys Occup Ther Pediatr 36(3):330–342. https://doi.org/10.3109/01942638.2015.1050150

Windham GC, Pearl M, Anderson MC et al (2019) Newborn vitamin D levels in relation to autism spectrum disorders and intellectual disability: a case–control study in California. Autism Res 12 (6):989–998. https://doi.org/10.1002/aur.2092

Wozniak JR, Fuglestad AJ, Eckerle JK et al (2013) Choline supplementation in children with fetal alcohol spectrum disorders has high feasibility and tolerability. Nutr Res N Y N 33 (11):897–904. https://doi.org/10.1016/j.nutres.2013.08.005

Wozniak JR, Fuglestad AJ, Eckerle JK et al (2015) Choline supplementation in children with fetal alcohol spectrum disorders: a randomized, double-blind, placebo-controlled trial. Am J Clin Nutr 102(5):1113–1125. https://doi.org/10.3945/ajcn.114.099168

Wozniak JR, Fink BA, Fuglestad AJ et al (2020) Four-year follow-up of a randomized controlled trial of choline for neurodevelopment in fetal alcohol spectrum disorder. J Neurodev Disord 12 (1):9. https://doi.org/10.1186/s11689-020-09312-7

Wu BTF, Dyer RA, King DJ, Richardson KJ, Innis SM (2012) Early second trimester maternal plasma choline and betaine are related to measures of early cognitive development in term infants. PLoS One 7(8):e43448. https://doi.org/10.1371/journal.pone.0043448

Wullschleger S, Loewith R, Hall MN (2006) TOR signaling in growth and metabolism. Cell 124 (3):471–484. https://doi.org/10.1016/j.cell.2006.01.016

Yu X, Ren T, Yu X (2013) Disruption of calmodulin-dependent protein kinase II α/brain-derived neurotrophic factor (α-CaMKII/BDNF) signalling is associated with zinc deficiency-induced impairments in cognitive and synaptic plasticity. Br J Nutr 110(12):2194–2200. https://doi.org/10.1017/S0007114513001657

Zeisel SH (2006) Choline: critical role during fetal development and dietary requirements in adults. Annu Rev Nutr 26:229–250. https://doi.org/10.1146/annurev.nutr.26.061505.111156

Zeisel SH (2012) A brief history of choline. Ann Nutr Metab 61(3):254–258. https://doi.org/10.1159/000343120

Zeisel S (2017) Choline, other methyl-donors and epigenetics. Nutrients 9(5). https://doi.org/10.3390/nu9050445

Zhou SJ, Skeaff SA, Ryan P et al (2015) The effect of iodine supplementation in pregnancy on early childhood neurodevelopment and clinical outcomes: results of an aborted randomised placebo-controlled trial. Trials 16:563. https://doi.org/10.1186/s13063-015-1080-8

Zimmermann MB (2020) Iodine supplements for mildly iodine-deficient pregnant women: are they worthwhile? Am J Clin Nutr 112(2):247–248. https://doi.org/10.1093/ajcn/nqaa116

Understanding Sensitive Period Effects in Musical Training

Virginia B. Penhune

Contents

Abstract Adult ability in complex cognitive domains, including music, is commonly thought of as the product of gene–environment interactions, where genetic predispositions influence and are modulated by experience, resulting in the final phenotypic expression. Recently, however, the important contribution of maturation to gene–environment interactions has become better understood. Thus, the timing of exposure to specific experience, such as music training, has been shown to produce long-term impacts on adult behaviour and the brain. Work from our lab and others shows that musical training before the ages of 7–9 enhances performance on musical tasks and modifies brain structure and function, sometimes in unexpected ways. The goal of this paper is to present current evidence for sensitive period effects for musical training in the context of what is known about brain maturation and to present a framework that integrates genetic, environmental and maturational influences on the development of musical skill. We believe that this framework can also

---

V. B. Penhune (✉)
Department of Psychology, Concordia University, Montreal, QC, Canada

Montreal Laboratory for Brain Music and Sound (BRAMS), Centre for Research in Brain, Language and Music, Montreal, QC, Canada
e-mail: virginia.penhune@concordia.ca

Curr Topics Behav Neurosci (2022) 53: 167–188
https://doi.org/10.1007/7854_2021_250
Published Online: 26 August 2021

be applied more broadly to understanding how predispositions, brain development and experience interact.

Keywords Brain maturation · Critical period · Development · Interactive specialization · Musical expertise · Musical training

1 Introduction

As with most complex abilities, adult musical skill is the product of genetic and environmental contributions that interact with each other in multiple ways. Genes encode a constellation of predispositions that contribute to specific abilities and these predispositions interact with the environment to passively and actively influence the development of skills (Sauce and Matzel 2018; Ullen et al. 2016). Genes also control neural and physical maturation, such that different brain networks and their related functions have developmental peaks at different ages (Fjell et al. 2019; Hensch and Quinlan 2018; Werker and Hensch 2015). For a function or ability to emerge, the underlying neural system must be not only adequately mature, but also have appropriate input (Hensch and Quinlan 2018). It is hypothesized that maturation in neural systems is primed to respond to relevant experience and that the rate of maturation is modulated by the type, duration and intensity of experience. Across all major neural systems it has been shown that relevant experience has greater effects during periods of peak maturational change, which have been termed sensitive or critical periods (Werker and Hensch 2015; Knudsen 2004; Hensch and Quinlan 2018; De Villers-Sidani and Merzenich 2011). Appropriate stimulation during sensitive periods results in short-term plasticity, but is also thought to promote long-term receptiveness to experience, thus forming a scaffold on which later experience can build (Hensch and Quinlan 2018; Kuhl 2010). The goal of this paper is to present current evidence for sensitive period effects for musical training in the context of what is known about brain maturation and to present a framework that integrates genetic, environmental and maturational influences on the development of musical skill. The ideas presented here build on a previous review and synthesis that proposed a gene-environment-maturation model for the development of musical skill (Penhune 2020).

2 Brain Maturation

Maturational changes in brain structure and function are continuous throughout life. Changes in brain structure and function across development are controlled by genes and modulated by environmental stimulation; typically proceeding from primary sensory and motor regions to frontal and parietal association areas (Fjell et al. 2019).

Brain regions do not develop in isolation, but rather in structurally and functionally connected networks (Ball et al. 2019) that may also share timelines of modulatory gene expression (Fjell et al. 2019).

After birth, the number of synapses, and therefore the volume of grey matter (GM), continues to increase for between 3 and 15 months (Huttenlocher and Dabholkar 1997). Following this period, synapses are reduced through pruning: a process thought to underlie early experience-dependent specialization, including perceptual narrowing for native language speech sounds (Werker and Hensch 2015). Studies of brain development based on structural magnetic resonance imaging (MRI) show that GM volume and cortical thickness (CT) typically peak in childhood (age 6–10) and then decrease across adolescence and into adulthood. At the same time, white-matter (WM) volume increases and then remains relatively stable into older adulthood (Mills et al. 2016; Brown and Jernigan 2012). Thus, GM thinning is largely the result of WM expansion, which shifts the GM/WM boundary as measured by MRI (Brown and Jernigan 2012). While the surface area or extent of cortical regions is thought to be largely under genetic control, individual variation in CT is thought to reflect a combination of accumulated genetic and environmental effects (Amlien et al. 2016).

Cortical maturation peaks first in sensory and motor regions and then proceeds across childhood and adolescence to frontal, parietal and temporal association areas. Some sub-cortical regions also have relatively late peaks of structural change, including the cerebellum and striatum. Cerebellar volume peaks in adolescence (Tiemeier et al. 2010), at a time that coincides with changes in its functional interaction with connected cortical regions (Amemiya et al. 2019). The cerebellum forms loop connections with many regions of the cortex and is thought to be important for optimization of sensory, motor and cognitive function (Sokolov et al. 2017; Bostan and Strick 2018). Later maturation of the cerebellum may thus be linked to parallel peaks in the fronto-parietal cortical networks to which it is connected. Striatal volumes also peak in adolescence (Goddings et al. 2014; Larsen and Luna 2015), and these nuclei also have strong interconnections with the frontal and parietal cortex (Bostan and Strick 2018). The ventral striatum, a region important for the processing of reward (including musical reward) also appears to peak in adolescence (Urosevic et al. 2012) at a time that coincides with changes in dopamine modulation of incentive reward (Luciana et al. 2012).

Thus, brain maturation shows a global pattern of rapid growth in the first 1–2 years of life that is largely genetically controlled, and which underpins development of basic sensory and cognitive functions. Across childhood and adolescence, maturational changes occur largely in WM density and connectivity within and between cortical networks that underlie complex sensorimotor and cognitive processes, changes that are thought to be more strongly modulated by experience. This pattern presents an intriguing contradiction. Peak cortical maturation of sensorimotor networks occurs in middle childhood, but basic perceptual and motor abilities develop much earlier. Additionally, sensorimotor abilities and their associated neurophysiological markers continue to develop after their childhood peak through adolescence and into early adulthood, (Eggermont and Moore 2012, Trainor 2012,

Ferronato et al. 2014). We can make sense of this contradiction by hypothesizing that development may rely on interactions between more mature, low-level sensory and motor processes and less mature, higher-level processes (Eggermont and Moore 2012, Johnson 2011). These interactions can thus bootstrap processing in less mature regions. This may first occur in infancy between early-developing brainstem regions and later-developing cortex, and in childhood between primary cortical areas and later-developing frontal-parietal areas. Interactions between earlier- and later-developing networks are likely an on-going feature of development and have been termed *interactive specialization* (Johnson 2011). These findings lead to a more complex picture of brain maturation where regional maturation depends on both local changes and network-level properties.

3 Sensitive and Critical Periods

A common feature of maturation in many neural systems is the presence of sensitive periods: windows in development when specific experience has long-term effects on behaviour and the brain (Knudsen 2004). Examples include better grammar and accent when a second language is acquired early (Hartshorne et al. 2018; Flege et al. 1999; Kuhl 2010) and improved restoration of hearing function with earlier cochlear implantation (Kral and Sharma 2012). This is in contrast to a critical period where specific experience is *required* for appropriate behaviour and brain function to develop (Hensch and Quinlan 2018). The classic example of a critical period is "lazy eye" syndrome in which early visual deprivation in one eye results in loss of acuity and responsiveness in visual cortex that cannot be fully remediated by intervention after about age 7 (Hensch and Quinlan 2018). In contrast, most complex cognitive functions are characterized by sensitive rather than critical periods: there are better and worse times to learn a second language, pick up a musical instrument or start playing chess, but these skills can be acquired to some degree at any time of life. This is because complex skills tap multiple, interacting abilities based on brain networks whose structure and function are more and less plastic at different stages of maturation, and because there are numerous routes to successful performance.

There is growing evidence that many neural systems exhibit sensitive or critical periods, each of which has a specific temporal window, and that within brain regions, different functions or features have distinct periods of heightened sensitivity (De Villers-Sidani and Merzenich 2011; Hensch and Quinlan 2018; Voss et al. 2017; Werker and Hensch 2015; Penhune 2020). It is thought that these windows are opened by a combination of maturational readiness – programmed peaks of plasticity controlled by genes – and exposure to the appropriate quantity and quality of experience (Hensch and Quinlan 2018; Voss et al. 2017; Werker and Hensch 2015). The opening and closing of sensitive periods are regulated by the balance of excitatory and inhibitory mechanisms that regulate plasticity, and they are also influenced by the quality, duration and intensity of experience (Hensch and Quinlan 2018; Voss et al. 2017). Further, it is important to keep in mind that brain regions and their associated functions do not develop in isolation, but within connected networks where both earlier and later maturing regions can influence development (Johnson 2011; Eggermont and Moore 2012; Emberson 2017). Finally, brain maturation also occurs in tandem with physical maturation of the body and is likely both primed and limited by it. A child cannot play the piano or learn to skate if he/she is unable to move his/her fingers independently or if his/her leg muscles are not sufficiently strong. The timing of sensitive periods will thus depend on the maturational timelines of relevant brain networks and on the physical systems they control.

Thus sensitive and critical periods can be seen as lying on a continuum where genetics, the mechanisms underlying opening and closing, and experience may contribute more or less strongly depending on the functional system and the maturational time-point. Across development we can imagine a cascade of sensitive periods that open and close at different times and which are sensitive to different types of experience (Werker and Hensch 2015; Hensch and Quinlan 2018; Voss et al. 2017; Penhune 2011, 2020). Further, because brain maturation occurs within connected networks, plasticity induced during one sensitive period may contribute to learning and plasticity in more or less mature regions (Eggermont and Moore 2012, Johnson 2011). This is consistent with the greater temporal variability of sensitive period effects for more complex functions. Basic sensory and motor functions that are represented in a single cortical or sub-cortical region may have earlier, more clearly defined and less malleable windows for plasticity. More complex abilities, such as language or music, which are represented across multiple functional regions – some of which develop into adulthood – may have broader, more variable windows for change, and multiple functional routes for plasticity.

4 Evidence for Sensitive Periods in Music Training

Anecdotal evidence from the lives of famous musicians suggests that early age of start (AoS) can promote the development of extraordinary skill in adulthood (Jorgensen 2011). (Mozart, who mastered the piano at age 4–5 and performed professionally throughout his childhood is probably the most famous example.) Early behavioural evidence for the impact of AoS came from the study of "perfect" or "absolute" pitch – the ability to name the pitch height of isolated tones. Two large-scale studies found that 78% of musicians with absolute pitch began their training before age 6 (Baharloo et al. 1998) and that pitch-naming accuracy was higher for those who began lessons before age 7 (Vanzella and Schellenberg 2010). Initial evidence for the effect of AoS on the brain came from work showing that the surface area of the anterior corpus callosum was larger in musicians who began training before age 7 (Schlaug et al. 1995) and the central sulcus was longer in those who started early (Amunts et al. 1997). However, none of these studies controlled for years of experience, thus the observed differences could be related to the fact that early starters had more years of experience than late starters of a similar age. Further, there were no measures of musical skill, and thus brain structural differences could

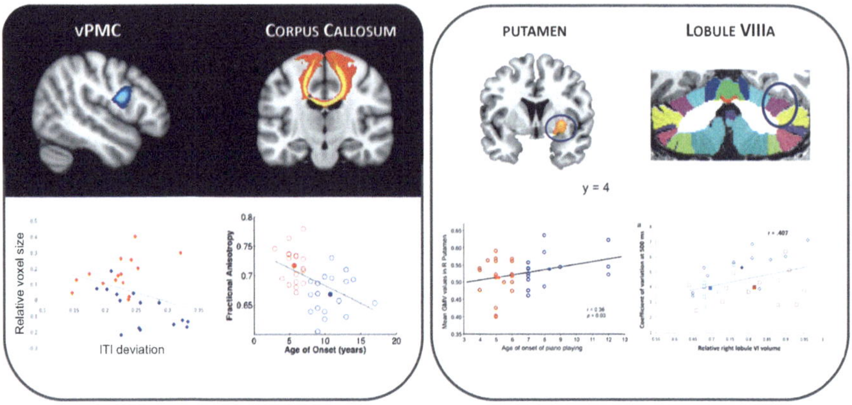

Fig. 1 Effects of early musical training on brain structure. (left to right): The first panel shows expansion of right vPMC in ET vs LT musicians and the relationship between vPMC volume and performance on a rhythm synchronization task (ET = red; LT = blue) (Bailey et al. 2014). The second panel shows greater fractional anisotropy (FA; blue) in the posterior mid-body of the corpus callosum and its connections to motor and premotor cortex (yellow-red). The graph below shows the relationship between FA in this region and AoS (Steele et al. 2013). The third panel shows enhanced volume of the right putamen in ET musicians and the relationship with AoS (Vaquero et al. 2016). The fourth panel shows the region of decreased cerebellar volume in lobule VI and its relationship to tapping variability (ET = light blue; LT = red; non-musicians = dark blue (Baer et al. 2015)

not be linked to relevant behaviour. Therefore, to systematically examine the effect of AoS, we conducted a series of studies comparing musical abilities and brain structure in adult musicians who began training before and after age 7 [Early-trained < age 7 (ET); Late-trained > age 7 (LT)]. Initially, the age-7 cut-off was based empirically on Schlaug's early findings, but has been partially validated in more recent studies (Bailey and Penhune 2013; van Vugt et al. 2021). Importantly, groups were matched for years of music experience, years of formal training and hours of current practice (Bailey and Penhune 2010, 2012, 2013; Bailey et al. 2014; Steele et al. 2013; Baer et al. 2015). We also assessed global cognitive function and working memory. Data from these and other studies described in this section have been reviewed elsewhere (Penhune 2011, 2020) and are summarized in Fig. 1.

Across all studies, we found that ET musicians out-perform LT musicians on measures of rhythm synchronization and melody discrimination. Further, ET musicians showed enlargement of ventral premotor cortex (vPMC) and the degree of enlargement was related to performance on the rhythm task (Bailey et al. 2014). We also found that ET musicians had enhanced WM integrity in the posterior mid-body of the corpus callosum (CC), the location of fibres connecting primary motor (M1) and the premotor cortex (PMC) in the two hemispheres (Steele et al. 2013). Intriguingly, in the same sample, we showed that ET musicians had *smaller* volumes of cerebellar lobules IV, V and VI, and that smaller volumes were related to reduced variability on an auditory-motor timing task (Baer et al. 2015). We have recently replicated this finding and shown that the cerebellar and cortical volumes are

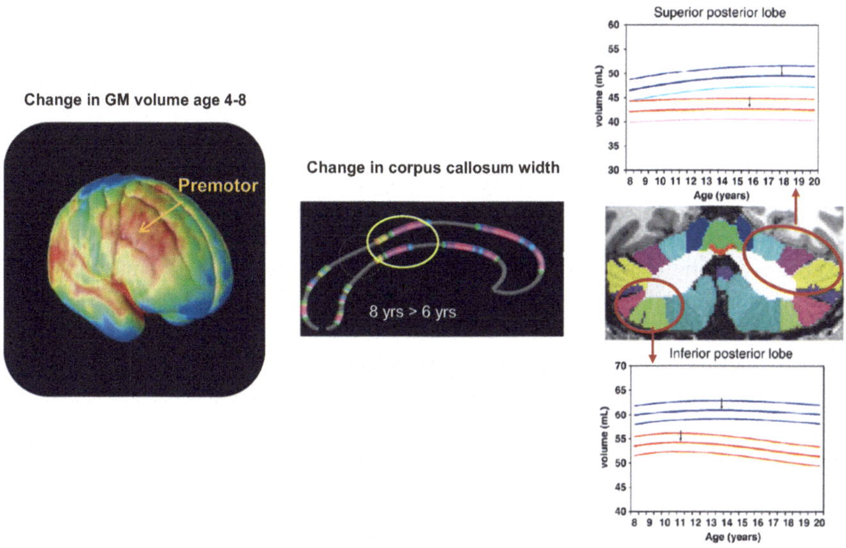

Fig. 2 Maturational changes in brain structure in regions associated with early musical training. (left to right): The first panel shows changes in whole brain GM volume between ages 4–8 (hot colours are regions of greater change) (Gogtay et al. 2004). The second panel shows changes in the width of the mid-sagittal corpus callosum between the ages of 6–8 (Westerhausen et al. 2011). Yellow circle indicates the region of the posterior mid-body where differences were found between ET and LT musicians (Steele et al. 2013). The third panel shows the developmental trajectories of GM volume in cerebellar regions (Tiemeier et al. 2010), including those that showed decreased volume in ET compared to LT musicians (Baer et al. 2015)

inversely correlated (Shenker et al. 2019). This suggests that plasticity in the two regions is interdependent and is consistent with evidence that connected regions change together across development (Johnson 2011; Fjell et al. 2019). Two other studies in an independent and well-matched sample of pianists showed that ET musicians had smaller GM volume in the right putamen, and lower timing variability when playing musical scales (Vaquero et al. 2016). Further, in ET musicians the putamen showed greater functional connectivity with premotor and occipital regions of the brain compared to LT musicians (van Vugt et al. 2021).

To interpret these findings we have considered data about the function of the regions involved, the timing of their normative maturation, and the relative contribution of genes and environment to their variability (See Fig. 2). Developmental data show that GM volume of anterior motor regions, including M1 and PMC has a peak rate of change between the ages of 6 and 8 (Giedd et al. 1999). The PMC is known to be involved in auditory-motor integration (Zatorre et al. 2007; Lega et al. 2016) and is thought to play a key role in motor timing (Merchant et al. 2013). The size of the posterior mid-body of the CC also has a maturational peak in middle childhood (Westerhausen et al. 2011), and adult variability in this region has a strong unique environmental contribution (Chiang et al. 2009). This part of the CC connects M1 and PMC across hemispheres, and its size is related to bimanual coordination in children and adolescents (Kurth et al. 2013). In contrast, peak maturation in the cerebellum occurs later, between the ages of 12 and 18 (Tiemeier et al. 2010). The cerebellum is structurally homogeneous, with sub-regions connected through feedforward and feedback loops to the rest of the brain (Kelly and Strick 2003). In motor control, cortico-cerebellar circuits are known to play a role in error-correction, optimization and instantiation of forward models. Because they are uniform and connect to all regions of the cortex, it is hypothesized that they may also perform the same role in optimizing a wide variety of sensory and cognitive functions (Sokolov et al. 2017). The cerebellar regions we found to be smaller in ET musicians are connected to frontal motor and association regions, including M1 and PMC (Kelly and Strick 2003). Because these regions are functionally connected, plasticity effects of early training on cortical regions may be partially mediated by cerebellar optimization mechanisms that result in greater efficiency and thus reduced volume. This is consistent with evidence from studies in mice where early environmental enrichment produced volumetric increases in cortical sensorimotor regions and decreases in the cerebellum (Scholz et al. 2015). Taken together, we hypothesize that early experience during periods of peak maturation promotes brain plasticity, and that regional differences in genetic permeability mean that some brain regions are more susceptible to the impact of training than others. Our findings emphasize that the impact of early experience occurs at a network level, with changes in one region influencing changes in connected regions.

Differential effects of AoS and type of training were revealed by a recent study in which we compared the effects of bilingualism and music training on the structure of the arcuate fasciculus (AF) (Vaquero et al. 2020). The AF is a white-matter pathway that connects posterior auditory cortex and frontal motor regions, and whose structure has been linked to language in the left hemisphere (Hamalainen et al. 2017), and to music in the right hemisphere (Halwani et al. 2011). All participants in our study exhibited a left-greater-than-right asymmetry in the volume of the long segment of the AF, consistent with previous findings (Powell et al. 2006). The long segment of the AF has been linked to left-hemisphere dominance for language, and it is hypothesized that its structure may be largely under genetic control (Budisavljevic et al. 2015). In an interesting dissociation, simultaneous bilinguals, who acquired their second language in the first year of life, showed an increased volume of the left long segment and a greater leftward asymmetry, while ET musicians showed an increased volume of the right long segment and reduced leftward asymmetry. These findings suggest that very early, intensive bilingual experience can modify a white-matter pathway that is typically under strong genetic control, whereas music training at a later period in childhood affects the right hemisphere, which may be more open to on-going experience. These results support the idea that brain plasticity depends on region-specific differences in malleability by experience, the age at which experience begins as well as the type of experience.

If the observed differences in behaviour and brain structure between adult ET and LT musicians result from the interaction between AoS and experience, an important question is whether sensitive period effects can be observed in childhood. To address this question we compared groups of children who started before and after age 7 on age-normed tests of musical skill (Ireland et al. 2018). Groups were matched for years of music lessons and hours of weekly practice, as well as working memory, global cognitive function and socio-economic status (Ireland et al. 2019). We found that ET children performed better than their LT counterparts on a simple melody discrimination task, and that across groups, AoS and global cognitive function independently predicted discrimination scores. There were no group differences or effects of AoS for the rhythm synchronization or a transposed melody discrimination task, but working memory ability predicted these scores. In addition, weekly practice predicted transposed melody discrimination. These results illustrate the combined effects of maturation, training and cognitive abilities on the development of musical skills. Further, they suggest that basic pitch abilities may mature earlier than more complex auditory and motor functions, and/or that children who begin early have advantages in pitch processing.

Integrating these results with those of adult ET musicians, we hypothesize that early training has an immediate impact on skills such as simple melody discrimination that rely on earlier-maturing regions including auditory cortex. Better simple melody discrimination skills may then promote acquisition of more complex skills, such as transposition or even rhythm synchronization. More crucially, we propose that early practice exerts a metaplastic effect in which experience promotes the long-term potential for learning and plasticity when exposed to additional training. The concept of metaplasticity originates from studies of hippocampal learning mechanisms (Abraham 2008) and denotes the idea that experience can change the potential for plasticity of a synapse (for review of this concept in the context of music training, see: Altenmuller and Furuya 2016, Herholz and Zatorre 2012). Thus, we hypothesize that abilities, such as transposition and rhythm synchronization, which do not show immediate gains in early-trained children, would be more readily acquired with later training because early experience has primed the underlying neural systems to learn. This is consistent with evidence that adult musicians show better learning of new sensory and motor skills (Herholz et al. 2011; Ragert et al. 2004; Rosenkranz et al. 2007), and greater increases in M1 activity during learning (Hund-Georgiadis and Von Cramon 1999). Finally, skills that are more complex might also require further cognitive and physical maturation as well as additional training.

5 Training-Related Plasticity in Childhood

The hypothesis that the observed effects of early AoS are partly the result of training-dependent plasticity is consistent with longitudinal studies of music training in childhood, which provide causal evidence that training produces changes in behaviour and the brain. Results show structural and functional changes in auditory and motor cortex, as well as the corpus callosum after 1–2 years of practice (Habibi et al. 2018; Hyde et al. 2009; Putkinen et al. 2014; Shahin et al. 2005; Fujioka et al. 2006). In these studies, no behavioural or neurophysiological differences were observed before the start of training, even though the groups self-selected to take music lessons, and thus might represent children with pre-existing skills or predilections. The fact that these changes occur in similar regions of the auditory-motor network that have been shown to differ after long-term training in adults supports the inference that these differences can be attributed in part to training. It is also consistent with a recent study comparing brain structure in monozygotic twins who were discordant for music practice. It found that the twins who played an instrument had greater cortical thickness in auditory and motor regions as well as WM enhancements in the corpus callosum compared to the other members of the twin pair who did not (De Manzano and Ullen 2018). Because these twins start life with an identical genetic makeup and can be supposed to have a very similar family environment, this study provides the strongest evidence thus far that observed differences in brain structure between musicians and non-musicians can be attributed to the effects of training.

6 Individual Differences: Predispositions and Possible Genetic Contributions

Although direct evidence for specific genetic contributions to musical skill is currently limited (Oikkonen et al. 2016), there is indirect evidence for possible structural or functional predispositions that are related to better performance and learning of musical tasks. A longitudinal study of 8- to 10-year-old children taking music lessons found that a larger auditory cortex volume was associated with measures of music ability, as well as behavioural and physiological measures of auditory processing (Seither-Preisler et al. 2014). Further, ability accounted for a greater proportion of the variance in auditory cortex volume than practice time. Based on this, and because no changes in auditory cortex volume were observed over the course of training, the authors suggest that larger volumes of auditory cortex might be a pre-existing anatomical feature that contributes to long-term development of musical skills.

Even in untrained adults, individual differences, particularly in the auditory-motor network, are linked to musical abilities and the capacity to learn. GM concentration and cortical thickness in auditory and parietal regions were found to be related to the ability to discriminate transposed melodies (Foster and Zatorre 2010), and WM integrity in the left arcuate fasciculus and the segment of the CC that connects the auditory cortices were found to predict individual differences in auditory-motor synchronization (Blecher et al. 2016). The ability to learn musical skills has also been linked to individual variation in auditory and motor regions of the brain. For example, in one longitudinal study non-musicians learned to play short piano melodies, and were scanned before and after training. Regression analyses revealed that greater activity in auditory cortex when listening to the melodies before training predicted how quickly a person learned during training (Herholz et al. 2016). Similarly, greater responsivity to pitch differences in auditory cortex was related to the rate of micro-melody learning (Zatorre et al. 2012). In a study of cello learning, greater activity in the supplementary motor area as well as its degree of connectivity with auditory cortex during passive listening before training was positively correlated with learning success (Wollman et al. 2018). Finally, WM connectivity between auditory and motor regions was also related to melody learning success (Engel et al. 2014). Taken together, these studies support the idea that individual variation in brain structure and function contributes to the ability to learn musical skills.

Genome-wide association studies have linked musical abilities to genes involved in development of the cochlea and auditory brainstem (Oikkonen et al. 2016). But, as with other complex abilities, such as global cognitive function (IQ), the likelihood that a small number of genes account for the broad range of individual differences is small (For review see, Sauce and Matzel 2018). A series of studies have examined genetic contributions to various aspects of musical skill in a large sample of mono- and dizygotic twins. As subset of these twins have some musical experience with an average of approximately 3,300 h of lifetime practice (Mosing et al. 2014). The findings show that genes appear to account for a moderate portion of the variance in music ability, as assessed by a test of melody, rhythm and pitch discrimination (Ullen et al. 2014), and that genes also contribute to the propensity to practice, represented by the number of hours of lifetime practice (Mosing et al. 2014). The researchers also found that personality variables such as "openness to experience" are associated with hours of lifetime practice (Butkovic et al. 2015), which is supported by correlational evidence that musical ability is linked to cognitive and personality variables in addition to practice (Swaminathan and Schellenberg 2018). In the same twin sample, the researchers also found evidence that an enriched musical environment in childhood is associated with greater professional achievement later in life, and that enrichment increases the relative contribution of genetic variability to outcome (Wesseldijk et al. 2019). These findings parallel those for global IQ, which show that higher SES and greater educational opportunity are associated with increased heritability estimates, likely because greater opportunity allows genetic variation to be fully expressed (Sauce and Matzel 2018).

A recent study again used the same sample of twins and a group of trained musicians to examine the impact of AoS on musical ability and professional outcomes (Wesseldijk et al. 2021). In both groups, adults who began training before age 8 showed better performance on a test of musical ability, even when controlling for hours of lifetime practice. Co-twin comparison and heritability analyses found that the interaction of AoS and ability was largely accounted for by genetic and familial factors. The authors interpret these results as providing little support for sensitive period effects in musical training, invoking a scenario in which children inherit musical skills from musically engaged parents who send their precocious offspring for lessons early. However, the data do not seem to support such a strong interpretation. First, the behavioural findings in both samples show that musical ability is significantly related to AoS after controlling for experience. Second, the claim that only genetic and familial factors contribute to ability is an inference about the effect of AoS in trained musicians based on the twin sample. These twins have limited lifetime hours of practice, or training and whether they currently play is not reported. The concept of a sensitive period effect is not that any amount of training in childhood will promote better performance later in life, but rather that AoS interacts with long-term experience. This is consistent with recent work in second-language learning that shows that proficiency plateaus only after about 30 years of experience, but that early AoS still predicts better outcome (Hartshorne et al. 2018).

Taken together, the evidence for genetic contributions to musical ability closely parallels that for global cognitive function (IQ), which has been found to be highly heritable, but which is also modulated by family and social environment, as well as personality and experience (Sauce and Matzel 2018). It appears logical that early AoS would be predicted by skills that might be genetically mediated, such as motor coordination, attentional control, or pitch and temporal discrimination ability. It also seems logical that families who value music would send their children for lessons or that those children with strong musical interests will ask their parents to take them. In addition, early AoS may contribute to skill acquisition in the context of long-term training. If we compare music training with other complex cognitive functions, we are reminded that the simple idea that either genes, family, or environment alone determine ability should give way to a more sophisticated appreciation of the range of gene–environment interactions that contribute to ultimate skill and achievement.

7 Why Is Music an Effective Driver of Sensitive-Period Plasticity?

We have argued that musical training that begins during a sensitive period promotes long-term changes in brain and behaviour. But what features of music training produce these effects? One obvious answer is practice – lots of practice. When children begin lessons they typically play only a few hours per week (Ireland et al. 2018; Seither-Preisler et al. 2014), but the average total duration of training for musicians in the studies reviewed here was 15–20 years. This is the equivalent of thousands of hours of practice across a large portion of a person's life. While the idea that simply practicing long enough will result in expertise has been debunked (Mosing et al. 2014; Hambrick et al. 2020), length of training is typically strongly related to both brain differences and performance (Herholz and Zatorre 2012). A recent study comparing musical prodigies to groups of ET and LT musicians found that one of the few differences between those who were extraordinarily successful at an early age and other highly trained musicians was the number of hours they practiced in childhood (Marion-St-Onge et al. 2020). This is consistent with the idea that early AoS, combined with intensive practice contributes to skill.

The nature of music training may also be particularly effective in promoting plasticity. A second reason that music training may be particularly effective in driving brain plasticity is the highly specific nature of practice. The majority of musicians are experts on single instrument; thus, they perform millions of repetitions of the same movements and listen attentively to an even larger number of associated sounds. When practicing, a musician imagines and plans a precise sequence of sounds and the movements required to produce them. Once the plan is set in motion, they use auditory and somatosensory information to detect subtle deviations in sound and movement, implementing adjustments to enhance performance. Thus, as a musician practices, he or she engages in a repeated cycle of prediction, feedback and error-correction (Patel 2011; Novembre and Keller 2014). Feedback and error-correction are key components of motor learning (Wolpert et al. 2011; Sokolov et al. 2017; Diedrichsen et al. 2010), and studies of both motor and sensory learning show that functional and structural changes in the brain are driven by decreases in error and improved precision. For example, learning to juggle (Scholz et al. 2009), balancing on a tilting board (Taubert et al. 2010) or performing a complex visuomotor task (Landi et al. 2011; Lakhani et al. 2016) have all been shown to produce changes in grey- or white-matter architecture that were related to decreases in error with learning. Thus, error-driven learning, particularly during periods of high developmental plasticity may be an important contributor to structural brain changes measured in adult musicians. Music training may also be particularly potent because it is inherently multi-sensory (combining hearing, sight and touch), and co-activation of circuits deriving from multiple senses may drive plasticity more strongly than input from a single sense (Lee and Noppeney 2011).

A final feature that is likely crucial in promoting plasticity is the rewarding nature of music and the pleasure of accurate performance. The intrinsic pleasure derived from music appears to be common to most people (Mas-Herrero et al. 2011) and is hypothesized to be based on the same dopamine-modulated, predictive systems that regulate reward in other domains (Salimpoor et al. 2015; Ferreri et al. 2019). Learning to produce a rewarding stimulus, such as music, is thus likely to be rewarding to the player. We also know that the reward value of what is learned strongly influences learning and plasticity. Animal studies show that brain plasticity associated with auditory learning is greater when the information to be learned is rewarded or behaviourally relevant (David et al. 2012). Importantly, dopamine has been shown to modulate motor learning in both humans and animals (Floel et al. 2005; Tremblay et al. 2009, 2010), possibly through the reinforcement and habit-formation circuitry of the striatum (Haith and Krakauer 2013; Graybiel and Grafton 2015). The interaction of music training and reward may be at its peak during late childhood and adolescence when brain structures and mechanisms associated with motivation are developing rapidly (Urosevic et al. 2012; Luciana et al. 2012).

8 What Sensitive Periods Are and What They Are Not

Having reviewed what we know about sensitive period effects for musical training it is important to pause and clarify what these effects encompass and what they do not. First, sensitive periods are windows of maturational change where relevant experience results in greater plasticity in behaviour and the brain than at other times during development. This means that evidence supporting a sensitive period must show specificity for both input and timing. It is not enough for a particular type of early experience to exert a long-term effect on behaviour or the brain, unless it can be shown that a different type of experience or the same experience later in life does not produce a similar effect. This specificity may be easier to demonstrate for narrower functions, such as binocular vision or phonemic learning, whereas functions that are more complex, such as music and language, may solicit similar brain networks, so that experience in either domain may produce overlapping effects. Sensitive periods for more complex experiences are also likely to have broader windows, with less abrupt onsets and offsets. Second, it is not the case that any amount of experience during a sensitive period is adequate to produce change (although this could be true for a critical period). Six months of piano lessons, French class or ice hockey practice at age 5 is unlikely to produce long-term effects on behaviour or the brain. Minimal experience during a sensitive period may however prime relearning later in life, as has been seen in studies of early second language exposure (Oh et al. 2010; Singh et al. 2011). Further, sensitive period effects likely diminish if exposure is not maintained, so even playing hockey every day from age 5 to 10 does not mean you can hop on the ice and execute a perfect slapshot at age 40 with no practice in between. Third, postulating a sensitive period in childhood for acquiring a particular skill does not necessarily mean that children will initially perform or learn better than adults. There is ample evidence that children learn most skills more slowly, less accurately and with greater variability than adults (Solum et al. 2020; Savion-Lemieux et al. 2009; Zwart et al. 2019), but there is also evidence that they may learn certain aspects of procedural skills more quickly (Juhasz et al. 2019). Sensitive period effects can be conceptualized as scaffolding plasticity, laying the groundwork on which later experience can build. Finally, demonstrating that early experience can alter behaviour and the brain is not the same as saying that it is the only determinant of later skill. An early start of musical training likely taps pre-existing skills and facilitates their development. But other factors, such as family and social environment, motivation and opportunity are clearly critical to long-term outcome.

9 Gene-Maturation-Environment Framework for Understanding Sensitive Periods

Based on current data and drawing on existing models (Ullen et al. 2016; Werker and Hensch 2015; Voss et al. 2017; Kuhl 2010), we propose a multidimensional gene-maturation-environment framework for understanding the development of musical skill (See Fig. 3). We have discussed this model in previous work (Penhune 2011, 2020) and have developed it further here. Importantly, we believe that this framework is applicable to the understanding of sensitive period effects for other complex abilities. Under this framework, genetic variation produces individual differences in musically relevant abilities such as auditory perception and motor control, as well as

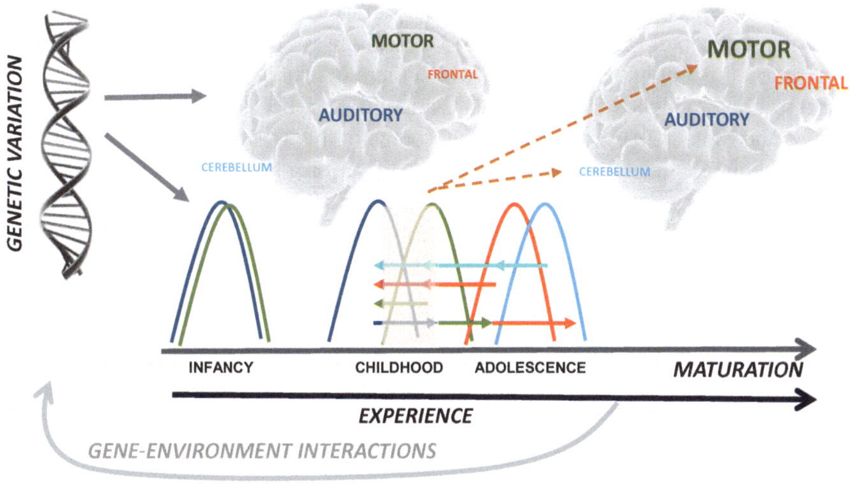

Fig. 3 Gene-maturation-environment interaction model. This Figure illustrates the gene-maturation-environment model (Penhune 2020), with genetic variation codes for individual differences in brain structure as well as maturational changes. Maturation is shown to have two distinct bursts, one during infancy (the first 2 years of life) and another during childhood and adolescence. Different regions such as auditory, motor, frontal and cerebellum have peaks of maturational change at different times. Experience during a sensitive period in childhood (orange bar) can have differential effects on structure in different regions. Plasticity in early and later maturing regions can have feedforward and feedback effects on connected regions. Experience can also modify plasticity through gene–environment interactions

in non-specific cognitive and personality factors including attention, memory, openness to experience and propensity to practice, that contribute to the potential for training. These individual differences interact with experience in passive and active ways to push or pull for development of skill. Genes also control maturation, and the timing of structural and functional changes differs across brain regions and networks, producing a cascade of maturational effects (Werker and Hensch 2015, Voss et al. 2017, Kuhl 2010). Early and late maturing regions interact, and experience can drive change through both bottom-up sensorimotor processes and through top-down cognitive influences, consistent with the concept of interactive specialization (Eggermont and Moore 2012, Johnson 2011). Thus, we propose that the timing of music experience interacts with both predispositions and maturation to influence long-term behavioural and brain plasticity. Early music training may initially affect earlier maturing, lower-level sensory processes such as pitch discrimination, but early training and its related plasticity may also have a metaplastic effect, laying the foundation for augmented development of later maturing sensorimotor integration abilities. Further, interactions within connected brain networks may produce different kinds of plastic effects: for example, structural enlargement in premotor cortex or the cerebellum, but structural reductions in connected regions of the cerebellum. Taken together, we propose that early training has a metaplastic effect where early training promotes brain plasticity that makes a region or network more receptive to future experience. Thus early training can be seen as a scaffold on which later experience can build (Steele et al. 2013). In this way, musical skill appears to be similar to other complex abilities, in which there is strong evidence for heritability, but also good evidence that individual variability can be moderated by experience (Sauce and Matzel 2018). One important source of this variability may be the age at which training begins. Stepping onto a stage to perform for others requires skill and practice, but also determination, poise and a fair dose of self-confidence. These characteristics likely contribute to the propensity to start lessons early, may be enhanced by early experience, are honed by training and depend on the influence of family, friends and opportunity.

Acknowledgements Funding for this work was provided by grants from the National Sciences and Engineering Research Council of Canada, the Canadian Institutes for Health Research and the Fonds de recherche du Québec.

References

Abraham WC (2008) Metaplasticity: tuning synapses and networks for plasticity. Nat Rev Neurosci 9:387. https://doi.org/10.1038/nrn2356

Altenmuller E, Furuya S (2016) Brain plasticity and the concept of metaplasticity in skilled musicians. Adv Exp Med Biol 957:197–208. https://doi.org/10.1007/978-3-319-47313-0_11

Amemiya K, Morita T, Saito DN, Ban M, Shimada K, Okamoto Y, Kosaka H, Okazawa H, Asada M, Naito E (2019) Local-to-distant development of the cerebrocerebellar sensorimotor network in the typically developing human brain: a functional and diffusion MRI study. Brain Struct Funct 224:1359–1375. https://doi.org/10.1007/s00429-018-01821-5

Amlien IK, Fjell AM, Tamnes CK, Grydeland H, Krogsrud SK, Chaplin TA, Rosa MGP, Walhovd KB (2016) Organizing principles of human cortical development – thickness and area from 4 to 30 years: insights from comparative primate neuroanatomy. Cereb Cortex 26:257–267. https://doi.org/10.1093/cercor/bhu214

Amunts K, Schlaug G, Jancke L, Steinmetz H, Schleicher A, Dabringhaus A, Zilles K (1997) Motor cortex and hand motor skills: structural compliance in the human brain. Hum Brain Mapp 5:206–215. https://doi.org/10.1002/(SICI)1097-0193(1997)5:3<206::AID-HBM5>3.0.CO;2-7

Baer L, Park M, Bailey J, Chakravarty M, Li K, Penhune V (2015) Regional cerebellar volumes are related to early musical training and finger tapping performance. NeuroImage 109:130–139. https://doi.org/10.1016/j.neuroimage.2014.12.076

Baharloo S, Johnston P, Service S, Gitschier J, Freimer N (1998) Absolute pitch: an approach for identification of genetic and nongenetic components. Am J Hum Genet 62:224–231. https://doi.org/10.1086/301704

Bailey JA, Penhune VB (2010) Rhythm synchronization performance and auditory working memory in early- and late-trained musicians. Exp Brain Res 204:91–101. https://doi.org/10.1007/s00221-010-2299-y

Bailey J, Penhune VB (2012) A sensitive period for musical training: contributions of age of onset and cognitive abilities. Ann N Y Acad Sci 1252:163–170. https://doi.org/10.1111/j.1749-6632.2011.06434.x

Bailcy JA, Penhune VB (2013) The relationship between the age of onset of musical training and rhythm synchronization performance: validation of sensitive period effects. Front Neurosci 7:227. https://doi.org/10.3389/fnins.2013.00227

Bailey JA, Zatorre RJ, Penhune VB (2014) Early musical training is linked to gray matter structure in the ventral premotor cortex and auditory-motor rhythm synchronization performance. J Cogn Neurosci 26:755–767. https://doi.org/10.1162/jocn_a_00527

Ball G, Beare R, Seal ML (2019) Charting shared developmental trajectories of cortical thickness and structural connectivity in childhood and adolescence. Hum Brain Mapp 40:4630–4644. https://doi.org/10.1002/hbm.24726

Blecher T, Tal I, Ben-Shachar M (2016) White matter microstructural properties correlate with sensorimotor synchronization abilities. NeuroImage 138:1–12. https://doi.org/10.1016/j.neuroimage.2016.05.022

Bostan AC, Strick PL (2018) The basal ganglia and the cerebellum: nodes in an integrated network. Nat Rev Neurosci 19:338–350. https://doi.org/10.1038/s41583-018-0002-7

Brown TT, Jernigan TL (2012) Brain development during the preschool years. Neuropsychol Rev 22:313–333. https://doi.org/10.1007/s11065-012-9214-1

Budisavljevic S, Dell'acqua F, Rijsdijk FV, Kane F, Picchioni M, Mcguire P, Toulopoulou T, Georgiades A, Kalidindi S, Kravariti E, Murray RM, Murphy DG, Craig MC, Catani M (2015) Age-related differences and heritability of the Perisylvian Language Networks. J Neurosci 35:12625–12634. https://doi.org/10.1523/JNEUROSCI.1255-14.2015

Butkovic A, Ullen F, Mosing MA (2015) Personality-related traits as predictors of music practice: underlying environmental and genetic influences. Personal Individ Differ 74:133–138. https://doi.org/10.1016/j.paid.2014.10.006

Chiang MC, Barysheva M, Shattuck DW, Lee AD, Madsen SK, Avedissian C, Klunder AD, Toga AW, Mcmahon KL, De Zubicaray GI, Wright MJ, Srivastava A, Balov N, Thompson PM (2009) Genetics of brain fiber architecture and intellectual performance. J Neurosci 29:2212–2224. https://doi.org/10.1523/JNEUROSCI.4184-08.2009

David SV, Fritz JB, Shamma SA (2012) Task reward structure shapes rapid receptive field plasticity in auditory cortex. Proc Natl Acad Sci U S A 109:2144–2149. https://doi.org/10.1073/pnas.1117717109

De Manzano O, Ullen F (2018) Same genes, different brains: neuroanatomical differences between monozygotic twins discordant for musical training. Cereb Cortex 28:387–394. https://doi.org/10.1093/cercor/bhx299

De Villers-Sidani E, Merzenich MM (2011) Lifelong plasticity in the rat auditory cortex: basic mechanisms and role of sensory experience. Prog Brain Res 191:119–131. https://doi.org/10.1016/B978-0-444-53752-2.00009-6

Diedrichsen J, Shadmehr R, Ivry RB (2010) The coordination of movement: optimal feedback control and beyond. Trends Cogn Sci 14:31–39. https://doi.org/10.1016/j.tics.2009.11.004

Eggermont JJ, Moore JK (2012) Morphological and functional development of the auditory nervous system. In: Werner L, Fay RR, Popper AN (eds) Springer handbook of human auditory development. Springer, New York

Emberson LL (2017) How does experience shape early development? Considering the role of top-down mechanisms. Adv Child Dev Behav 52:1–41. https://doi.org/10.1016/bs.acdb.2016.10.001

Engel A, Hijmans BS, Cerliani L, Bangert M, Nanetti L, Keller PE, Keysers C (2014) Inter-individual differences in audio-motor learning of piano melodies and white matter fiber tract architecture. Hum Brain Mapp 35:2483–2497. https://doi.org/10.1002/hbm.22343

Ferreri L, Mas-Herrero E, Zatorre RJ, Ripolles P, Gomez-Andres A, Alicart H, Olive G, Marco-Pallares J, Antonijoan RM, Valle M, Riba J, Rodriguez-Fornells A (2019) Dopamine modulates the reward experiences elicited by music. Proc Natl Acad Sci U S A 116:3793–3798. https://doi.org/10.1073/pnas.1811878116

Ferronato PA, Domellof E, Ronnqvist L (2014) Early influence of auditory stimuli on upper-limb movements in young human infants: an overview. Front Psychol 5:1043. https://doi.org/10.3389/fpsyg.2014.01043

Fjell AM, Chen CH, Sederevicius D, Sneve MH, Grydeland H, Krogsrud SK, Amlien I, Ferschmann L, Ness H, Folvik L, Beck D, Mowinckel AM, Tamnes CK, Westerhausen R,

Haberg AK, Dale AM, Walhovd KB (2019) Continuity and discontinuity in human cortical development and change from embryonic stages to old age. Cereb Cortex 29:3879–3890. https://doi.org/10.1093/cercor/bhy266

Flege JE, Yeni-Komshian G, Liu S (1999) Age constraints on second-language acquisition. J Mem Lang 41:78–104. https://doi.org/10.1006/jmla.1999.2638

Floel A, Breitenstein C, Hummel F, Celnik P, Gingert C, Sawaki L, Knecht S, Cohen LG (2005) Dopaminergic influences on formation of a motor memory. Ann Neurol 58:121–130. https://doi.org/10.1002/ana.20536

Foster NE, Zatorre RJ (2010) Cortical structure predicts success in performing musical transformation judgments. NeuroImage 53:26–36. https://doi.org/10.1016/j.neuroimage.2010.06.042

Fujioka T, Ross B, Kakigi R, Pantev C, Trainor LJ (2006) One year of musical training affects development of auditory cortical-evoked fields in young children. Brain 129:2593–2608. https://doi.org/10.1093/brain/awl247

Giedd J, Blumenthal J, Jeffries N, Castellanos F, Liu H, Zijdenbos A, Paus T, Evans A, Rapoport J (1999) Brain development during childhood and adolescence: a longitudinal MRI study. Nat Neurosci 2:861–863. https://doi.org/10.1038/13158

Goddings AL, Mills KL, Clasen LS, Giedd JN, Viner RM, Blakemore SJ (2014) The influence of puberty on subcortical brain development. NeuroImage 88:242–251. https://doi.org/10.1016/j.neuroimage.2013.09.073

Gogtay N, Giedd J, Lusk L, Hayashi K, Greenstein D, Vaituzis A, Nugent T, Merman D, Clasen L, Toga A, Rapoport J, Thompson P (2004) Dynamic mapping of human cortical development during childhood and through early adulthood. Proc Natl Acad Sci U S A 101:8174–8179

Graybiel AM, Grafton ST (2015) The striatum: where skills and habits meet. Cold Spring Harb Perspect Biol 7:a021691. https://doi.org/10.1101/cshperspect.a021691

Habibi A, Damasio A, Ilari B, Veiga R, Joshi AA, Leahy RM, Haldar JP, Varadarajan D, Bhushan C, Damasio H (2018) Childhood music training induces change in micro and macroscopic brain structure: results from a longitudinal study. Cereb Cortex 28(12):4336–4347. https://doi.org/10.1093/cercor/bhx286

Haith AM, Krakauer JW (2013) Model-based and model-free mechanisms of human motor learning. Adv Exp Med Biol 782:1–21. https://doi.org/10.1007/978-1-4614-5465-6_1

Halwani GF, Loui P, Ruber T, Schlaug G (2011) Effects of practice and experience on the arcuate fasciculus: comparing singers, instrumentalists, and non-musicians. Front Psychol 2:156. https://doi.org/10.3389/fpsyg.2011.00156

Hamalainen S, Sairanen V, Leminen A, Lehtonen M (2017) Bilingualism modulates the white matter structure of language-related pathways. NeuroImage 152:249–257. https://doi.org/10.1016/j.neuroimage.2017.02.081

Hambrick DZ, Macnamara BN, Oswald FL (2020) Is the deliberate practice view defensible? A review of evidence and discussion of issues. Front Psychol 11:1134. https://doi.org/10.3389/fpsyg.2020.01134

Hartshorne JK, Tenenbaum JB, Pinker S (2018) A critical period for second language acquisition: evidence from 2/3 million English speakers. Cognition 177:263–277. https://doi.org/10.1016/j.cognition.2018.04.007

Hensch TK, Quinlan EM (2018) Critical periods in amblyopia. Vis Neurosci 35:E014. https://doi.org/10.1017/S0952523817000219

Herholz S, Zatorre R (2012) Musical training as a framework for brain plasticity: behavior, function, and structure. Neuron 76:486–502. https://doi.org/10.1016/j.neuron.2012.10.011

Herholz S, Boh B, Pantev C (2011) Musical training modulates encoding of higher-order regularities in the auditory cortex. Eur J Neurosci 34:524–529. https://doi.org/10.1111/j.1460-9568.2011.07775.x

Herholz SC, Coffey EB, Pantev C, Zatorre RJ (2016) Dissociation of neural networks for predisposition and for training-related plasticity in auditory-motor learning. Cereb Cortex 26:3125–3134. https://doi.org/10.1093/cercor/bhv138

Hund-Georgiadis M, Von Cramon D (1999) Motor-learning-related changes in piano players and non-musicians revealed by functional magnetic-resonance signals. Exp Brain Res 125:417–425. https://doi.org/10.1007/s002210050698

Huttenlocher P, Dabholkar A (1997) Regional differences in synaptogenesis in human cerebral cortex. J Comp Neurol 387:167–178. https://doi.org/10.1002/(sici)1096-9861(19971020)387:2<167::aid-cne1>3.0.co;2-z

Hyde KL, Lerch J, Norton A, Forgeard M, Winner E, Evans AC, Schlaug G (2009) Musical training shapes structural brain development. J Neurosci 29:3019–3025. https://doi.org/10.1523/JNEUROSCI.5118-08.2009

Ireland K, Parker A, Foster N, Penhune V (2018) Rhythm and melody tasks for school-aged children with and without musical training: age-equivalent scores and reliability. Front Audit Cogn Neurosci 9:426. https://doi.org/10.3389/fpsyg.2018.00426

Ireland K, Iyer TA, Penhune VB (2019) Contributions of age of start, cognitive abilities and practice to musical task performance in childhood. PLoS One 14:e0216119. https://doi.org/10.1371/journal.pone.0216119

Johnson MH (2011) Interactive specialization: a domain-general framework for human functional brain development? Dev Cogn Neurosci 1:7–21. https://doi.org/10.1016/j.dcn.2010.07.003

Jorgensen H (2011) Instrumental learning: is an early start a key to success? Br J Music Educ 18:227–239

Juhasz D, Nemeth D, Janacsek K (2019) Is there more room to improve? The lifespan trajectory of procedural learning and its relationship to the between- and within-group differences in average response times. PLoS One 14:e0215116. https://doi.org/10.1371/journal.pone.0215116

Kelly R, Strick P (2003) Cerebellar loops with motor cortex and prefrontal cortex of a non-human primate. J Neurosci 23:8432–8444. https://doi.org/10.1523/JNEUROSCI.23-23-08432.2003

Knudsen E (2004) Sensitive periods in the development of the brain and behavior. J Cogn Neurosci 16:1412–1425. https://doi.org/10.1162/0898929042304796

Kral A, Sharma A (2012) Developmental neuroplasticity after cochlear implantation. Trends Neurosci 35:111–122. https://doi.org/10.1016/j.tins.2011.09.004

Kuhl PK (2010) Brain mechanisms in early language acquisition. Neuron 67:713–727. https://doi.org/10.1016/j.neuron.2010.08.038

Kurth F, Mayer EA, Toga AW, Thompson PM, Luders E (2013) The right inhibition? Callosal correlates of hand performance in healthy children and adolescents callosal correlates of hand performance. Hum Brain Mapp 34:2259–2265. https://doi.org/10.1002/hbm.22060

Lakhani B, Borich MR, Jackson JN, Wadden KP, Peters S, Villamayor A, Mackay AL, Vavasour IM, Rauscher A, Boyd LA (2016) Motor skill acquisition promotes human brain myelin plasticity. Neural Plast 2016:7526135. https://doi.org/10.1155/2016/7526135

Landi SM, Baguear F, Della-Maggiore V (2011) One week of motor adaptation induces structural changes in primary motor cortex that predict long-term memory one year later. J Neurosci 31:11808–11813. https://doi.org/10.1523/JNEUROSCI.2253-11.2011

Larsen B, Luna B (2015) In vivo evidence of neurophysiological maturation of the human adolescent striatum. Dev Cogn Neurosci 12:74–85. https://doi.org/10.1016/j.dcn.2014.12.003

Lee H, Noppeney U (2011) Long-term music training tunes how the brain temporally binds signals from multiple senses. Proc Natl Acad Sci U S A 108:E1441–E1450. https://doi.org/10.1073/pnas.1115267108

Lega C, Stephan M, Zatorre R, Penhune V (2016) Testing the role of dorsal premotor cortex in auditory-motor association learning using transcranical magnetic stimulation (TMS). PLoS One 11:e0163380. https://doi.org/10.1371/journal.pone.0163380

Luciana M, Wahlstrom D, Porter JN, Collins PF (2012) Dopaminergic modulation of incentive motivation in adolescence: age-related changes in signaling, individual differences, and implications for the development of self-regulation. Dev Psychol 48:844–861. https://doi.org/10.1037/a0027432

Marion-St-Onge C, Weiss MW, Sharda M, Peretz I (2020) What makes musical prodigies? Front Psychol 11:566373. https://doi.org/10.3389/fpsyg.2020.566373

Mas-Herrero E, Marco-Pallares J, Lorenzo-Seva U, Zatorre RJ, Rodriguez-Fornells A (2011) Individual differences in music reward experiences. Music Percept 31:118–138. https://doi.org/10.1525/MP.2013.31.2.118

Merchant H, Harrington DL, Meck WH (2013) Neural basis of the perception and estimation of time. Annu Rev Neurosci 36:313–336. https://doi.org/10.1146/annurev-neuro-062012-170349

Mills KL, Goddings AL, Herting MM, Meuwese R, Blakemore SJ, Crone EA, Dahl RE, Guroglu B, Raznahan A, Sowell ER, Tamnes CK (2016) Structural brain development between childhood and adulthood: convergence across four longitudinal samples. NeuroImage 141:273–281. https://doi.org/10.1016/j.neuroimage.2016.07.044

Mosing MA, Madison G, Pedersen NL, Kuja-Halkola R, Ullen F (2014) Practice does not make perfect: no causal effect of music practice on music ability. Psychol Sci 25:1795–1803. https://doi.org/10.1177/0956797614541990

Novembre G, Keller PE (2014) A conceptual review on action-perception coupling in the musicians' brain: what is it good for? Front Hum Neurosci 8:603. https://doi.org/10.3389/fnhum.2014.00603

Oh JS, Au TK, Jun SA (2010) Early childhood language memory in the speech perception of international adoptees. J Child Lang 37:1123–1132. https://doi.org/10.1017/S0305000909990286

Oikkonen J, Kuusi T, Peltonen P, Raijas P, Ukkola-Vuoti L, Karma K, Onkamo P, Jarvela I (2016) Creative activities in music – a genome-wide linkage analysis. PLoS One 11:e0148679. https://doi.org/10.1371/journal.pone.0148679

Patel AD (2011) Why would musical training benefit the neural encoding of speech? The OPERA hypothesis. Front Psychol 2:142. https://doi.org/10.3389/fpsyg.2011.00142

Penhune VB (2011) Sensitive periods in human development: evidence from musical training. Cortex 47:1126–1137. https://doi.org/10.1016/j.cortex.2011.05.010

Penhune V (2020) A gene-maturation-environment model for understanding sensitive period effects in musical training. Curr Opin Behav Neurosci 36:13–22. https://doi.org/10.1016/j.cobeha.2020.05.011

Powell HW, Parker GJ, Alexander DC, Symms MR, Boulby PA, Wheeler-Kingshott CA, Barker GJ, Noppeney U, Koepp MJ, Duncan JS (2006) Hemispheric asymmetries in language-related pathways: a combined functional MRI and tractography study. NeuroImage 32:388–399. https://doi.org/10.1016/j.neuroimage.2006.03.011

Putkinen V, Tervaniemi M, Saarikivi K, Ojala P, Huotilainen M (2014) Enhanced development of auditory change detection in musically trained school-aged children: a longitudinal event-related potential study. Dev Sci 17:282–297. https://doi.org/10.1111/desc.12109

Ragert P, Schmidt A, Altenmuller E, Dinse H (2004) Superior tactile performance and learning in professional pianists: evidence for meta-plasticity in musicians. Eur J Neurosci 19:473–478. https://doi.org/10.1111/j.0953-816x.2003.03142.x

Rosenkranz K, Williamon A, Rothwell JC (2007) Motorcortical excitability and synaptic plasticity is enhanced in professional musicians. J Neurosci 27:5200–5206. https://doi.org/10.1523/JNEUROSCI.0836-07.2007

Salimpoor VN, Zald DH, Zatorre RJ, Dagher A, Mcintosh AR (2015) Predictions and the brain: how musical sounds become rewarding. Trends Cogn Sci 19:86–91. https://doi.org/10.1016/j.tics.2014.12.001

Sauce B, Matzel LD (2018) The paradox of intelligence: heritability and malleability coexist in hidden gene-environment interplay. Psychol Bull 144:26–47. https://doi.org/10.1037/bul0000131

Savion-Lemieux T, Bailey JA, Penhune VB (2009) Developmental contributions to motor sequence learning. Exp Brain Res 195:293–306. https://doi.org/10.1007/s00221-009-1786-5

Schlaug G, Jancke L, Huang Y, Staiger JF, Steinmetz H (1995) Increased corpus callosum size in musicians. Neuropsychologia 33:1047–1055. https://doi.org/10.1016/0028-3932(95)00045-5

Scholz J, Klein MC, Behrens TE, Johansen-Berg H (2009) Training induces changes in white-matter architecture. Nat Neurosci 12:1370–1371. https://doi.org/10.1038/nn.2412

Scholz J, Allemang-Grand R, Dazai J, Lerch JP (2015) Environmental enrichment is associated with rapid volumetric brain changes in adult mice. NeuroImage 109:190–198. https://doi.org/10.1016/j.neuroimage.2015.01.027

Seither-Preisler A, Parncutt R, Schneider P (2014) Size and synchronization of auditory cortex promotes musical, literacy, and attentional skills in children. J Neurosci 34:10937–10949. https://doi.org/10.1523/JNEUROSCI.5315-13.2014

Shahin A, Roberts LE, Pantev C, Trainor LJ, Ross B (2005) Modulation of P2 auditory-evoked responses by the spectral complexity of musical sounds. Neuroreport 16:1781–1785. https://doi.org/10.1097/01.wnr.0000185017.29316.63

Shenker J, Zatorre RJ, Penhune VB (2019) The relationship between cerebellar volume and cortical sensorimotor areas in early-trained musicians. Organization for Human Brain Mapping, Rome

Singh L, Liederman J, Mierzejewski R, Barnes J (2011) Rapid reacquisition of native phoneme contrasts after disuse: you do not always lose what you do not use. Dev Sci 14:949–959. https://doi.org/10.1111/j.1467-7687.2011.01044.x

Sokolov AA, Miall RC, Ivry RB (2017) The cerebellum: adaptive prediction for movement and cognition. Trends Cogn Sci 21:313–332. https://doi.org/10.1016/j.tics.2017.02.005

Solum M, Loras H, Pedersen AV (2020) A golden age for motor skill learning? Learning of an unfamiliar motor task in 10-year-olds, young adults, and adults, when starting from similar baselines. Front Psychol 11:538. https://doi.org/10.3389/fpsyg.2020.00538

Steele CJ, Bailey JA, Zatorre RJ, Penhune VB (2013) Early musical training and white-matter plasticity in the corpus callosum: evidence for a sensitive period. J Neurosci 33:1282–1290. https://doi.org/10.1523/JNEUROSCI.3578-12.2013

Swaminathan S, Schellenberg EG (2018) Musical competence is predicted by music training, cognitive abilities, and personality. Sci Rep 8:9223. https://doi.org/10.1038/s41598-018-27571-2

Taubert M, Draganski B, Anwander A, Muller K, Horstmann A, Villringer A, Ragert P (2010) Dynamic properties of human brain structure: learning-related changes in cortical areas and associated fiber connections. J Neurosci 30:11670–11677. https://doi.org/10.1523/JNEUROSCI.2567-10.2010

Tiemeier H, Lenroot RK, Greenstein DK, Tran L, Pierson R, Giedd JN (2010) Cerebellum development during childhood and adolescence: a longitudinal morphometric MRI study. NeuroImage 49:63–70. https://doi.org/10.1016/j.neuroimage.2009.08.016

Trainor LJ (2012) Musical experience, plasticity, and maturation: issues in measuring developmental change using EEG and MEG. Ann N Y Acad Sci 1252:25–36. https://doi.org/10.1111/j.1749-6632.2012.06444.x

Tremblay PL, Bedard MA, Levesque M, Chebli M, Parent M, Courtemanche R, Blanchet PJ (2009) Motor sequence learning in primate: role of the D2 receptor in movement chunking during consolidation. Behav Brain Res 198:231–239. https://doi.org/10.1016/j.bbr.2008.11.002

Tremblay PL, Bedard MA, Langlois D, Blanchet PJ, Lemay M, Parent M (2010) Movement chunking during sequence learning is a dopamine-dependant process: a study conducted in Parkinson's disease. Exp Brain Res 205:375–385. https://doi.org/10.1007/s00221-010-2372-6

Ullen F, Mosing M, Holm L, Eriksson H, Madison G (2014) Psychometric properties and heritability of a new online test for musicality, the Swedish musical discrimination test. Personal Individ Differ 63:87–93. https://doi.org/10.1016/j.paid.2014.01.057

Ullen F, Hambrick DZ, Mosing MA (2016) Rethinking expertise: a multifactorial gene-environment interaction model of expert performance. Psychol Bull 142:427–446. https://doi.org/10.1037/bul0000033

Urosevic S, Collins P, Muetzel R, Lim K, Luciana M (2012) Longitudinal changes in behavioral approach system sensitivity and brain structures involved in reward processing during adolescence. Dev Psychol 48:1488–1500. https://doi.org/10.1037/a0027502

van Vugt FT, Hartmann K, Altenmüller E, Mohammadi B, Margulies DS (2021) The impact of early musical training on striatal functional connectivity. NeuroImage 238:118251. https://doi.org/10.1016/j.neuroimage.2021.118251

Vanzella P, Schellenberg EG (2010) Absolute pitch: effects of timbre on note-naming ability. PLoS One 5:e15449. https://doi.org/10.1371/journal.pone.0015449

Vaquero L, Hartmann K, Ripolles P, Rojo N, Sierpowska J, Francois C, Camara E, Van Vugt FT, Mohammadi B, Samii A, Munte TF, Rodriguez-Fornells A, Altenmuller E (2016) Structural neuroplasticity in expert pianists depends on the age of musical training onset. NeuroImage 126:106–119. https://doi.org/10.1016/j.neuroimage.2015.11.008

Vaquero L, Rousseau PN, Vozian D, Klein D, Penhune V (2020) What you learn & when you learn it: impact of early bilingual & music experience on the structural characteristics of auditory-motor pathways. NeuroImage 213:116689. https://doi.org/10.1016/j.neuroimage.2020.116689

Voss P, Thomas ME, Cisneros-Franco JM, De Villers-Sidani E (2017) Dynamic brains and the changing rules of neuroplasticity: implications for learning and recovery. Front Psychol 8:1657. https://doi.org/10.3389/fpsyg.2017.01657

Werker JF, Hensch TK (2015) Critical periods in speech perception: new directions. Annu Rev Psychol 66:173–196. https://doi.org/10.1146/annurev-psych-010814-015104

Wesseldijk LW, Mosing MA, Ullen F (2019) Gene–environment interaction in expertise: the importance of childhood environment for musical achievement. Dev Psychol 55:1473–1479. https://doi.org/10.1037/dev0000726

Wesseldijk LW, Mosing MA, Ullen F (2021) Why is an early start of training related to musical skills in adulthood? A genetically informative study. Psychol Sci 32:3–13. https://doi.org/10.1177/0956797620959014

Westerhausen R, Luders E, Specht K, Ofte SH, Toga AW, Thompson PM, Helland T, Hugdahl K (2011) Structural and functional reorganization of the corpus callosum between the age of 6 and 8 years. Cereb Cortex 21:1012–1017. https://doi.org/10.1093/cercor/bhq165

Wollman I, Segado M, Charpentier T, Penhune V, Zatorre R (2018) Neural network retuning and neural predictors of learning success associated with cello training. Proc Natl Acad Sci U S A 115:E6056–E6064. https://doi.org/10.1073/pnas.1721414115

Wolpert DM, Diedrichsen J, Flanagan JR (2011) Principles of sensorimotor learning. Nat Rev Neurosci 12:739–751. https://doi.org/10.1038/nrn3112

Zatorre R, Chen J, Penhune V (2007) When the brain plays music: sensory-motor interactions in music perception and production. Nat Rev Neurosci 8:547–558

Zatorre RJ, Delhommeau K, Zarate JM (2012) Modulation of auditory cortex response to pitch variation following training with microtonal melodies. Front Psychol 3:544. https://doi.org/10.3389/fpsyg.2012.00544

Zwart FS, Vissers C, Kessels RPC, Maes JHR (2019) Procedural learning across the lifespan: a systematic review with implications for atypical development. J Neuropsychol 13:149–182. https://doi.org/10.1111/jnp.12139

Sensitive Periods for Recovery from Early Brain Injury

Bryan Kolb

Contents

Abstract The developing brain is remarkably plastic as it changes in response to a wide range of experiences including sensory and motor experience, psychoactive drugs, peer relationships, parent–infant interactions, gonadal hormones, intestinal flora, diet, and injury. There are sensitive periods for many of these experiences, including cerebral injury. Comparisons across mammalian species (humans, monkeys, cats, rats, mice) show a sensitive period for good outcomes from cerebral injury around the time of intense synaptogenesis. This period is postnatal in humans, cats, and rats, but prenatal in monkeys, reflecting the differences in neuronal development at birth across species. In addition, there appears to be a sensitive period prenatally during the time of maximum cortical neurogenesis and possibly during adolescence as well, although these periods are not as well studied as the period related to synaptogenesis and to date only examined in rats. Here we review

B. Kolb (✉)
Department of Neuroscience, University of Lethbridge, Lethbridge, AB, Canada
e-mail: kolb@uleth.ca

Curr Topics Behav Neurosci (2022) 53: 189–212
https://doi.org/10.1007/7854_2021_296
Published Online: 14 February 2022

the evidence for sensitive periods related to brain injury across species and propose mechanisms that may underlie the plasticity during these periods.

Keywords Brain development · Early brain injury · Prefrontal cortex · Recovery of function

1 Background

Neural circuits are shaped by experience during sensitive periods of development, during which time the brain is especially plastic and can be changed by many different experiences. Although sensitive periods are most often associated with the development of specific behaviors, such as language or binocular vision (e.g., Werker and Hensch 2015), or the vulnerability to certain perturbations such as stress or drug exposure (e.g., Andersen 2003; Jordan and Andersen 2017), there are also sensitive periods in which the brain is able to turn on plastic processes that allow for significant compensation in both the brain and behavior at certain ages following brain injury during brain development. Thus, there are sensitive periods during brain development where the brain is capable of remarkable repair after injury and these periods contrast to other developmental periods in which the outcome of injury is dramatically poorer (see also Oberman and Pascual-Leone 2013). These contrasting periods reveal stark differences in developmental processes at different developmental times and these windows are common across mammalian species. I will refer to good functional outcomes as "recovery" but recognize that the behaviors in question often have not yet developed and thus they have been "spared" but this word is seldom used today so I will stick with "recovery."

There is a long history of interest in the effect of age on severity of chronic behavioral symptoms after brain injury. The most dramatic evidence of an age-related difference comes from observations that damage to language areas of the cortex in young children rarely leads to persisting aphasia. According to Lennenberg (1967), it was Paul Broca who first noted that children were not aphasic after damage to the third frontal convolution on the left (i.e., Broca's area). Studies in the 1970s by Rasmussen and Milner (1977) used unilateral sodium Amytal injections to locate language after early left hemisphere injury and found that although injury prior to age 5 would allow a shift of language to the right hemisphere, after 6 years language could move within the left hemisphere but not shift to the right. This intrahemispheric reorganization appears to be restricted to injuries at about age 6–10 years and allows less recovery than injuries from 1 to 5 years. Although Rasmussen and Milner did not specifically address injury in the first year of life, other studies have shown that injuries in the first year of life produce more severe impairments in IQ than those occurring after 1 year (e.g., Riva and Cazzangia 1986). A more recent study of a large number of children suggests that the period of severe impairment is longer, including the second year of life (Anderson et al. 2009).

The first laboratory studies of the effects of early brain injury were conducted by Margaret Kennard (1938, 1940) who did behavioral studies of rhesus monkeys and chimpanzees with motor cortex lesions as infants. Kennard reported that the monkeys with early lesions (at 3 weeks of age) appeared to have a better behavioral outcome than those with adult injuries and she proposed that this was due to some type of compensatory change(s) after the early injury. Kennard's findings later led Hans Lukas Teuber (1975) to name this phenomenon the "Kennard Principle," which basically says that if you are going to have a brain injury, have it early. Thus, Teuber believed that there was a sensitive period early in development during which the brain was able to compensate for injury. This idea had intuitive appeal because it is a common observation that infants and young children appear to recover more quickly from various maladies than adults do.

Although there is considerable support for the Kennard Principle (e.g., Payne and Lomber 2003; Whishaw and Kolb 1988), there are three fundamental problems with the Kennard studies. First, her work was early days in the development of behavioral analyses and thus her behavioral examinations of motor functions were relatively primitive compared to current behavioral methods. Later studies by others such as Passingham et al. (1983) did more thorough behavioral analyses and found much greater behavioral deficits than Kennard had reported. Second, monkeys are born more mature than humans and much more mature than lab animals such as cats and rats, so it is difficult to generalize from monkeys to species whose brains are much more immature, such as humans, cats, or rats. The brain goes through many developmental stages (see Fig. 1) and we now know that the consequences of early brain injury vary with the developmental stage. These stages begin with neurogenesis, followed by neuronal migration, neuronal maturation, synaptic formation, gliogenesis, and finally cell death and synaptic pruning. The sensitive periods for good recovery from brain injury are related to the stage of brain development at the time of injury, and not related to age at birth, per se. For example, brain perturbation during the period of neuronal migration is likely to have a different (and worse) effect than a similar perturbation during synaptogenesis. Third, whereas Kennard was studying monkeys, Donald Hebb was studying children with early frontal injuries and he concluded that early damage to the frontal lobes actually has worse behavioral sequelae than damage later in life (Hebb 1949). Hebb suggested that damage to certain structures during sensitive periods of development could compromise cognitive development and the children are not able to compensate adequately. We could call this the "Hebb Principle." It is important to note, however, that whereas Kennard mostly was studying motor functions, Hebb was studying cognitive functions.

The difference between the outcomes of the Kennard and Hebb studies could be related to many factors including developmental age at injury, location of injury, age at the time of behavioral assessment, nature of the behavioral assessment, sex of the individuals, and whether the injuries are diffuse, unilateral, or bilateral. We need to turn to studies of laboratory animals to sort this out.

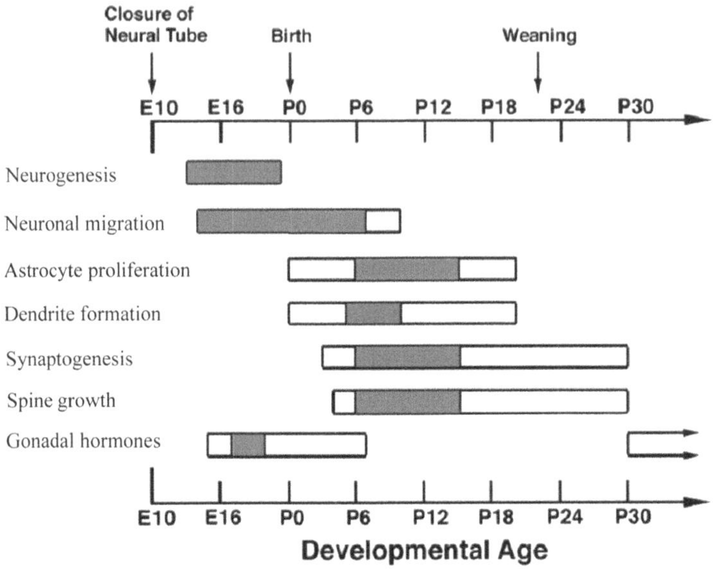

Fig. 1 Stages of cortical development in the rat. Gray areas in each box represent the maximum activity in each stage. The white areas represent the temporal range of activity. *E* embryonic, *P* postnatal. Timing is based on Andersen (2003), Berry (1974), Hicks and D'Amato (1968), and Ignacio et al. (1995)

2 Effects of Frontal Lobe Injuries in Young Monkeys and Cats

Harry Harlow and colleagues were the first to make prefrontal cortical lesions in young monkeys and reported that the animals had much smaller deficits in cognitive tests than monkeys with later lesions (Harlow et al. 1964), a finding that was consistent with the Kennard studies on the effects of motor cortex lesions. Patricia Goldman (1974) confirmed Harlow et al. findings but over the ensuing years it became clear to her that she and others had overestimated the extent of recovery because the animals were tested behaviorally when they were young. Thus, Goldman-Rakic et al. (1983) were able to show that monkeys with early dorsolateral prefrontal injuries became progressively more impaired at cognitive tasks such as delayed alternation as they grew older. By adulthood they were severely impaired, showing that deficits emerge as the brain develops, a finding that has also been reported for children with early brain injuries (e.g., Banich et al. 1990; Johnson et al. 2003), and hamsters with medial prefrontal injuries (Kolb and Whishaw 1985a). The emergence of deficits during development suggests that during development certain brain areas, such as the prefrontal cortex, become increasingly important to the solution of cognitive problems. Goldman-Rakic et al. (1983) demonstrated this in normal brains by developing a technique for reversible lesions by implanting cooling probes into the prefrontal cortex. When the tissue is cooled, it does not function

Fig. 2 Performance on test of delayed alternation by prenatally prefrontal cortex operated monkey (E106), postnatally operated monkeys (P50), and adult operated monkeys (P540–720). Adapted from Goldman and Galkin (1978)

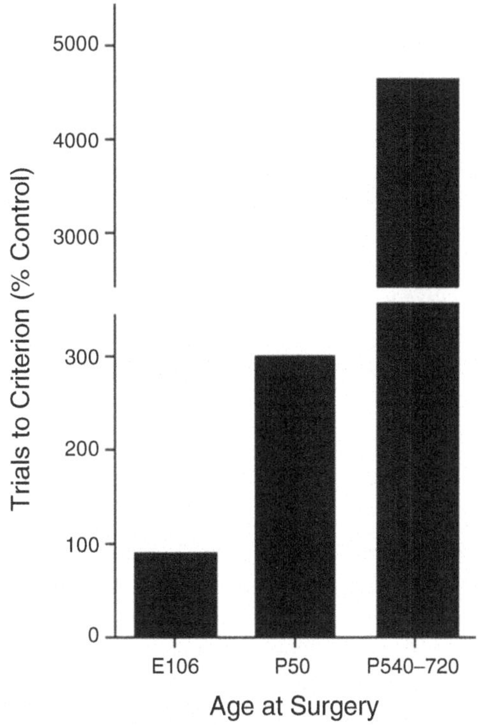

normally leading to cognitive deficits when cooled at about 3 years of age but no deficits, or only mild deficits, up until about 2 years of age. Thus, if a monkey was given a prefrontal lesion in infancy, and tested behaviorally at 2 years of age, there would be no deficit because the prefrontal cortex is not necessary to perform the task at 2 years. In contrast if the animal were tested at 3 years, there would be obvious impairment.

One problem with the Goldman-Rakic lesion studies was that the prefrontal lesions were made in relatively well-developed animals. To address this issue Goldman and Galkin (1978) performed a prenatal prefrontal lesion in a monkey, at a developmental time more like 1–2 year-old humans, and found dramatically better functional outcomes in this animal, suggesting that there may be prenatal sensitive period in monkey brain development in which the brain can compensate more effectively than after later injuries (see Fig. 2). Clearly, the developmental stage of brain development is an important variable in functional outcome after injury and there is at least one sensitive period for injury in which there can be remarkable recovery of function. Although Goldman's studies on monkeys with prefrontal lesions only examined cognitive functions, studies by Bachevalier's group (e.g., Bachevalier and Mishkin 1994; Malkova et al. 1997) have shown that whereas there is recovery of memory functions after neonatal temporal lesions in monkeys, there is no recovery of socioemotional behaviors (Malkova et al. 2010). The difference in the outcomes for cognitive and species-typical behaviors is important and true for other species, including humans, as well.

Cats are born much younger than monkeys so the sensitive period equivalent to the prenatal period in monkeys is likely to be postnatal. Villablanca et al. (1984, 1993a, b) made frontal lesions in newborn cats and showed results similar to the prenatal lesions in monkeys: the cats had good recovery of motor and cognitive functions later in life relative to similar lesions in adult cats (Olmstead and Villablanca 1979; Villablanca et al. 1978). But Villablanca et al. also made prenatal lesions in cats and unlike the monkey study, these animals were severely impaired relative to the cats with postnatal injuries (Villablanca et al. 1993a, b). This study is important because it shows that the sensitive period for good recovery is constrained to a specific developmental period.

3 Effects of Frontal Injuries in Young Rats

The advantage of studying rodents such as rats and mice is that they are born developmentally younger than cats, which allows a further investigation of the boundaries of the sensitive period(s) in brain development. We began investigating the effects of neonatal frontal lesions in rats by making medial prefrontal lesions (mPFC) at about 7 days of age (P7), which is during a period of active synaptogenesis (e.g., Kolb and Nonneman 1976, 1978). These animals were virtually indistinguishable from normal animals on cognitive and motor tasks sensitive to frontal lesions in adulthood. When we later investigated the effects of lesions as early as P1 the results were clear: lesions from P1-P5 resulted in a dismal outcome relative to the injuries at P7-P15, which again allowed good recovery on cognitive tests (e.g., Kolb 1987; Kolb and Whishaw 1981, 1985a, b; see also Nonneman et al. 1984; Vicedomini et al. 1982) (see Fig. 3). When we did more extensive behavioral testing, including motor and species-typical (e.g., nest building, food hoarding), we found that whereas the cognitive functions appeared normal in the P7–10 frontal rats, motor and species-typical behaviors showed much less recovery (e.g., Kolb and Whishaw 1981). These functions also showed no recovery in the P1-5 operates (Kolb 1987). The end of the postnatal sensitive period is not firmly established but appears to extend to about P15 for cognitive and motor tasks (Kolb et al. 1996; Kolb and Whishaw 1985b). But there may be an additional sensitive period during adolescence. Nemati and Kolb (2012) compared the effects of P35 and P55 mPFC lesions on various motor tasks and found good recovery at P35 but poor performance at P55, suggesting a sensitive period in early adolescence. Unfortunately, they did not study cognitive behavior.

When we examined the effects of prenatal lesions at embryonic day 18 (E18) in the presumptive prefrontal cortex the brains were grossly abnormal when examined in adulthood, but with the exception of a few animals that had hydrocephalus, the behavior of the E18 lesion animals was indistinguishable from controls on both cognitive and motor tests (Kolb et al. 1998a). This unexpected result suggests that

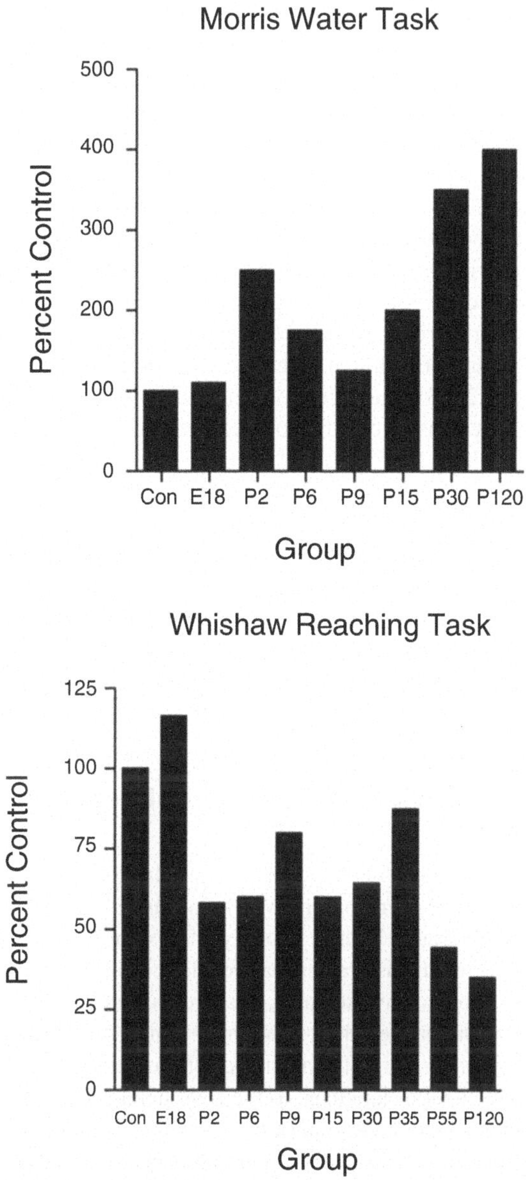

Fig. 3 Performance of adult rats with medial prefrontal lesions at various ages as a percent of control groups. The latency to find the hidden platform summed over 20 trials is illustrated for the Morris Water Task in which a low score reflects better performance. The success in reaching for food pellets through bars is illustrated for the Whishaw Reaching Task in which a high score reflects better performance. Data from Kolb (1987), Kolb et al. (1996, 1998b), Kolb and Whishaw (1985b)

there is an early sensitive period occurring during cortical neurogenesis, which begins in the cortex around E12. The finding of a prenatal sensitive period is consistent with an earlier demonstration by Sam Hicks (Hicks 1954; Hicks et al. 1984) in which he irradiated fetuses in utero on E12, which would kill the population of newly formed cortical neurons. These animals grew a new, albeit peculiar, brain after the irradiation-induced damage! And, like the Kolb et al. prenatal study, the motor behavior of Hicks' animals was nearly normal in spite of a weird brain.

In sum, taking the monkey, cat, and rat studies together there appear to be three sensitive periods for recovery from early brain injury: one during neurogenesis and another during synaptogenesis and a third during early adolescence. Similar injury during neuronal migration (~P1-P6 in the rat) produces a very poor functional outcome. The presence of three sensitive periods for recovery from brain injury suggests that these periods would also be plastic in response to other types of experiences and they are (see below). Although most of the monkey and cat studies were done with animals given frontal lesions, studies of rats with neonatal lesions of motor, parietal, orbital frontal, cingulate, temporal, or visual areas show a pattern of results that is similar to the results of the frontal cortex studies (e.g., Kolb 1995; Kolb et al. 2013), one exception being the posterior parietal lesions, in which P10, but not P7, lesions allowed good recovery of motor and cognitive functions (Kolb and Whishaw 1985b; Kolb 1987), suggesting that the sensitive period may be slightly later for posterior parietal cortex, which is slightly later to mature than more anterior cortex (Kolb 1987).

There have been few studies of the effect of unilateral focal cortical injuries at different developmental ages but rats with unilateral motor cortex lesions at P1, P10, or adulthood reveal a similar pattern of better recovery after P10 than P1 or adult lesions (Kolb et al. 2000). We have also seen a similar pattern of recovery in mice with bilateral mPFC or posterior parietal lesions (Calder et al. 2001).

4 Effects of Diffuse Cortical Lesions in Children

It is difficult to generalize the age-related sensitivity to brain injury from laboratory animals to humans for several reasons. First, the brain injuries in children are much more variable both in etiology and extent than in the laboratory animal studies and are often diffuse rather than focal. Second, whereas children with early brain injury are liable to have seizures, this rarely occurs in lab animals. Third, the human brain is more lateralized than nonhuman brains so the laterality of the injuries in children may have a significant impact on outcomes, as we saw earlier for language. The movement of language to the right hemisphere can significantly impact right hemisphere functions because of "crowding of functions" in the right hemisphere (e.g., Aram and Eisele 1994; Vargha-Khadem et al. 1985).

But the question remains as to whether there is sensitive period for good recovery in children. To address the question Anderson et al. (2009) divided 164 children with early brain injury into several age groups: congenital (first and second trimester), perinatal (third trimester to 1 month), infancy (2 months to 2 years), preschool (3–6 years), mid childhood (7–9 years), and late childhood (after 10 years). In order to provide sufficient children across different ages, the etiologies were diverse including stroke, contusion from falls, penetrating brain injury, tumor, malformation, dysplasias, cyst, and abscess. Nonetheless, there were some clear findings in

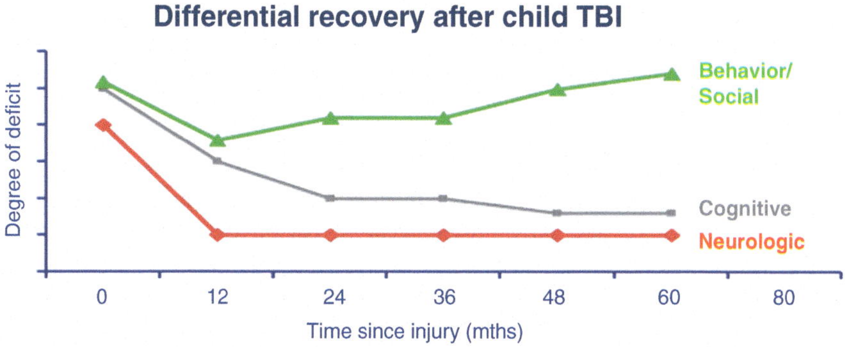

Fig. 4 Differential recovery on measures of neurologic, cognitive, and behavior/social behavior after child traumatic brain injury. Adapted from a figure courtesy of Vicki Anderson

behavioral assessments done in adolescence. Children sustaining an early brain injury before 2 years of age had severe cognitive deficits relative to children in the later groups. Children in the preschool group (3–6 years) had much better outcomes whereas those in the mid childhood group performed worse and more like the early group. Thus, this study is consistent with the increased vulnerability of the young brain followed by sensitive period for better recovery. With the limited data available for children, it is difficult to identify the duration of the sensitive period in early childhood, but the data do show that age and recovery are not linearly related but are associated with underlying developmental processes such as synaptogenesis, dendritic arborization, and myelination (Greenham et al. 2018). There is no evidence to date of an earlier sensitive period, which would likely be in the latter part of the first trimester and early in the second trimester when neurogenesis is still in progress. Although about half of the children in the congenital group had focal injuries the timing of these injuries would be difficult to identify exactly when the injuries occurred and whether or not they affected the subventricular zone, which could interfere with neurogenesis.

Finally, one additional finding from the Anderson group is that socioemotional behaviors show the worst outcomes regardless of age at injury (see Fig. 4). This result parallels similar findings in monkeys and rats described earlier.

5 Effects of Hemispherectomy in Children

During the course of treatment for severe and often life-threatening neurological disorders it has sometimes proven necessary to remove an entire cerebral hemisphere in children. For example, the standard treatment for Rasmussen's encephalitis, which is an inflammatory disease characterized by severe seizures originating in one hemisphere and producing a wide range of cognitive and motor symptoms, is to remove the affected hemisphere (Rasmussen 1983). And hemispherectomy is also used to treat other pre/perinatal disease such as Sturge-Weber syndrome or seizures resulting from a cerebral vascular accident. The age at surgery can vary widely from early childhood until adolescence with the age at onset of the disease being a major consideration. Liegeois et al. (2008) compared the cognitive outcomes in children with pre/perinatal vs postnatal disease onset with the general finding that those with pre/perinatal onset fare more poorly following hemispherectomy. These authors see this result as consistent with the idea that there is a cost with brain injury sustained before the emergence and early development of cognitive abilities, which was Hebb's conclusion regarding his children with early frontal lobe lesions discussed earlier.

One important feature of the outcome of hemispherectomy is that even with removal of the left hemisphere, most children are still capable of some language, which contrasts with the dense aphasia in adults with large left hemisphere injuries, and many children develop very good language skills, which allows some to graduate from university. One reason for this is that in many cases language has already partly shifted to the right hemisphere and removal of the dysfunctional hemisphere removes the inhibition on the right hemisphere, allowing the right hemisphere to support language.

But is there a sensitive period for recovery from hemispherectomy? Owing to the neurological disease that eventually leads to the surgery, this is a difficult question as the presurgical pathology alone produces significant cognitive symptoms (e.g., Tavares et al. 2020). Laboratory animal studies do not have this problem.

6 Effects of Hemispherectomy or Hemidecortication in Laboratory Animals

Maurice Ptito and his colleagues (e.g., Burke et al. 2010, 2012; Boire et al. 2001) developed a nonhuman primate model of hemispherectomy in vervet monkeys. In these studies one hemisphere was removed at 9 weeks or 4 years of age and the animals were studied intensively for 4 years. The results showed significant functional recovery of motor and visual functions in the early operates, although some motor functions took over 2 years to resolve. The behavioral recovery was correlated with extensive anatomical reorganization involving both the intact hemisphere and residual subcortical structures. These studies are important because they show that in the absence of pre-existing neurological dysfunction seen in children, there is an early sensitive period for recovery from hemispherectomy.

Jamie Villablanca and colleagues (e.g., Burgess and Villablanca 1986; Burgess et al. 1986; Hovda et al. 1990; Villablanca and Hovda 2000) performed hemispherectomies in 5- to 20-day old kittens and in adult cats. After waiting until the monkeys were at least 6 months of age, the authors measured a wide spectrum of motor, sensory, and cognitive tests. The early-lesioned cats showed remarkable recovery on 23 of 25 behavioral tests, and the authors note that on casual inspection the animals were indistinguishable from their littermates. As in the case of the Ptito monkeys, postmortem anatomical analyses revealed extensive morphological changes in the early operates relative to the adults. Villablanca et al. (1993b) also tried to produce viable fetal-hemispherectomized animals but they were all stillborn. When they made incomplete lesions many animals survived, but just as in the case of prenatal frontal lesions in cats, their performance on behavioral tests was significantly worse than kittens with hemispherectomies on postnatal day 10. Thus, the sensitive period for recovery again appears to be postnatal in the cats, although the postnatal boundary of this period was not determined.

The first studies of newborn hemidecortication in rats were done by Hicks and D'Amato (1970, 1975). They removed the neocortex of one hemisphere at P0 (12–28 h old) or in adulthood and showed that although the early operates showed motor impairments, they were significantly better than the adult operates. My colleagues and I later varied the age of hemidecortication (P1, P5, P10, adult) and found that although hemidecortication at all ages produced deficits in motor and cognitive behaviors at all ages, the deficits were much smaller in the neonatal operates. To our surprise, however, earlier was better: hemidecortication at P1 allowed the best recovery (Kolb and Tomie 1988; Whishaw and Kolb 1988). As in the monkey and cat studies, the neonatal hemidecortications resulted in extensive morphological changes (Kolb et al. 1992) as the intact hemisphere had increased cortical thickness, enhanced dendritic arborization in cortical pyramidal neurons in motor and parietal cortex, and changes in cortical connectivity of the prefrontal and sensorimotor regions, with the effects being largest in the P1 animals. One obvious difference between the effects of bilateral focal lesions versus hemidecortication is that the brains of animals with hemidecortications have an intact hemisphere, which has extensive anatomical reorganization. We later showed that even a slight perturbation (a stab wound) of the "intact" hemisphere at the time of hemidecortication completely blocked the recovery from P1 hemidecortications (Kolb 1995). Animals with just a stab wound were indistinguishable from control animals. The age-dependent effects of hemidecortication need not impugn the sensitive periods theory proposed earlier, however. If we assume that the intact hemisphere is especially plastic around P10, the advantage of the P1 lesion may be that at P10 the shock (diaschisis) following the loss of one hemisphere has resolved and there is an opportunity for extensive anatomical changes in an intact hemisphere. Such changes after P10 lesions may be reduced as the sudden removal of callosal connections is likely to produce significant diaschisis, which would interfere with the plastic changes occurring shortly after the P10 lesion.

7 Adolescence and Sensitive Periods for Recovery

The postnatal sensitive period in rats begins to decline by about by P15, but the question arises as to when the plastic changes occur after P7-P15 lesions. There is strong evidence that the effect of early experiences such as drug exposure or stress often does not show up until adolescence (e.g., Andersen 2016; 2018; Tottenham 2020), which is an extremely plastic period. To investigate the possibility that plastic changes following P10 lesions are delayed until adolescence, Kolb and Gibb (1993) removed the medial prefrontal cortex at P1 or P10 and trained the animals in the Morris Water task at either P21 or P56. Rats with lesions at either age were severely impaired at P21 but by P56, animals with P10 lesions did not differ from controls. Anatomical analyses showed no differences in pyramidal neurons in adjacent sensorimotor cortex at P22 but by P56 there was hypertrophy of dendritic length, branching, and spine density in the P10 group but atrophy in the P1 group. Thus, the plastic period for the plasticity supporting recovery appears to be sometime in adolescence, although the exact timing has not been demonstrated. A subsequent study by de Brabander and Kolb (1997) showed that the dendritic hypertrophy in P10 animals was not present at P25 but it was present at P35. Thus, the plastic period for recovery appears to be sometime in adolescence although this may vary depending upon when and where the injury is. Thus, the plastic period for recovery appears to be sometime in adolescence although this may vary depending upon when and where the injury is. One remaining question, however, is why it is only the P7-15 age that allows for the morphological plasticity to occur later.

Further evidence for a sensitive period in adolescence comes from a study by Nemati and Kolb (2010) that compared the effect of unilateral motor cortex lesions at P35 and P55, reflecting injury early in adolescence versus late in adolescence. Our expectation was that because P35 was early in adolescence there would be a larger plastic response, and possibly a sensitive period, around P35 but it would be over by P55, just as we saw in rats with mPFC lesions. In fact, we found the opposite: the P35 animals were as impaired as P1 animals whereas the P55 fared somewhat better, although they still had some impairments. The behavioral recovery in the P55 group was correlated with enhanced dendritic branching in adjacent sensorimotor cortex, whereas the P35 brains showed reduced dendritic branching and reduced spine density. It would appear that there may be a sensitive period in adolescence but that the actual timing might vary depending upon the region damaged.

Although there was no recovery after P35 lesions in the Nemati and Kolb (2010) study, it is possible to stimulate plasticity after injury at that age. Nemati and Kolb (2011) gave P35 rats subcutaneous injections of Fibroblast Growth Factor-2 (FGF-2) or vehicle and at P65 the FGF-treated lesion rats had virtually total recovery, which was correlated with increased dendritic length and spine density in adjacent pyramidal neurons. The FGF-2 effect suggests that plasticity can be turned on in adolescence although we do not know when the plastic changes actually occurs after P35 lesions.

Finally, one plastic change that has not yet been studied in adolescent animals with brain injury is neurogenesis. Using the isotropic fractionator method, Mortera and Herculano-Houzel (2012) found that during adolescence there is a major increase in the number of cortical neurons that peaks between 1 and 2 months of age, followed by significant neuronal loss by 3 months. Thus, it is possible that there may be neurogenesis in adolescence that is stimulated by the early brain injuries. And, the FGF-2 effect in the P35 rats could be related to increased neurogenesis as we have shown FGF-2 stimulated neurogenesis in P10 rats (e.g., Monfils et al. 2005) (see below), although the lesion cavities did not show any filling in P35 rats.

8 Other Examples of Extensive Plasticity in the Sensitive Periods for Brain Injury

In the course of studying the effects of P10 lesions on brain morphology we were able to demonstrate that there was spontaneous neurogenesis following midline telencephalic lesions including the olfactory bulb, medial frontal cortex, and posterior cingulate cortex (see Fig. 5) (Kolb et al. 1998b; Gonzalez et al. 2002), but more

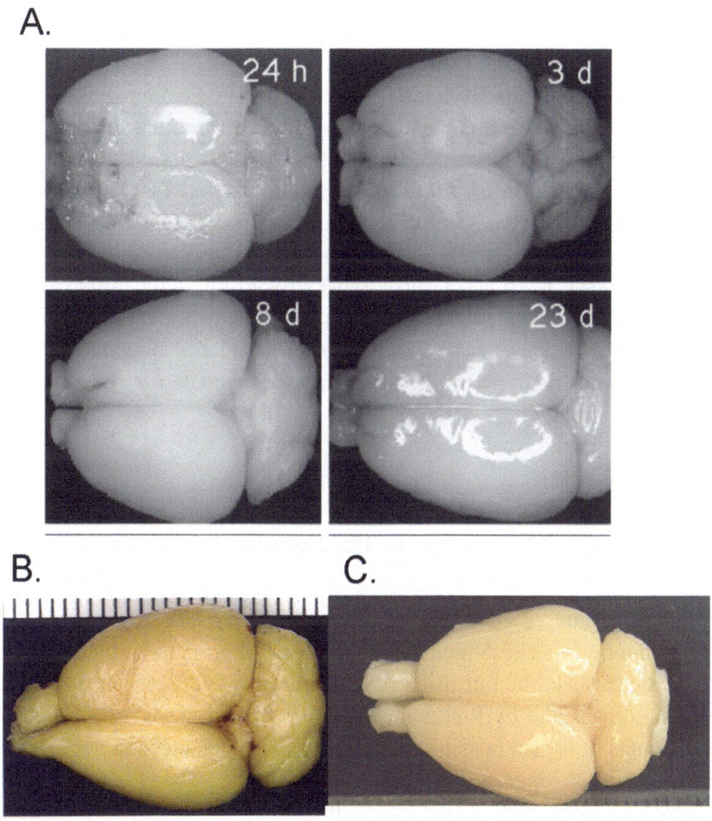

Fig. 5 Photographs of brains of rats with medial prefrontal lesions (mPFC) (**a**) or olfactory bulbectomy at P1 (**b**) or P10 (**c**). Brains from littermate rats with mPFC lesions were collected at different ages to illustrate the regeneration of the mPFC region (Adapted from Kolb et al. 1998b). Brains from the littermate rats with olfactory bulb removals were collected in adulthood. Note that the left hemisphere of the P1 olfactory bulbectomy is visibly smaller than the right hemisphere. (Adapted from Goldsbury et al. 2006)

lateral lesions such as motor or parietal cortex, do not result in regeneration. Driscoll et al. (2007) examined the neurophysiological properties of the neurons in the regenerated frontal cortex and found them to be essentially normal. We further showed that the P10 regeneration and behavioral recovery could be blocked by embryonic injections of Bromodeoxyuridine (BrdU) (Kolb et al. 1998b), which we had accidently discovered altered stem cell activity when given prenatally but not postnatally (Kolb et al. 1999).

P10 motor cortex lesions do not induce spontaneous neurogenesis but administration of FGF-2 following P10 motor cortex lesions, but not P3 lesions, does (e.g., Monfils et al. 2005). This neurogenesis of motor cortex is associated with virtually complete recovery of motor functions, which is supported by the regeneration of corticospinal projections and functions (Monfils et al. 2008) (Fig. 6). Figure 6 also shows the morphology of the regrown cortex, which is thinner than normal and lacks normal cortical lamination. Like the spontaneous regeneration of medial frontal cortex, the FGF-2 stimulated neurogenesis is blocked by embryonic injections of BrdU.

In view of the Nemati and Kolb (2011) finding that FGF-2 can stimulate recovery after P35 lesions, it is reasonable to ask if FGF-2, or other treatments, might stimulate recovery if administered during the sensitive period around P10 after lesions at P1-5. Comeau et al. (2007, 2008) administered FGF-2 after P3 mPFC lesions and found recovery of cognitive functions when the animals were tested as adults. FGF-2 can also be increased endogenously by tactile stimulation in young animals (Gibb 2004). Kolb and Gibb (2010) stroked rats with mPFC or posterior parietal lesions at P3 by using a small brush for 15 min three times daily for 2 weeks following the surgery. There was virtually complete functional recovery, which was correlated with dendritic hypertrophy and increased spine density in adjacent cortex in adulthood (Kolb and Gibb 2010).

9 Mechanisms of Plasticity in the Sensitive Periods

There are several putative mechanisms of the sensitive periods for recovery (or not) from early brain injury. These include: (1) rewiring of cortical inputs and outputs as well as modifications to the intrinsic cortical connectivity; (2) changes in functional connectivity as measured by functional MRI; (3) neurogenesis; and (4) changes in perineuronal nets. It is also possible that injury triggers changes in gene expression, although to my knowledge this has only been studied extensively after traumatic brain injury (e.g., Hehar et al. 2017; Rao et al. 2002; Zamani et al. 2020) and there is no systematic investigation of change in gene expression after early brain injuries. It is quite likely, however, that there will be changes in gene expression related to the onset of sensitive periods. Many, or even all, of the putative mechanisms are at play in the sensitive periods, the details depending upon the precise age at injury and the precise locus of injury.

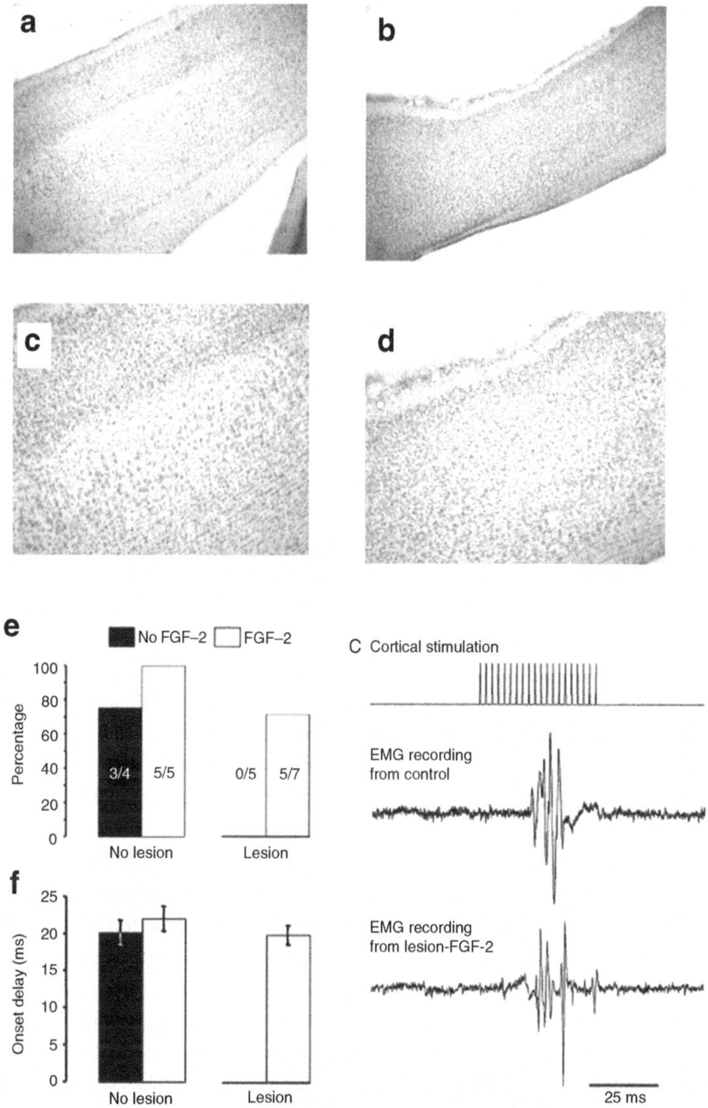

Fig. 6 Top. Photomicrographs of coronal sections through brains from a control (**A**, **C**) and P10 + FGF-2 motor cortex lesion (**B**, **D**). The regenerated cortex of the lesion rats is visibly thinner and lacks the normal lamination. Bottom. EMG recordings from wrist extensors following cortical stimulation in P10 motor cortex lesion rats. (**A**) Number of rats from each group in which EMG activity could be recorded from wrist extensors. (**B**) Delay of movement onset in the rats that in which EMGs were recorded. (**C**) Representative EMG recordings. (Adapted from Monfils et al. 2008)

9.1 Rewiring of Cortical Inputs and Outputs and Alterations in Neuronal Morphology

Margaret Kennard proposed that the recovery she observed in her young monkeys with motor cortex injury was related to some type of reparative process in the developing brain. The first studies looking for such processes were done by Schneider (1973) who damaged the superior colliculus unilaterally in newborn hamsters, which caused striking anomalies in the distribution of the optic tract as it terminated in the contralateral lateral superior colliculus and the ipsilateral thalamic nucleus lateralis posterior. These anomalous connections did not support normal behavior, however, as the animals responded to visual stimulation in the contralateral visual field as if the stimulation was in the ipsilateral field. Nonetheless, the study was a proof of principle that the brain could rewire significantly after early injury. Following up on Schneider's work, Cornwell et al. (1978) showed that kittens with lesions on P3 had partial sparing of visual functions and in later studies showed that this resulted from system-wide changes in neural circuitry that included the retina, thalamus, midbrain, and extrastriate cortex (Payne and Cornwell 1994). These results paralleled the results by Villablanca et al. (1993b) discussed earlier showing widespread rewiring in kittens with frontal lesions in the neonatal period.

Based on our demonstration that the outcome of P1 and P10 cortical lesions in rats was so different, we predicted that rats with P10 lesions would show similar extensive rewiring whereas those with P1 lesions would not. We were wrong (Kolb et al. 1994). In fact, the P1 rats had extensive abnormalities in thalamo-cortical, amygdalo-cortical, and nigro-cortical connections whereas the P10 rats had no obvious abnormalities in those long connections. But the effects of P1 lesions did not result from rewiring but rather a failure to prune abnormal connections as the same pattern of connectivity was present in 4-day-old normal rats. These anomalous projections were not pruned in the P1 operates and the anomalous connections acted to interfere with functional recovery. Thus, the functional recovery in the P10 animals did not result from extensive rewiring of long connections. Rather, it resulted in part from changes in dendritic organization of cortical pyramidal neurons: adult rats with P10 frontal lesions showed enhanced dendritic branching, dendritic length, and spine density whereas the P1 rats showed atrophy on those measures, as we discussed earlier. It is possible that Schneider's anomalous connections in the hamster would not have occurred with later lesions and like our rats, the anomalous connections may have been at least partly a result of a failure to prune aberrant connections.

9.2 Functional Connectivity

There may also be changes in functional connectivity without obvious widespread changes on anatomical connectivity. Li et al. (2021) used resting-state functional

MRI (fMRI) to examine functional connectivity in the brains of rhesus monkeys with neonatal hippocampal lesions. The results showed that the functional connectivity was significantly altered, especially with respect to connections with the dorsolateral prefrontal cortex. The connectivity changes were not observed in monkeys with adult lesions. The extent of change was correlated with the extent of deficits in tests of working memory. Although they did not examine whether changes occur after neocortical lesions, it is likely that cortical insult during development would also alter resting-state fMRI. One advantage of using rs-fMRI is that it is not invasive and could be used repeatedly to track the timeline of the network changes following early brain injury. Although the rat brain is small, Bajic et al. (2017) showed that rs-fMRI can be used with rats to identify large-scale networks, thus opening the door for future studies using this method in laboratory animals to look for sensitive periods (see also Pan et al. 2018; Whitesell et al. 2021). The earliest age that this method could be used developmentally remains to be determined.

9.3 Neurogenesis

Another mechanism discussed earlier is neurogenesis (see Bandeira et al. 2009; Mortera and Herculano-Houzel 2012). Thus, we have seen spontaneous neurogenesis after P10 lesions of midline cortex or olfactory bulb and FGF-2 induced neurogenesis following P10 motor cortex injury. It is important to note that there is some spontaneous neurogenesis in the injured adult brain (e.g., Altman 1962; see review by Altman 2011) but new neocortical neurons appear to be relatively rare after cortical injury. It is possible to stimulate large numbers of new neurons in adulthood following cortical injury, however, by intraventricular infusions of Epidermal Growth Factor (EGF) and erythropoietin together, but the new neurons do not integrate into existing cortical circuits (Kolb et al. 2007). There is clearly something special about the neurogenesis in the postnatal sensitive period.

There are likely several mechanisms in play in the early postnatal period neurogenesis. We have shown, for example, that precursor cells in the subventricular zone are the source of many of the new neurons that migrate to the site of injury (Kolb et al. 1998b). The new neurons appear to migrate only to the site of the midline injury and if the lesion is unilateral they are only mobilized in the ipsilateral hemisphere. These observations suggest that there is some sort of signal generated after P7-P12 lesions that attracts the new cells to the injury site. Given that there are precursor cells in the subventricular zone in the first week of life, and after the sensitive period, this putative signal either must only be present during the sensitive period or the precursor cells are only responsive to it during that period.

Another putative mechanism of the injury-stimulated neurogenesis is related to astrocytes in the cortex or striatum adjacent to lesions that may convert into neurons. This astrocyte-neuron conversion has not been shown after injury in the infant brain but it has been shown after adult lesions (e.g., Guo et al. 2014; Niu et al. 2013; Wu et al. 2020). In addition, we have seen increased production of astrocytes both in the regenerating tissue and in the proximal intact cortex (B. Kolb and R. Gibb, unpublished observations), so it is quite likely that the conversion could be occurring in the infant brain.

9.4 Perineuronal Nets

Sensitive period plasticity may be related to the relationship between neurons and perineuronal nets (PNNs) (Drzewieki et al. 2020; Reichelt et al. 2019). In the cortex PNNs surround inhibitory parvalbumin+ (PV+) neurons that are implicated in sensitive period plasticity in sensory cortex (e.g., Takesian and Hensch 2013). As these neurons develop they are surrounded by PNNs, which provide scaffolding for the cells and in addition, the PNNs play a significant role in neurogenesis, synaptogenesis, and synaptic and homeostatic plasticity (e.g., Dityatev et al. 2014). The initiation of the sensitive period in sensory system plasticity requires the maturation of specific inhibitory circuitry involving the PV+ neurons, which regulates the timing of the sensitive periods through the generation of an optimal ratio of excitatory and inhibitory (E/I) circuit activity in the PV+ circuits. This E/I balance triggers a sequence of molecular events, including second messenger molecules such as CREB, CaMKII, ERK, TNFa, which induce structural changes such as spine pruning, regrowth, and axonal rewiring (Takesian and Hensch 2013) (see

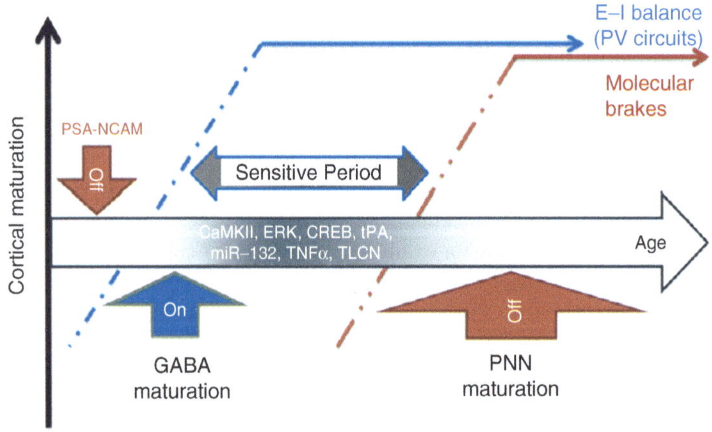

Fig. 7 Putative mechanism controlling onset and closure of sensitive periods. Precocious plasticity is blocked by factors such as polysialic acid (PSA) acting on neural cell adhesion molecule (NCAM). The sensitive period onset is triggered once factors promote GABA and PV cell maturation, which leads to an optimal excitatory/inhibitory balance. This stimulates a sequence of molecular events, which induce structural changes such as spine pruning, regrowth, and axonal wiring. The sensitive period closes as molecular brakes such as PNNs dampen plasticity. (Adapted from Takesian and Hensch 2013)

Fig. 7). The sensitive period closes as molecular brakes related to the PNNs emerge to dampen plasticity and put brakes on plasticity to end the sensitive period (e.g., Hensch 2018). The E/I balance can be stimulated to develop earlier by enhancing GABAergic activity, such as by exposure to GABAergic drugs (e.g., prenatal benzodiazepines) (e.g., Reh et al. 2020; Weikum et al. 2012; Werker and Hensch 2015). Just as exposure to GABAergic drugs can shift the sensitive period earlier in sensory systems, such GABAergic drug exposure also shifts the sensitive period for recovery from mPFC injury earlier as well (Kolb et al. 2008), although it does not lead to neurogenesis. Thus, the timing of the injury-related plasticity in the postnatal sensitive period may be related to the PNNs in the cortex. It is also worth noting here that removal of PNNs can enhance plasticity so it is possible that during the sensitive period an injury could slow down the development of the molecular brakes related to PNNs. In contrast, injury before the sensitive period may slow down the initial development of the PNNs, which could reduce plasticity.

Piekarski et al. (2017) note that estrogen receptors in the frontal cortex are specifically localized to PV+ interneurons in both males and females. The rise of estrogens or androgens, which are converted to estradiol via aromatase enzyme present in the frontal cortex, can act as a transcription factor to alter gene expression that can regulate the physiological properties of the PV+ interneurons and change the network inhibition in adolescence. This would provide a mechanism for a sensitive period for plasticity after injury in adolescence. Neonatal gonadectomy in both sexes blocks recovery from P10 mPFC lesions (Kolb and Stewart 1995). Recall that the plastic changes in neuronal morphology after P10 lesions do not occur until adolescence and thus could be related to the effect of the production of gonadal hormones. The intensity of PNNs in prefrontal cortex has been found to increase across adolescence between P30 and P60 (Drzewieki et al. 2020) in both male and female rats. This would correlate with the closing of the sensitive period during adolescence.

10 Conclusions

The developing injured brain shows a remarkable plasticity for plastic change, which occurs especially during sensitive periods in development. Looking across laboratory animal species (monkeys, cats, rats, mice) there is a sensitive period for recovery after neuronal migration in the cortex ends and rapid synaptogenesis begins. Studies on rats have also shown a sensitive period sometime between E12 and E19 and possibly another one during adolescence. A variety of mechanisms are proposed including changes in neuronal morphology and local connectivity, neurogenesis and astrocytosis, and perineuronal nets. All of these are likely related to injury-induced changes in gene expression.

References

Altman J (1962) Are new neurons formed in the brains of adult mammals? Science 135:1127–1128

Altman J (2011) Memoir: the discovery of adult mammalian neurogenesis. In Seki T (ed) Neurogenesis in the adult brain I: neurobiology. Springer, New York, p 1. https://doi.org/10.1007/978-4-431-53933-9_1

Andersen SL (2003) Trajectories of brain development: point of vulnerability or window of opportunity? Neurosci Biobehav Rev 27:3–18

Andersen SL (2016) Commentary on the special issue on the adolescent brain: adolescence, trajectories, and the importance of prevention. Neurosci Biobehav Rev 70:329–333

Andersen SL (2018) Stress, sensitive periods, and substance abuse. Neurobiol Stress 10:100140. https://doi.org/10.1016/j.ynstr.2018.100140

Anderson V, Spencer-Smith M, Leventer R, Coleman L, Anderson P, Williams J, Greenham M, Jacobs R (2009) Childhood brain insult: can age at insult help us predict outcome? Brain 132: 45–56

Aram DM, Eisele JA (1994) Intellectual stability in children with unilateral brain lesions. Neuropsychologia 32:85–95

Bachevalier J, Mishkin M (1994) Effects of selective neonatal temporal lobe lesions on visual recognition memory in rhesus monkeys. J Neurosci 14:2128–2139

Bajic D, Craig MM, Mongerson CRL, Borsook D, Becerra L (2017) Identifying rodent resting-state brain component networks with independent component analysis. Front Neurosci. https://doi.org/10.3389/fnins.2017.00685

Bandeira F, Lent R, Herculano-Houzel S (2009) Changing numbers of neuronal and non-neuronal cells underlie postnatal brain growth in the rat. Proc Natl Acad Sci 106:14108–14113

Banich MT, Cohen-Levine S, Kim H, Huttenlocher P (1990) The effects of developmental factors on IQ in hemiplegic children. Neuropsychologia 28:35–47

Berry MJ (1974) Development of the cerebral neocortex of the rat. In: Gottlieb G (ed) Aspects of neurogenesis. Academic Press, New York, pp 7–67

Boire D, Theoret H, Ptito M (2001) Visual pathways following cerebral hemispherectomy. Prog Brain Res 134:379–397

Burgess JW, Villablanca JR (1986) Recovery of function after neonatal or adult hemispherectomy in cats. I. Motor deficits, limb bias and development, paw usage, locomotion and rehabilitative effects of motor training. Behav Brain Res 20:1–18

Burgess JW, Villablanca JR, Levine MS (1986) Recovery of function after neonatal or adult hemispherectomy in cats. II. Complex functions: open field exploration, social interactions, maze and holeboard performance. Behav Brain Res 20:177–230

Burke MW, Zangenehpour S, Ptito M (2010) Partial recovery of hemiparesis following hemispherectomy in infant monkeys. Neurosci Lett 469:243–247.wqasdfx

Burke MW, Kupers R, Ptito M (2012) Adaptive neuroplastic responses in early and late hemispherectomied monkeys. Neural Plast. https://doi.org/10.1155/2012/852423

Calder S, Hastings E, Kindt R, Gibb R, Kolb B (2001) The pattern of functional recovery and morphological change after neonatal prefrontal or posterior parietal cortex lesions is similar in mice and rats. Soc Neurosci Abstr 27(476):5

Comeau W, Hastings E, Kolb B (2007) Differential effect of pre and postnatal FGF-2 following medial prefrontal cortical injury. Behav Brain Res 180:18–27

Comeau W, Gibb R, Hastings E, Cioe J, Kolb B (2008) Therapeutic effects of complex rearing or bFGF after perinatal frontal lesions. Dev Psychobiol 50:134–146

Cornwell P, Overman W, Ross C (1978) Extent of recovery from neonatal damage to the cortical visual system of cats. J Comp Physiol Psychol 92:255–270

de Brabander J, Kolb B (1997) Development of pyramidal cells in medial frontal cortex following neonatal lesions of anterior midline cortex. *Restor Neurol Neurosci* 11:91–97

Dityatev A, Wehrle-Hadller B, Pitkanen A (2014) Preface. Brain extracellular matrix in health and disease. Prog Brain Res 214:xiii-xvii

Driscoll I, Monfils M-H, Flynn C, Teskey GC, Kolb B (2007) Neurophysiological properties of cells filling the neonatal medial prefrontal cortex lesion cavity. Brain Res 1178:1209–1218
Drzewieki CM, Willing J, Juraska JM (2020) Influences of age and pubertal status on number and intensity of perineuronal nets in the rat medial prefrontal cortex. Brain Struct Funct 225:2495–2507
Gibb R (2004) Perinatal experience and recovery from brain injury. Unpublished PhD thesis. University of Lethbridge
Goldman PS (1974) An alternative to developmental plasticity: heterology of CNS structures in infants and adults. In: Stein DG, Rosen JJ, Butters N (eds) Plasticity and recovery of function in the central nervous system. Academic, New York, pp 149–174
Goldman PS, Galkin TW (1978) Prenatal removal of frontal association cortex in the fetal rhesus monkey: anatomical and functional consequences in postnatal life. Brain Res 152:451–485
Goldman-Rakic PS, Isseroff AM, Schwartz ML, Bugbee NM (1983) The neurobiology of cognitive decelopment. In: Mussen PH (ed) Handbook of child psychology: biology and infancy development. Wiley, New York, pp 311–344
Goldsbury RP, Pearce SM, Gibb R, Kolb B (2006) Unilateral neonatal bulbectomy alters motor behaviour and overall cortical organization in a sexually dimorphic manner. Soc Neurosci Abstr 620:1
Gonzalez CLR, Gibb R, Kolb B (2002) Functional recovery and dendritic hypertrophy after posterior and complete cingulate lesions on postnatal day 10. Dev Psychobiol 40:138–146
Greenham M, Ryan NP, Anderson V (2018) Injury. In: Gibb R, Kolb B (eds) The neurobiology of brain and behavioral development. Elsevier, New York, pp 413–438
Guo ZY, Zhang L, Wu Z, Chen UY, Wang F, Chen G (2014) In vivo direct reprogramming of reactive glial cells into functional neurons after brain injury and in an Alzheimer's disease model. Cell Stem Cell 14:188–202
Harlow HF, Akert K, Schlitz KA (1964) The effects of bilateral lesions of prefrontal cortex on the learned behavior of neonatal, infant, and preadolescent monkeys. In: Warren JM, Akert K (eds) The frontal granular cortex and behavior. McGraw-Hill, New York, pp 126–148
Hebb DO (1949) The organization of behavior. McGraw-Hill, New York
Hehar H, Ma I, Mychasiuk R (2017) Intergenerational transmission of paternal epigenetic marks: mechanisms influencing susceptibility to post-concussion symptomology in a rodent model. Sci Rep 7:7171. https://doi.org/10.1038/s41598-017-07784-7
Hensch TK (2018) Critical periods in brain development. In: Gibb R, Kolb B (eds) The neurobiology of brain and behavioral development. Elsevier, New York, pp 133–152
Hicks SP (1954) Mechanisms of radiation anencephaly, anopthalamia, and pituitary anomalies. Repair in the mammalian embryo. Arch Pathol 54:363–378
Hicks S, D'Amato C (1968) Cell migrations to the isocortex of the rat. Anat Rec 160:619–634
Hicks S, D'Amato C (1970) Motor-sensory and visual behavior after hemispherectomy in newborn and mature rats. Exp Neurol 29:416–438
Hicks S, D'Amato C (1975) Motor-sensory cortex corticospinal system and developing locomotion and placing in rats. Am J Anat 143:1–42
Hicks S, D'Amato C, Glover RA (1984) Recovery or malformation after fetal radiation and other injuries. In: Finger S, Almli CR (eds) Early brain damage, vol 1. Academic, New York, pp 127–147
Hovda DA, Villablanca JR, Chugani HT, Phelps ME (1990) Cerebral metabolism following adult or neonatal hemineodecortication in cats: I. Effects on glucose metabolism using [14C]2-deoxy-d-glucose autoradiography. J Cereb Blood Flow Metab 16:134–146
Ignacio MP, Kimm EJ, Kageyama GH, Yu J, Robertson RT (1995) Postnatal migration of neurons and formation of laminae in rat cerebral cortex. Anat Embryol 191(2):89–100
Johnson DA, Rose FD, Brooks BM, Eyers S (2003) Age and recovery from brain injury: legal opinions, clinical beliefs and experimental evidence. Pediatr Rehabil 6:103–109
Jordan CJ, Andersen SL (2017) Sensitive periods of substance abuse: early risk for the transition to dependence. Dev Cogn Neurosci 25:29–44

Kennard M (1938) Reorganization of motor function in the cerebral cortex of monkeys deprived of motor and premotor areas in infancy. J Neurophysiol 1:477–496
Kennard M (1940) Relation of age to motor impairment in man and in subhuman primates. Arch Neurol Psychiatry 44:377–397
Kolb B (1987) Recovery from early cortical damage in rats. I. Differential behavioral and anatomical effects of frontal lesions at different ages of neural maturation. Behav Brain Res 25:205–220
Kolb B (1995) Brain plasticity and behavior. Lawrence Erlbaum Associates, Mahwah
Kolb B, Gibb R (1993) Possible anatomical basis of recovery of spatial learning after neonatal prefrontal lesions in rats. Behav Neurosci 107:799–811
Kolb B, Gibb R (2010) Tactile stimulation facilitates functional recovery and dendritic change after neonatal medial frontal or posterior parietal lesions in rats. Behav Brain Res 214:115–120
Kolb B, Nonneman AJ (1976) Functional development of prefrontal cortex in rats continues into adolescence. Science 193(4250):335–336
Kolb B, Nonneman AJ (1978) Sparing of function in rats with early prefrontal cortex lesions. Brain Res 151:135–148
Kolb B, Stewart J (1995) Changes in neonatal gonadal hormonal environment prevent behavioral sparing and alter cortical morphogenesis after early frontal cortex lesions in male and female rats. Behav Neurosci 109:285–294
Kolb B, Tomie J (1988) Recovery from early cortical damage in rats. IV. Effects of hemidecortication at 1, 5, or 10 days of age. Behav Brain Res 28:259–274
Kolb B, Whishaw IQ (1981) Neonatal frontal lesions in the rat: sparing of learned but not species-typical behavior in the presence of reduced brain weight and cortical thickness. J Comp Physiol Psychol 95:863–879
Kolb B, Whishaw IQ (1985a) Neonatal frontal lesions in hamsters impair species-typical behaviors and reduce brain weight and cortical thickness. Behav Neurosci 99:691–704
Kolb B, Whishaw IQ (1985b) Earlier is not always better: behavioural dysfunction and abnormal cerebral morphogenesis following neonatal cortical lesions in the rat. Behav Brain Res 17:25–43
Kolb B, Gibb R, van der Kooy D (1992) Neonatal hemidecortication alters cortical and striatal structure and connectivity. J Comp Neurol 322:311–324
Kolb B, Gibb R, van der Kooy D (1994) Neonatal frontal cortical lesions in rats alter cortical structure and connectivity. Brain Res 645:85–97
Kolb B, Petrie B, Cioe J (1996) Recovery from early cortical damage in rats. VII. Comparison of the behavioural and anatomical effects of medial prefrontal lesions at different ages of neural maturation. Behav Brain Res 79:1–13
Kolb B, Cioe J, Muirhead D (1998a) Cerebral morphology and functional sparing after prenatal frontal cortex lesions in rats. Behav Brain Res 91:143–155
Kolb B, Gibb R, Gorny G, Whishaw IQ (1998b) Possible brain regrowth after cortical lesions in rats. Behav Brain Res 91:127–141
Kolb B, Pedersen B, Ballerman M, Gibb R, Whishaw IQ (1999) Embryonic exposure to BrdU produces chronic changes in brain and behavior *in rats*. J Neurosci 19:2337–2346
Kolb B, Cioe J, Whishaw IQ (2000) Is there an optimal age for recovery from motor cortex lesions? I. Behavioural and anatomical sequelae of bilateral motor cortex lesions in rats on postnatal days 1, 10, and in adulthood. Brain Res 882:62–74
Kolb B, Morshead C, Gonzalez C, Kim N, Shingo T, Weiss S (2007) Growth factor-stimulated generation of new cortical tissue and functional recovery after stroke damage to the motor cortex of rats. J Cereb Blood Flow Metab 27:983–997
Kolb B, Gibb R, Pearce S, Tanguay R (2008) Prenatal exposure to prescription medications alters recovery following early brain injury in rats. Soc Neurosci Abstr 349:5
Kolb B, Mychasiuk R, Muhammad A, Gibb R (2013) Brain plasticity in the developing brain. Prog Brain Res 207:35–64
Lennenberg E (1967) Biological foundations of language. Wiley, New York

Li CX, Li Z, Hu X, Zhang X, Bachevalier J (2021) Altered hippocampal-prefrontal functional network integrity in adult macaque monkeys with neonatal hippocampal lesions. Neuroimage 227:117645

Liegeois F, Cross JH, Polkey C, Harkness W, Vargha-Khadem F (2008) Language after hemispherectomy in childhood: contributions from memory and intelligence. Neuropsychologia 46: 3101–3107

Malkova L, Mishkin M, Suomi SJ, Bachevalier J (1997) Socioemotional behavior in adult rhesus monkeys after early versus late lesions of the medial temporal lobe. Ann N Y Acad Sci 807:538–540

Malkova L, Mishkin M, Suomi SJ, Bachevalier J (2010) Long-term effects of neonatal medial temporal ablations on socioemotional behavior in monkeys (*Macaca mulatta*). Behav Neurosci 124:742–760

Monfils MH, Driscoll I, Vandenberg PM, Thomas NJ, Danka D, Kleim JA, Kolb B (2005) Basic fibroblast growth factor stimulates functional recovery after neonatal lesions of motor cortex in rats. Neuroscience 134:1–8

Monfils M-H, Driscoll I, Vavrek R, Kolb B, Fouad K (2008) FGF-2 induced functional improvement from neonatal motor cortex injury via corticospinal projections. Exp Brain Res 185:453–460

Mortera P, Herculano-Houzel S (2012) Age-related neuronal loss in the rat brain starts at the end of adolescence. Front Neuroanat. https://doi.org/10.3389/fnana.2012.00045

Nemati F, Kolb B (2010) Motor cortex injury has different behavioral and anatomical effects in juvenile and adolescent rats. Behav Neurosci 24:612–622

Nemati F, Kolb B (2011) FGF-2 induces behavioral recovery after early adolescent injury to the motor cortex of rats. Behav Brain Res 225:184–191

Nemati F, Kolb B (2012) Recovery from medial prefrontal cortex injury during adolescence: implications for age-dependent plasticity. Behav Brain Res 229:168–175

Niu W, Zang T, Zou Y, Fang S, Smith DK, Bachoo R, Zhang C-L (2013) In vivo reprogramming of astrocytes to neuroblasts in the adult brain. Nat Cell Biol 15:1164–1175

Nonneman AJ, Corwin JV, Sahley CL, Vicedomini JP (1984) Functional development of the prefrontal system. In: Finger S, Almli CR (eds) Early brain damage, vol 2. Academic, New York, pp 139–153

Oberman L, Pascual-Leone A (2013) Changes in plasticity across the lifespan: cause of disease and target for intervention. Prog Brain Res 207:91–120

Olmstead CE, Villablanca JR (1979) Effects of caudate nuclei or frontal cortical ablations in kittens: maze learning performance. Exp Neurol 63:244–256

Pan W-J, Billings J, Nezafati M, Abbas A, Keilholz S (2018) Resting state fMRI in rodents. Curr Protoc Neurosci https://doi.org/10.1002.cpns.45

Passingham RE, Perry VH, Wilkinson F (1983) The long-term effects of removal of sensorimotor cortex in infant and adult rhesus monkeys. Brain 106:675–705

Payne BR, Cornwell P (1994) System-wide repercussions of damage to the immature visual cortex. Trends Neurosci 17:126–130

Payne BR, Lomber SG (2003) Quantitative analyses of principal and secondary compound parieto-occipital feedback pathways in cat. Exp Brain Res 152:420–433

Piekarski DJ, Johnson C, Boivin JR, Thomas AW, Lin WC, Delevich K, Galarce E, Wilbrecht L (2017) Does puberty mark a transition in sensitive periods for plasticity in the association cortex? Brain Res 1654(Pt B):123–144

Rao VLR, Dhodda VK, Song G, Bowen KK, Dempsey RJ (2002) Traumatic brain injury-induced acute gene expression changes in rat cerebral cortex identified by gene chip analysis. J Neurosci Res. https://doi.org/10.1002/jnr.10486

Rasmussen T (1983) Hemispherectomy for seizures revisited. Can J Neurol Sci 10(2):71–78

Rasmussen T, Milner B (1977) Clinical and surgical studies of speech areas in man. In: Zulch KJ, Creutzfeldt O, Galbraith GC (eds) Cerebral localization. Springer, Berlin, pp 238–257

Reh RK, Dias BG, Nelson CA, Kaufer D, Werker JF, Kolb B, Levine JD, Hensch TK (2020) Critical period regulation across multiple timescales. Proc Natl Acad Sci 117:23242–23251
Reichelt AC, Hare DJ, Bussey TJ, Saksida LM (2019) Perineuronal nets: plasticity, protection, and therapeutic potential. Trends Neurosci 42:458–470
Riva D, Cazzangia L (1986) Late effects of unilateral brain lesions sustained before and after age one. Neuropsychologia 24:423–428
Schneider GE (1973) Early lesions of superior colliculus: factors affecting the formation of abnormal retinal projections. Brain Behav Evol 8:73–90
Takesian AE, Hensch TK (2013) Balancing plasticity/stability across brain development. Prog Brain Res 207:3–34
Tavares TP, Kerr EN, Smith ML (2020) Memory outcomes following hemispherectomy in children. Epilepsy Behav 112:107360
Teuber H-L (1975) Recovery of function after brain damage in man. In: CFS (ed) Outcome of severe damage to the nervous system. Elsevier, Amsterdam
Tottenham N (2020) Early adversity and the neonotenous human brain. Biol Psychiatry 87:350–358
Vargha-Khadem F, Watters G, O'Gorman AM (1985) Development of speech and language following bilateral frontal lesions. Brain Lang 37:167–183
Vicedomini JP, Corwin JV, Nonneman AJ (1982) Role of residual anterior neocortex in recovery from neonatal prefrontal lesions in the rat. Physiol Behav 28:797–806
Villablanca JR, Hovda DA (2000) Developmental neuroplasticity in a model of cerebral hemispherectomy and stroke. Neuroscience 95:625–637
Villablanca JR, Olmstead CE, Levine MS, Marcus RJ (1978) Effects of caudate nuclei or frontal cortical ablations in kittens: neurology and gross behavior. Exp Neurol 61:615–634
Villablanca JR, Burgess JW, Sonnier BJ (1984) Neonatal cerebral hemispherectomy: a model for postlesion reorganization of the brain. In: Finger S, Almli CR (eds) Early brain damage, vol 2. Academic, New York, pp 179–210
Villablanca JR, Hovda DA, Jackson GF, Gayek R (1993a) Neurological and behavioral effects of a unilateral frontal cortical lesion in fetal kittens. I. Brain morphology, movements, posture and sensorimotor tests. Behav Brain Res 57:63–77
Villablanca JR, Hovda DA, Jackson GF, Infante C (1993b) Neurological and behavioral effects of a unilateral frontal cortical lesion in fetal kittens. II. Visual system tests, and proposing an "optimal developmental period" for lesion effects. Behav Brain Res 57:79–92
Weikum WM, Oberlander TF, Hensch TK, Werker JF (2012) Prenatal exposure to antidepressants and depressed maternal mood alter trajectory of infant speech perception. Proc Natl Acad Sci 109(Suppl 2):17221–17227
Werker JF, Hensch TK (2015) Critical periods in speech perception: new directions. Annu Rev Psychol 66:173–196
Whishaw IQ, Kolb B (1988) Sparing of skilled forelimb reaching and corticospinal projections after neonatal motor cortex removal or hemidecortication in the rat: support for the Kennard Doctrine. Brain Res 451:97–114
Whitesell JD, Liska A, Coletta L, Hirokawa KE, Bohn P, Williford A et al (2021) Regional, layer, and cell-type connectivity of the mouse default mode network. Neuron 109:1–15
Wu Z, Parry M, Hou XY, Liu M-H, Wang H, Cain R, Pei Z-F, Chen Y-C, Guo Z-Y, Abhijeet S, Chen G (2020) Gene therapy conversion of striatal astrocytes into GABAergic neurons in mouse models of Huntington's disease. Nat Commun 11. https://doi.org/10.1038/s41467-020-14855-3
Zamani A, Powell KL, May A, Semple BD (2020) Validation of reference genes for gene epression alaysis following experimental traumatic brain injury in a pediatric mouse model. Brain Res Bull. https://doi.org/10.1016/j.brainresbull.2020.09.021

Part III
Community Level Impact of Sensitive Periods: Analyses and Applications

Statistical Modeling of Sensitive Period Effects Using the Structured Life Course Modeling Approach (SLCMA)

Brooke J. Smith, Andrew D. A. C. Smith, and Erin C. Dunn

Contents

Abstract Sensitive periods are times during development when life experiences can have a greater impact on outcomes than at other periods during the life course. However, a dearth of sophisticated methods for studying time-dependent

B. J. Smith
Psychiatric and Neurodevelopmental Genetics Unit, Center for Genomic Medicine, Massachusetts General Hospital, Boston, MA, USA

A. D. A. C. Smith
Mathematics and Statistics Research Group, University of the West of England, Bristol, UK

E. C. Dunn (✉)
Psychiatric and Neurodevelopmental Genetics Unit, Center for Genomic Medicine, Massachusetts General Hospital, Boston, MA, USA

Department of Psychiatry, Harvard Medical School, Boston, MA, USA

Stanley Center for Psychiatric Research, The Broad Institute of Harvard and MIT, Cambridge, MA, USA

Harvard Center on the Developing Child, Harvard University, Cambridge, MA, USA
e-mail: edunn2@mgh.harvard.edu; http://www.thedunnlab.com

Curr Topics Behav Neurosci (2022) 53: 215–234
https://doi.org/10.1007/7854_2021_280
Published Online: 23 April 2022

exposure-outcome relationships means that sensitive periods are often overlooked in research studies in favor of more simplistic and easier-to-use hypotheses such as ever being exposed, or the effect of an exposure accumulated over time. The structured life course modeling approach (SLCMA; pronounced "slick-mah") allows researchers to model complex life course hypotheses, such as sensitive periods, to determine which hypothesis best explains the amount of variation between a repeated exposure and an outcome. The SLCMA makes use of the least angle regression (LARS) variable selection technique, a type of least absolute shrinkage and selection operator (LASSO) estimation procedure, to yield a parsimonious model for the exposure-outcome relationship of interest. The results of the LARS procedure are complemented with a post-selection inference method, called selective inference, which provides unbiased effect estimates, confidence intervals, and *p*-values for the final explanatory model. In this chapter, we provide a brief overview of the genesis of this sensitive period modeling approach and provide a didactic step-by-step user's guide to implement the SLCMA in sensitive- period research. R code to complete the SLCMA is available on our GitHub page at: https://github.com/thedunnlab/SLCMA-pipeline.

Keywords ALSPAC · Childhood adversity · Life course modeling · Sensitive periods · SLCMA

1 The Structured Life Course Modeling Approach (SLCMA) as an Aide to Building Targeted Policy Intervention and Prevention Approaches

To achieve optimal effectiveness and make the best use of limited resources, public health interventions need to be based on statistical models that are specific and targeted to the goals of the intervention. It is incumbent on researchers to scrutinize the role timing plays in an exposure-outcome relationship so that we may tailor interventions if appropriate to the periods in development when they are most needed to promote better health outcomes across the life course.

Until recently, life course statistical modeling – commonly used in the field of life course epidemiology, a subspecialty field in public health (Kuh and Ben-Shlomo 2004) – focused on testing simplistic hypotheses. For example, researchers often grouped study participants into ever-exposed groups (i.e., participants who had a lifetime history of a given exposure versus those that never had) or created variables to count the number of exposures over time (building from accumulation-of-risk models). The downside of crude approaches like these is that information about the timing of the exposure is lost. The better our methods are for comparing theories on how and when exposures have their greatest impact, the more we can understand how these diseases operate. Whether or not timing plays a role in the exposure-outcome relationship provides us with information about the etiology of a disease. Once we understand etiology better, the more targeted and precise we can make our policy interventions (or prevention strategies) at key points in development (Zhu et al. 2021). Thus, to do meaningful research that results in concrete action in the public health sphere, more must be known about if, and if so how the timing of an exposure can affect a given health outcome.

Increasingly, epidemiologists and other researchers have expanded their portfolio of statistical modeling approaches to incorporate sensitive period effects. Sensitive periods are "high-risk, high-reward" windows of time during the course of development when people are more susceptible to a given exposure (Knudsen 2004). However, until recently, few methods were available to analyze repeated exposure measurements in order to identify sensitive periods in development.

Herein, we describe the structured life course modeling approach (SLCMA; pronounced "slick-mah") an analytic procedure developed to address this need for methods that test sensitive periods and other effects. The SLCMA is a powerful modeling tool that enables the user to examine how sensitive periods relate to health or other outcomes. The SLMCA also has applications in life course epidemiology beyond investigation of sensitive periods.

In this chapter, our aim is to provide a thorough overview of the SLCMA as well as a didactic guide for executing this method. We hope this chapter will encourage researchers to implement the SLCMA in their own sensitive periods research so that they can explore nuanced, time-dependent hypotheses. In the next section (Sect. 1.1), we describe what the SLCMA is and why it is advantageous to use in sensitive periods research. We also describe, more broadly, research studying life course and developmental hypotheses. In Sect. 2 we briefly describe the origins of life course modeling and the history of its development. In Sect. 3, we take the reader step-by-step through the setup and execution of the SLCMA, testing how exposure to parental intimate partner violence (IPV) during childhood could influence depressive symptoms that emerge at age 13 as an example. Finally, we conclude with a section (Sect. 4) that describes the limitations and future directions of the SLCMA.

1.1 What Is the SLCMA and How Is Using It Advantageous?

The SLCMA is an analytic approach to life course modeling that permits users to propose, a priori, a set of life course hypotheses that potentially explain the relationship between exposure and outcome over the life course, and then identify the life course hypothesis (or set of hypotheses) with the greatest explanatory power between the exposure and outcome (Smith et al. 2015).

SLCMA has many advantages over other life course modeling approaches. First, it allows researchers to simultaneously investigate more than one potential life course hypothesis and find which of these hypotheses are best supported by the data. The SLCMA offers the user the option to analyze various life course hypotheses, providing researchers with the opportunity to go beyond simple dualities, such as ever versus never-exposed or accumulation hypotheses, and explore more nuanced life course hypotheses, e.g., sensitive periods.

Second, the SLCMA has versatility in that it can be applied to situations with both small and extremely large datasets that feature continuous outcomes, bringing new levels of detail to life course research in fields such as epigenomics. Finally, because SLCMA can be used in conjunction with a post-selection inference method, called selective inference, it is possible to estimate unbiased effect estimates, confidence intervals, and p-values using this method. This helps evaluate the statistical significance of the findings without incurring bias from the model selection process.

Now that readers are more familiar with the SLCMA and how it can help with their sensitive periods research, we transition to an overview of the hypotheses as we examine them in our example. We also provide a brief description of other commonly employed life course hypotheses.

1.2 Commonly Examined Life Course Hypotheses

Before implementing the SLCMA, care must be taken to propose only those potential hypotheses with carefully considered explanations for the relationship between the exposure and the outcome. Traditional hypotheses for modeling the life course include critical periods/sensitive periods and accumulation. We also provide a brief description of other common life course hypotheses: recency, effect modification, change/mobility, threshold, and ever-exposed (Dunn et al. 2018).

The critical and sensitive period hypotheses postulate that there are specific windows of time when exposure has an outsized impact on outcome. The terms critical period and sensitive period are sometimes used interchangeably in the literature. It is important to understand, however, that critical and sensitive periods have distinct field-specific meanings that may contradict those in other contexts. We use the descriptor sensitive period to mean a period during which a given exposure has a greater impact on outcome than at other periods in development; we use sensitive period, rather than critical period, in acknowledgment that even when the sensitive period window closes for humans, there is likely to be some degree of malleability or plasticity. This definition of sensitive period is equivalent to the usage of the term critical period in the life course epidemiology literature (Kuh and Ben-Shlomo 2004; Smith et al. 2015).

The accumulation hypothesis treats exposures cumulatively. The expectation is that the effect of a negative exposure intensifies with each reoccurrence of the exposure (or increasing dosing of the exposure) and similarly, an increasingly positive effect with every positive exposure. Many authors have tested the effects of such accumulation-of-risk models in the field of childhood development research (Evans et al. 2013).

Other life course hypotheses include recency, which considers an exposure cumulatively but inversely weights the exposure by the length of time passed since it first occurred. For example, it has been applied to studying whether the recency of risk factors for psychosocial stress predicts depression onset in younger individuals (Shanahan et al. 2011). The change hypothesis proposes that the degree of change in an exposure is the most relevant factor to its later effect, rather than the frequency or timing of occurrence. Hallqvist et al. (2004) investigated such a hypothesis in a study of the effects of socioeconomic status (SES), showing how changes in SES from childhood to adulthood influence adult risk for myocardial infarction. Changes in SES relate to upward or downward social mobility, hence in such a context the change hypothesis may be referred to as a mobility hypothesis. The effect modification hypothesis explores how an early-life factor can lead to different outcomes if the factor interacts with the exposure. For instance, examining whether low birthweight infants of high SES experience fewer negative health outcomes later in life than those of infants with low SES (Currie and Hyson 1999).

Finally, the ever-exposed model is a commonly used, straightforward hypothesis that evaluates whether there is a higher risk of a given outcome if the participant *ever* experienced exposure to the specific situation. This translates to a binary variable of "ever" or "never" exposed. While simple to quantify and evaluate, this simplicity can mean specificity in the data is lost. For reasons that will become even clearer later, it is strongly suggested that an ever-exposed hypothesis not be included in the SLCMA, because of its correlation with all other life course hypotheses.

2 SLCMA Origins and Where It Is Today

The life course modeling approach proposed by Mishra and colleagues in 2009 challenged the practice of solely examining accumulation hypotheses. The authors suggested a nested model testing approach whereby each hypothesized life course model was compared to a saturated model that contained all combinations of the exposure measurements. Analysts could then use an F test to determine whether the hypothesized model was supported by the observed data; this could be achieved by evaluating the hypothesized (reduced) model versus the saturated (full) model. If the resulting *p*-value from this test was sufficiently high, the conclusion was that the hypothesized model was a good fit for the data. If more than one hypothesized model surpassed the *p*-value threshold, then the model with the higher *p*-value or, alternatively, lower Akaike information criterion (AIC) was chosen as the most likely explanation for the observed data. One limitation of the Mishra modeling approach is it requires fitting the saturated model as a starting point for comparison (the "full model first" approach). However, this saturated model cannot be fitted if some combinations of the exposures are unobserved or if the exposures are continuous.

In 2015, Smith and colleagues proposed a new method to address some of the shortcomings of the Mishra method. Their approach used a null model for comparison (the "null model first" approach) and instead applied *variable selection* to determine the best-fitting hypothesized model. Variable selection methods aim to identify the best possible subset of regression model predictors among all potential predictors. In this specific setting, variable selection methods aim to identify the predictors for the best-fitting hypothesized model among all potential predictors from all proposed hypothesized models. In Smith's approach, the variable selection procedure employed was least angle regression (LARS), a type of least absolute shrinkage and selection operator (LASSO) estimation that shrinks small coefficient estimates to zero to encourage sparse (i.e., parsimonious) models with reduced dimensionality. With LARS, variable selection occurs before regression parameters are estimated. LARS was used because of its advantages over other variable selection procedures, including the exclusive (at the time) availability of post-selection inference (Lockhart et al. 2014). Post-selection inference is a methodology that allows researchers to calculate effect estimates, confidence intervals, and p-values after the best-fitting hypothesized life course model has been selected without incurring bias from the variable selection procedure.

Like all LASSO estimators, LARS has the advantage of promoting sparse models, so that the simplest life course hypothesis may be discovered. The LASSO also allows selection between collinear variables, a frequent characteristic when choosing between more than one life course hypothesis. Moreover, LARS can identify the best-fitting hypothesized model without using p-values, a distinct advantage over the Mishra method, which relies on p-values to validate model selection. Note that neither the full model first nor the null model first approaches can identify the true underlying life course model if it is not within the a priori hypothesized models. However, these modeling approaches provide support for the best-fitting hypothesized model over all other a priori models proposed.

A 2016 update to the SLCMA method by Smith and colleagues provides several additional benefits over Mishra's nested modeling approach, which include suggesting methods for handling continuous exposures, confounders, missing data, and measurement error (Smith et al. 2016). These adaptations vastly improve the applicability and reliability of SLCMA to various life course modeling analyses. In 2018, Madathil and colleagues proposed a method that departs from the variable selection approach characteristic of the SLCMA and instead uses Bayesian inference to select the most likely life course hypothesis (Madathil et al. 2018).

Prior studies have focused on non-omics data, meaning data that does not include genetic, epigenetic, or other 'omic measures. However, a paper by Zhu et al. (2021) explored the application of the SLCMA to high-dimensional omics outcomes, investigating the best practices researchers should deploy when using the SLCMA with DNA methylation data, a type of epigenetic modification shown to be influenced by early life experiences, perhaps during sensitive periods in development (Dunn et al. 2019). Zhu's simulation (and empirical) paper was the first to advise users on methods to control for confounding that improve statistical power when analyzing hundreds of thousands of outcomes. Their paper also advises users of the SLCMA to apply the selective inference method for post-selection inference, as this approach outperformed other post-selection inference methods tested. It is with these recommendations that we execute the SLCMA example in the following section.

3 How to Use the SLCMA

In this third section, we discuss the steps to follow when using the SLCMA to identify sensitive periods. We illustrate the SLCMA using a low-dimensional data example (i.e., a sample with a relatively low number of variables and observations) focusing on the relationship between exposure to parental intimate partner violence (IPV) from infancy to age 11 and the outcome of the child's level of depressive symptoms at age 13. The steps we follow below are listed here in the Table 1 checklist, which readers can follow during their own life course modeling work with the SLCMA. Readers can also consult our GitHub page at: https://github.com/thedunnlab/SLCMA-pipeline for detailed code on how to conduct the SLCMA in R.

3.1 Background to Motivate the Example

Intimate partner violence (IPV) can have detrimental effects on children in both the short and long term (Olofsson 2013; Roustit et al. 2009; Russell et al. 2010). Negative effects from exposure to IPV may include depressive symptoms (McIntosh 2003; Russell et al. 2010), with some studies suggesting these depressive symptoms may be more pronounced in females than in males (Buckner et al. 2004; Cummings et al. 1999). Less is known about at which point during the life course exposure to IPV may have the greatest effect or if the effect is heightened as the occurrence of IPV exposure accumulates over time, an important consideration for the creation of effective intervention strategies to forestall IPV in the home and its negative consequences for children. SLCMA enables us to consider multiple life course hypotheses at once, making it an effective tool for broadening our understanding of how the timing of IPV exposure may influence adolescent depression.

3.2 Analytic Sample

We analyzed data from the Avon Longitudinal Study of Parents and Children (ALSPAC) cohort (Boyd et al. 2013; Fraser et al. 2013). ALSPAC is composed of pregnant women who resided in Avon, England, with expected dates of delivery from April 1, 1991, to December 31, 1992. Initially, the number of pregnancies enrolled in the study was 14,541, which resulted in 14,676 fetuses (adding in births of multiple fetuses, e.g., twins) and 14,062 live births, 13,988 of which were alive at 1 year. When the children were about 7 years old, 913 children were added to the study. The resulting sample size for data collected after the age of seven is 15,454 pregnancies resulting in 15,589 fetuses. The study website contains details of all the data that is available through a fully searchable data dictionary and variable search tool (http://www.bristol.ac.uk/alspac/researchers/our-data/).

Table 1 Checklist for conducting SLCMA

1. Define your exposure-outcome relationship of interest Define the life course relationship you are trying to model
2. Identify the life course hypotheses and covariates to include **2A.** Propose the hypotheses you will consider for the model based on current literature and personal expertise **2B.** Identify any possible confounders to this exposure-outcome relationship
3. Select and inspect your data **3A.** Identify your data sample **3B.** Identify your repeated exposure variable and its format (binary, continuous, or categorical) **3C.** List the frequency and time points when this exposure is measured **3D.** Identify your continuous outcome variable **3E.** Check if there are missing data in your exposure, outcome, or covariates. If so, evaluate whether imputation is a possibility for your data or if complete-case analysis is a better fit. Then, identify your final sample size
4. Establishing your model inputs **4A.** Encode the proposed hypotheses into key variables to use during model selection. Examples for how common hypotheses are encoded are provided in Table 2. Sensitive periods are coded as binary indicator variables for each time point when measurement has been taken **4B.** Check the correlation among confounders and exposure hypotheses using a correlation matrix. Those with very high correlation (e.g., over 0.9 in absolute value) may need to be removed or condensed
5. Reducing confounding bias in your analysis Use the Frisch-Waugh-Lovell (FWL) theorem and regress the key variables on the confounders and retain the residuals to use in the subsequent analysis, then do the same for the outcome
6. Performing the first stage of method Run least angle regression (LARS) on your exposure hypotheses and outcome If analyzing data in a low-dimensional context, create an elbow plot modeling the R^2 values for each increasingly complex model. Select the number of hypotheses that coincide with where the elbow occurs. In a high-dimensional context, typically only the first hypothesis is selected for feasibility and ease of interpretation
7. Performing the second stage of method Use selective inference to find effect estimates, confidence intervals, and p-values for the specified number of hypothesis variables selected by LARS
8. Check SLCMA implemented correctly using a naïve comparison (optional) Build a traditional linear model of the selected life course hypothesis and see if the effect estimates match those in the SLCMA model. Effect estimates should be the same, but confidence intervals and p-values may be larger in the SLCMA model because selective inference accounts for the multiple testing and automatic selection of the best-fitting model
9. Checking model assumptions are met Check that your final model meets the four basic assumptions for linear regression: a linear relationship between the independent and dependent variables, independent residuals, homoscedasticity of the variance, and normally distributed residuals. Keep in mind that any deviance from these assumptions could have implications for the interpretation of your results

Table 2 Construction of life course hypothesis variables

Life course hypothesis	Time frame	Binary measurement points of physical or emotional abuse
Sensitive periods		
Very early childhood	0–3 years	8 weeks, 8 months, 21 months, 33 months
Early childhood	4–5 years	47 months, 61 months
Middle childhood	6–8 years	73 months
Late childhood	9–11 years	9 years, 11 years
Accumulation	0–11 years	Sum of all binary sensitive period measures

In our example, we defined a child as exposed to IPV if the mother answered affirmatively to either of the following statements: "Your partner hurt you physically" and/or "Your partner was emotionally cruel to you." At each time point when these questions were asked, we created a binary exposure variable given a value of 1 for exposed and 0 for unexposed. See Table 2 for a list of time points.

Because of data constraints, we included only those cases of IPV in which the partner abused the mother, not where the mother abused the partner. The outcome of depressive symptoms was measured using the Short Mood and Feelings Questionnaire (SMFQ) score (ranging from 0 to 26 points), which was completed by the child at age 13. We focused on SMFQ score at this age, because it is the closest subsequent time point to our last measurement point for IPV at age 11, allowing us to maintain temporality in our exposure-outcome association. Age 13 also corresponds to a pivotal period in adolescent development and falls within the age range when children start having romantic interests and more complex peer relationships, which may be influenced by exposure to IPV.

3.3 Conducting the SLCMA

Step 1. Define your exposure-outcome relationship of interest.

The first step in conducting a SLCMA analysis is to define the exposure-outcome relationship to investigate. As mentioned, in this example we were particularly interested in examining the association between exposure to IPV from birth through 11 years of age on the outcome of depressive symptoms at age 13.

Step 2. Identify the life course hypotheses and covariates to include.

2A. *Propose the hypotheses you will consider for the model.*

Using the current literature and your own scientific knowledge of the possible life course hypotheses for the relationship you are attempting to model, decide which hypotheses should be investigated and included in your analysis. In other words, ask yourself: what is your specific theory (or set of theories) about how your exposure relates to your outcome? And what are any competing theories (to your primary one)? The number of life course hypotheses proposed may depend on the goals of the analysis (Smith et al. 2015). If the analysis is more exploratory in nature, then a small number of simple hypotheses may be proposed with the goal of establishing whether there is an association between exposure and outcome over the life course. If the analysis is confirmatory, with the goal of establishing support for one life course hypothesis over all others, then a larger number of hypotheses may be proposed to confirm that the hypothesis of interest is more supported than any other. For establishing sensitive periods, we recommend proposing several sensitive period hypotheses and an accumulation hypothesis. The proposed sensitive periods should cover as much of the life course as possible, in order to explore a wide range of potential sensitive periods. The accumulation hypothesis is included in order to confirm that, if a sensitive period is selected, the sensitive period is more supported than the hypothesis that the outcome does not depend on the timing of the exposure.

For our regression model, we tested four potential sensitive periods: very early childhood (0–3 years), early childhood (4–5 years), middle childhood (6–8 years), and late childhood (9–11 years), timeframes similar to what has been proposed in the literature (Dunn et al. 2018; Kaplow and Widom 2007). We also allowed for the possibility of an alternative model with no sensitive periods in the relationship between IPV and depressive symptoms, by including an accumulation of risk model.

2B. *Identify any possible confounders to this exposure-outcome relationship.*

Possible confounders of the exposure-outcome relationship should be considered during this step of your analysis preparation. The SLCMA can accommodate continuous, categorical, and binary covariates. For our analysis, we included the following confounders: child's sex, child's race, mother's education level at gestation, mother's age at child's birth, financial stability in the household during gestation, history of childhood abuse for the mother, and history of childhood abuse for the partner where in these cases childhood abuse was defined as either emotional abuse, physical abuse, or physical neglect (Bevan and Higgins 2002; Bowen 2015; Gartland et al. 2019; Roustit et al. 2009).

Step 3. Select and inspect your data.

3A. *Identify your data sample.*

Ideally, you want to use a data sample that is longitudinal in nature and as comprehensive as possible. A good data sample will include a large number of individuals, include all covariates of interest, and have a continuous outcome, because SLCMA can only currently accommodate this form of outcome data, though as we share later, extensions to accommodate binary outcomes are forthcoming. Also important is to have a sample with a good amount of variability across time. This way, if you are interested in testing sensitive period hypotheses, you have variability in when people are exposed, which allows sensitive period testing to occur. In this example, to conduct our analysis we used data from the previously described ALSPAC cohort.

3B. *Identify your repeated exposure variable and its format.*

To inform how you will encode your key variables, you should note the format of your variables in the data (binary, continuous, or categorical). In this example, to identify whether exposure to IPV was present, we will use a repeated binary exposure variable indicating whether IPV occurred during each given time point.

3C. *List the frequency and time points when this exposure is measured.*

Take stock of what sensitive period or other life course hypotheses you can construct by creating a list of all measurement points available in your data. In this example we have nine measures of IPV: at 8 weeks, 8 months, 21 months, 33 months, 47 months, 61 months, 73 months, 9 years, and 11 years.

3D. *Identify your continuous outcome variable.*

The SLCMA can currently only model continuous outcomes. Therefore, to use this modeling approach your outcome of interest must be in a continuous format or be able to translate over to a continuous format if deemed a scientifically sound approach to your analysis. As mentioned, for our example, we selected SMFQ score at age 13 as our outcome of interest.

3E. *Evaluating the need for imputation.*

After you have identified the data source for your analysis, you want to assess the extent of missingness in your data. If there is substantial missing data (e.g., > 5%) and it is believed that the data are missing at random (MAR) or missing completely at random (MCAR), you normally can impute the data to bolster the sample size and thereby increase statistical power, preventing the loss of information, and reducing bias in your effect estimates (Azur et al. 2011; Hughes et al. 2019).

For our example, we chose to use multiple imputation by chained equations (MICE) to impute our missing exposure and covariate data. When using MICE, it is important that you include the outcome, confounders, and exposures to diminish potential bias (Azur et al. 2011). Based on the level of missingness in our data (at its highest, 46.5% missing for covariate partner's history of abuse), we imputed our dataset 20 times per the recommendation in the literature (Graham et al. 2007). The imputation process left us with a final sample size of 6,022 individuals.

Further details of the MICE procedure and how it works can be found in the helpful 2011 paper, "Multiple imputation by chained equations: what is it and how does it work?" by Azur et al. (2011).

Step 4. Establishing your model inputs.

4A. *Encoding your key variables.*

Once the regression model inputs and data source are identified, you can start constructing your dataset for analysis. Each life course hypothesis proposed can be represented by one or more variables. For example, a sensitive period hypothesis is represented simply with a binary variable indicating the presence (versus absence) of that exposure during a given time period. Conversely, an accumulation variable can be represented by the sum of the total occurrences of the exposure over the life course. These *key variables*, which are combinations or transformations of the exposure measurements, are the means by which life course hypotheses are represented in the LARS variable selection procedure. Of importance, the inclusion of a given hypothesis in the SLCMA is contingent on there being sufficient exposure data available to encode it.

In Table 2, we show a breakdown of how we constructed each life course hypothesis before running the SLCMA. We collapsed our nine observations into four sensitive periods as proposed in Step 2A. For each sensitive period time frame, we created a new binary variable indicating whether IPV exposure occurred at any point during that window of time. You may decide in your own analysis, however, to leave measurements ungrouped and model each time point separately, especially because sensitive period timing could operate at more granular timescales.

Sensitive periods and accumulation are the only hypotheses we use in our demonstration; however, examples for the construction of additional common life course hypotheses can be found in Table 3 below.

4B. *Check the correlation among key variables.*

There is no limit to the number of hypotheses that can be included in the SLCMA selection process. However, strongly correlated key variables diminish the ability of the SLCMA to detect the hypothesis best supported by the data (Smith et al. 2015). For this reason, it is important to check the bivariate correlation among your key

Table 3 Common life course hypotheses translated into algebra with examples of how they could be coded

Hypothesis	Key variable	Example of coding
First sensitive period	x_1	Exposed during first sensitive period = 1 No exposure during first sensitive period = 0
Second sensitive period	x_2	Exposed during second sensitive period = 1 No exposure during second sensitive period = 0
Third sensitive period	x_3	Exposed during third sensitive period = 1 No exposure during third sensitive period = 0
Fourth sensitive period	x_4	Exposed during fourth sensitive period = 1 No exposure during fourth sensitive period = 0
Accumulation	$x_1 + x_2 + x_3 + x_4$	Sum of exposures in first through fourth sensitive periods (equal to the number of times exposed). Ranges from 0 to 4
Recency	$t_1x_1 + t_2x_2 + t_3x_3 + t_4x_4$	Weighted sum of exposures in first through fourth sensitive periods, the weights being proportional to the age in each sensitive period, so that more recent periods have a higher weighting
Change (e.g., between first and last sensitive period)	$x_4 - x_1$	Difference in exposure between fourth sensitive period and first sensitive period
Effect modification (e.g., effect of change modified by exposure in first sensitive period)	$x_1 (x_4 - x_1)$	Difference in exposure multiplied by the exposure in the first sensitive period.

Key: x_1–x_4 represent the exposure status in each of the sensitive periods 1–4; t_1–t_4 represent the average age in each of the sensitive periods 1–4 variables. Those key variables with very high correlation (e.g., absolute value greater than 0.9) should be considered for removal or for condensing.

Note that the statistical power of your model will depend on the number of hypotheses considered. There is no established way to determine the statistical power with the SLCMA other than performing simulations from your data.

Step 5. Reducing confounding bias in your analysis.

The Frisch-Waugh-Lovell (FWL) theorem is recommended as the most suitable way to adjust for confounders when using variable selection because it can be used in tandem with any post-selective inference method and using it improves statistical power (Zhu et al. 2021). You can implement the FWL theorem by regressing the key variables on the confounders and then regressing the outcome on the confounders. The residuals from both these regressions can then be used in the analysis in place of the original values for the key variables and the outcome.

Step 6. Performing the first stage of method.

You are now ready to engage in the first stage of the SLCMA. In this stage, we use the LARS algorithm to find the key variable, or combination of key variables, that explain the most variation in the outcome. Thus, the first stage of the SLCMA will identify the life course hypothesis, or combination of life course hypotheses, that is most supported by the data.

When operating in the low-dimensional context as we are in our example, you can create an elbow plot to visualize the optimal number of key variables to include in the model based on the coefficient of determination, R^2, which is the proportion of variation in the outcome explained by the fitted regression model. The optimal number is at the arc (or the elbow) of the plot. See Fig. 1 below as an example, which shows a modest elbow shape at the second variable, indicating that two is the preferred number of key variables to use in your model. After two selected key variables, the fitted model becomes more difficult to interpret while giving little improvement in the amount of outcome variation explained. Selecting two key

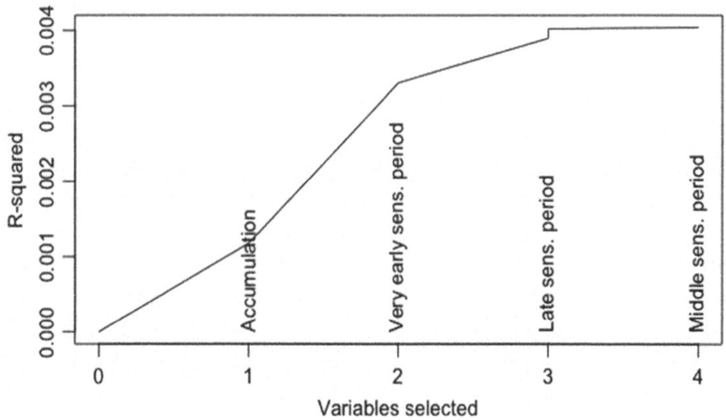

Fig. 1 Elbow plot of life course hypotheses for effect of exposure to childhood IPV on age 13 SMFQ scores using imputed data

variables means you have a *compound hypothesis*, and thus you combine two proposed life course hypotheses. In this instance, we combine an accumulation hypothesis and a very early sensitive period hypothesis, indicating that age 13 depressive symptoms vary with accumulated exposure to childhood IPV with an additional effect when exposure occurs during very early life.

In a high-dimensional context, for instance in the case of examining genome-wide DNA methylation, hundreds of thousands of regression models are tested, making examination of individual elbow plots impossible. In that context, the best solution may be to focus on selecting the model with the *first* key variable selected by LARS. Depending on the situation, it may be that focusing on the first key variable does not result in a loss of information. For instance, in a 2019 paper analyzing the effect of childhood adversity on DNA methylation, Dunn and colleagues demonstrated through sensitivity analyses of the top CpG site hits that including more than one theoretical life course model did not significantly improve the amount of methylation variability explained.

Step 7. Performing the second stage of method.

The second stage of the SLCMA applies the post-selection inference method, called selective inference, to calculate effect estimates, confidence intervals, and *p*-values for the model selected in the first stage. A naïve approach to calculate these estimates would be to take the selected model and calculate these values in the usual way, ignoring the fact that the linear model was selected from several possibilities by an automatic procedure. But such an approach would introduce bias due to the fact that SLCMA automatically selects the best-fitting model. Instead, post-selection inference methods are available to remove such bias. Selective inference has been recommended as the best post-selection inference method in both low- and high-dimensional contexts due to its ability to control family-wise error rate – meaning the probability of making at least one Type I error when conducting multiple hypotheses testing – as well as provide optimal statistical power and confidence interval coverage (Tibshirani et al. 2016, 2019; Zhu et al. 2021).

Step 8. Check whether SLCMA was implemented correctly using a naïve comparison (optional).

One way to check if the SLCMA was executed correctly is to build a traditional linear model of the selected life course hypothesis and see if the effect estimates match those in the SLCMA model. Whereas effect estimates should be the same, confidence intervals and *p*-values may be larger in the SLCMA model because selective inference accounts for the automatic selection of the best-fitting model. Accounting for these characteristics may result in larger *p*-values, but importantly, more accurate estimates. The life course hypothesis included in the model is now more robustly supported because it has been selected, over the other considered hypotheses, as the one that explains the most variance in the data. We demonstrate dissonance in estimates below in Table 4 where we compare the results of our own example using a naïve approach (not accounting for hypothesis selection) and the SLCMA results.

Step 9. Checking model assumptions are met.

Table 4 Comparison of naïve linear regression results and SLCMA results ($n = 6{,}022$)

	β (SE)	CI	p-value
Naïve approach			
Accumulation (0–11 years)	0.1765 (0.0876)	(0.0048, 0.3481)	0.0439
Very early childhood sensitive period (0–3 years)	0.3102 (0.2165)	(−0.1143, 0.7346)	0.1521
SLCMA			
Accumulation (0–11 years)	0.1765 (0.0876)	(−0.0831, 0.4889)	0.0776
Very early childhood sensitive period (0–3 years)	0.3102 (0.2166)	(−0.7005, 0.8287)	0.2777

It is important to realize that at the end of the SLCMA process, SLCMA users are left with a linear regression model. Therefore, to produce reliable estimates from the SLCMA, the regression model must meet the four assumptions needed for linear modeling, namely: (1) a linear relationship between the independent and dependent variables; (2) independent residuals; (3) homoscedasticity of the variance; and (4) normally distributed residuals. We encourage you to check the assumptions of the best-fitting hypothesized model in the same way that you would when analyzing any linear regression model. Be aware of how assumptions *not* being met could affect your model inference. For instance, heteroscedasticity of the variance or skewed residuals could yield imprecise estimates or inaccurate statistical inference. In either case, a transformed (e.g., log, square root) version of your variable may be preferable.

3.4 Results of Our Example Analysis Using the SLCMA

Now that we've completed all of the prior stages, what did we find? Was there evidence in support of sensitive periods, as we hypothesized?

The results from our LARS procedure indicate that a compound hypothesis, combining a sensitive period of very early childhood as well as accumulation, best explains the relationship between exposure to IPV during childhood and levels of depressive symptoms at age 13. Both the accumulation and sensitive period coefficients are positive (Table 4), indicating that depressive symptoms increase with increasing accumulation of exposure and there is an additional increase when exposure occurs in very early childhood. However, the results after applying selective inference suggest that we fail to reject the null; there does not appear to be potential sensitive periods in the data. Using selective inference, we calculated the estimated effect of very early childhood exposure to IPV [$\beta(SE) = 0.31(0.22)$] as well as the effect of each additional instance of exposure to IPV throughout childhood via the accumulation coefficient [$\beta(SE) = 0.18(0.09)$]. Given that the confidence interval for a very early childhood sensitive period spans zero and the p-value for this estimate is above 0.05, we conclude that there is no evidence that very early childhood is a sensitive period in the relationship between exposure to IPV and later adolescent depressive symptoms. Full details of our results, including p-values and effect estimate confidence intervals can be found in Table 4.

3.5 Sensitivity Analyses

Regardless of whether or not your hypotheses are supported by your observed data, it might be important to understand the stability of results derived from your application of the SLCMA. To demonstrate how our SLCMA results compare to different analysis conditions, we performed two sets of sensitivity analyses. First, we investigated what the analyses would show if we had focused on a complete-case analysis instead of using imputed data. We found that with a complete case reduced sample (n = 2,315), a late sensitive period combined with a very early sensitive period model was selected. Given that we think our data are MAR and the sample size for the complete-case analysis is approximately a third the size of the imputed data (n = 2,315 versus n = 6,022), we favor the results from the imputed data rather than the complete-case analysis. Despite the fact that our results were nonsignificant, the discrepancy between imputed and complete-case analysis results is important to note. The differences in hypotheses selected highlight the importance of ensuring your data are imputed when appropriate.

We also examined the potential impact of not grouping our time periods into wider sensitive period windows, but instead including each separate measurement occasion in the SLCMA procedure. Using the imputed data to perform this analysis we found that the SLCMA selected an 8-week postnatal sensitive period and an accumulation hypothesis as the best fit for the data. This finding is unsurprising, because it closely mirrors our results with grouped sensitive periods. In this situation, grouping our sensitive periods did not significantly change our results. However, based on your data and how your measurement points are clustered, grouping could affect your results. Therefore, it is wise to pick sensitive periods time frames that closely align with the literature and the current understanding of the biology.

4 Limitations of and Future Directions for the SLCMA

Although the strengths of the SLCMA method are numerous, this modeling approach still has room for improvement. For instance, the SLCMA is not yet able to model multiple time-varying exposures simultaneously. In addition, because the SLCMA looks for the exposure with the strongest association with the outcome, the ability to select the strongest association is weakened when highly correlated life course hypotheses are included in the procedure.

The SLCMA is only as strong as the association between exposure and outcome. If there is a weak relationship between exposure and outcome, then the researcher has little hope of finding compelling evidence for the selected life course hypothesis, even though it is most supported by the data. As is always the case when there is a weak relationship between exposure and outcome, the effect estimates in the final statistical model will likely be small. Similarly, as stated earlier in the chapter, if the true underlying life course hypothesis is not included in the proposed life course hypotheses, then the SLCMA is unable to identify the true hypothesis. Just as in cooking, your final dish is only as good as the ingredients you put in it. However, even if the true underlying hypothesis has not been included in the procedure, it does not mean that the SLCMA is a wasted exercise. The SLCMA can combine key variables from more than one life course hypothesis, potentially identifying a new combination of the proposed hypotheses that is closer to the true underlying relationship. Viewed differently, requiring researchers to propose a priori the life course hypotheses is a strength, preventing speculative generation of hypotheses after observing results and avoiding generation of hypotheses that are unique to a particular set of data. In this way, SLCMA can be viewed as possibly having the best of both worlds, allowing for exploratory and confirmatory analyses. If the true underlying life course hypothesis is not included in the proposed life course hypotheses, the SLCMA is unable to identify the true hypothesis and also the potential for such was not considered plausible – or perhaps not even known to the researcher at the start.

Next steps for improving the SLCMA will include allowing for *time-varying* covariates (also known as *time-dependent* covariates) in the modeling process. As the name suggests, time-varying covariates are covariates that may change over time. For instance, in our example, it would have been preferable to include the confounder financial stability as a time-varying covariate rather than being limited to the gestational measurement point because changes in this covariate could affect the prevalence of IPV. If a family feels greater financial security, the stress of daily life is diminished, which may in turn diminish the potential for conflict in the home. Additionally, this feeling of greater security could improve the mental wellbeing of the child, thus reducing their risk for depressive symptoms. But as it stands, to avoid contemporality issues, a covariate can only be included as it is measured at the start of the study period. Future updates to this method that allow for time-varying covariates will yield even more insightful models.

In addition, the SLCMA currently allows only for the study of *continuous* outcomes. As we hinted at before, methods research is currently underway to extend the SLCMA to binary outcomes to improve its versatility in sensitive periods research.

4.1 Conclusion

The deployment of science-based policies and interventions begins with targeted and precise research methods. These methods must allow for comparison between specific life course hypotheses to understand the etiology of disease and thus provide better guidance for prevention and intervention. The SLCMA helps researchers go beyond the study of simplistic, broad hypotheses to instead discover whether sensitive periods or other more complex relationships underlie the exposure-outcome relationship. The steps and strategies outlined here in this chapter enrich the potential of the SLCMA to provide scientifically supported evidence of the mechanisms by which diseases develop and persist. By homing in on these more nuanced exposure-outcome relationships, we can promote interventions that target negative exposures from the start and thus, promote healthier lives across the life course.

Acknowledgements We are extremely grateful to all the families who took part in the Avon Longitudinal Study of Parents and Children (ALSPAC) study, the midwives for their help in recruiting them, and the whole ALSPAC team, which includes interviewers, computer and laboratory technicians, clerical workers, research scientists, volunteers, managers, receptionists, and nurses. The UK Medical Research Council and the Wellcome Trust (Grant ref.: 102215/2/13/2) and the University of Bristol provide core support for ALSPAC. A comprehensive list of grants funding is available on the ALSPAC website (http://www.bristol.ac.uk/alspac/external/documents/grant-acknowledgements.pdf). This publication is the work of the authors, each of whom serves as guarantors for the contents of this paper. The authors thank Rebecca Lacey, PhD and Tochukwu Nweze for their suggestions on how to improve this manuscript as well as Susan T. Landry for her assistance in copy-editing the chapter.

Funding: The UK Medical Research Council and Wellcome (Grant ref.: 217065/Z/19/Z) and the University of Bristol provide core support for ALSPAC. This publication is the work of the authors and BJS, ADACS, and ECD will serve as guarantors for the contents of this paper. This work was supported by the National Institute of Mental Health of the National Institutes of Health [grant number R01MH113930 awarded to ECD]. The content is solely the responsibility of the authors and does not necessarily represent the official views of the NIH.

Ethical Approval and Informed Consent: Ethical approval for the study was obtained from the ALSPAC Ethics and Law Committee and the Local Research Ethics Committees. Informed consent for the use of data collected via questionnaires and clinics was obtained from participants following the recommendations of the ALSPAC Ethics and Law Committee at the time.

Conflict of Interest: None declared.

References

Azur MJ, Stuart EA, Frangakis C, Leaf PJ (2011) Multiple imputation by chained equations: what is it and how does it work? Int J Methods Psychiatr Res 20:40–49

Bevan E, Higgins D (2002) Is domestic violence learned? The contribution of five forms of child maltreatment to men's violence and adjustment. J Fam Violence 17:223–245

Bowen E (2015) The impact of intimate partner violence on preschool children's peer problems: an analysis of risk and protective factors. Child Abuse Negl 50:141–150

Boyd A, Golding J, Macleod J, Lawlor DA, Fraser A, Henderson J, Molloy L, Ness A, Ring S, Davey Smith G (2013) Cohort profile: the 'children of the 90s' – the index offspring of the Avon longitudinal study of parents and children. Int J Epidemiol 42:111–127

Buckner JC, Beardslee WR, Bassuk EL (2004) Exposure to violence and low-income children's mental health: direct, moderated, and mediated relations. Am J Orthopsychiatry 74:413–423

Cummings JG, Pepler DJ, Moore TE (1999) Behavior problems in children exposed to wife abuse: gender differences. J Fam Violence 14:133–156

Currie J, Hyson R (1999) Is the impact of health shocks cushioned by socioeconomic status? The case of low birthweight. Am Econ Rev 89(2):245–250

Dunn EC, Soare TW, Raffeld MR, Busso DS, Crawford KM, Davis KA, Fisher VA, Slopen N, Smith A, Tiemeier H, Susser ES (2018) What life course theoretical models best explain the relationship between exposure to childhood adversity and psychopathology symptoms: recency, accumulation, or sensitive periods? Psychol Med 48:2562–2572

Dunn EC, Soare TW, Zhu Y, Simpkin AJ, Suderman MJ, Klengel T, Smith A, Ressler KJ, Relton CL (2019) Sensitive periods for the effect of childhood adversity on DNA methylation: results from a prospective, longitudinal study. Biol Psychiatry 85:838–849

Evans GW, Li D, Whipple SS (2013) Cumulative risk and child development. Psychol Bull 139:342–396

Fraser A, Macdonald-Wallis C, Tilling K, Boyd A, Golding J, Davey Smith G, Henderson J, Macleod J, Molloy L, Ness A, Ring S, Nelson SM, Lawlor DA (2013) Cohort profile: the Avon longitudinal study of parents and children: ALSPAC mothers cohort. Int J Epidemiol 42:97–110

Gartland D, Giallo R, Woolhouse H, Mensah F, Brown SJ (2019) Intergenerational impacts of family violence – mothers and children in a large prospective pregnancy cohort study. EClinicalMedicine 15:51–61

Graham JW, Olchowski AE, Gilreath TD (2007) How many imputations are really needed? Some practical clarifications of multiple imputation theory. Prev Sci 8:206–213

Hallqvist J, Lynch J, Bartley M, Lang T, Blane D (2004) Can we disentangle life course processes of accumulation, critical period and social mobility? An analysis of disadvantaged socio-economic positions and myocardial infarction in the Stockholm Heart Epidemiology Program. Soc Sci Med 58(8):1555–1562

Hughes RA, Heron J, Sterne JAC, Tilling K (2019) Accounting for missing data in statistical analyses: multiple imputation is not always the answer. Int J Epidemiol 48:1294–1304

Kaplow JB, Widom CS (2007) Age of onset of child maltreatment predicts long-term mental health outcomes. J Abnorm Psychol 116:176–187

Knudsen E (2004) Sensitive periods in the development of the brain and behavior. J Cogn Neurosci 16:1412–1425

Kuh D, Ben-Shlomo Y (2004) Life course approaches to socioeconomic differentials in health a life course approach to chronic disease epidemiology. Oxford University Press, Oxford

Lockhart R, Taylor J, Tibshirani RJ, Tibshirani R (2014) A significance test for the LASSO. Ann Stat 42:413–468

Madathil S, Joseph L, Hardy R, Rousseau M, Nicolau B (2018) A Bayesian approach to investigate life course hypotheses involving continuous exposures. Int J Epidemiol 47:1623–1635

McIntosh J (2003) Children living with domestic violence: research foundations for early intervention. J Fam Stud 9:219–234

Olofsson N (2013) A life course model of self-reported violence exposure and ill- health with a public health problem perspective. AIMS Public Health 1:9–24

Roustit C, Renahy E, Guernec G, Lesieur S, Parizot I, Chauvin P (2009) Exposure to interparental violence and psychosocial maladjustment in the adult life course: advocacy for early prevention. J Epidemiol Community Health 63:563–568

Russell D, Springer KW, Greenfield EA (2010) Witnessing domestic abuse in childhood as an independent risk factor for depressive symptoms in young adulthood. Child Abuse Negl 34:448–453

Shanahan L, Copeland WE, Costello EJ, Angold A (2011) Child-, adolescent- and young adult-onset depressions: differential risk factors in development? Psychol Med 41:2265–2274

Smith AD, Hardy R, Heron J, Joinson CJ, Lawlor DA, Macdonald-Wallis C, Tilling K (2016) A structured approach to hypotheses involving continuous exposures over the life course. Int J Epidemiol 45:1271–1279

Smith AD, Heron J, Mishra G, Gilthorpe MS, Ben-Shlomo Y, Tilling K (2015) Model selection of the effect of binary exposures over the life course. Epidemiology 26:719–726

Tibshirani RJ, Taylor J, Lockhart R, Tibshirani R (2016) Exact post-selection inference for sequential regression procedures. J Am Stat Assoc 111:600–620

Tibshirani RJ, Tibshirani R, Taylor J, Loftus J, Reid S (2019) selectiveInference: tools for post-selection inference, R package version 1.2.0. https://cran.r-project.org/web/packages/selectiveInference/selectiveInference.pdf

Zhu Y, Simpkin AJ, Suderman MJ, Lussier AA, Walton E, Dunn EC, Smith ADAC (2021) A structured approach to evaluating life course hypotheses: moving beyond analyses of exposed versus unexposed in the omics context. Am J Epidemiol. https://academic.oup.com/aje/article-abstract/190/6/1101/5943467

Capitalizing on Neuroplasticity Across Development to Redirect Pathways from Juvenile Justice Involvement

Shannon Chaplo and Diana Fishbein

Contents

Abstract Adolescence is an exquisitely sensitive period of development during which pathways branch toward success in school and prosocial pursuits or, conversely, toward behavior problems and involvement in high-risk activities and systems, such as juvenile justice (JJ). Adverse childhood experiences (ACEs) such as poverty, family dysfunction, and child maltreatment, have been strongly and repeatedly associated with JJ involvement. A significant body of research from neuroscience has established that ACEs can alter facets of neurodevelopment that undergird self-regulation throughout childhood and adolescence, thereby increasing

S. Chaplo
FPG Child Development Institute, University of North Carolina at Chapel Hill, Chapel Hill, NC, USA

D. Fishbein (✉)
FPG Child Development Institute, University of North Carolina at Chapel Hill, Chapel Hill, NC, USA

Department of Human Development and Family Studies, The Pennsylvania State University, State College, PA, USA

National Prevention Science Coalition to Improve Lives, Chapel Hill, NC, USA
e-mail: dvf5211@psu.edu

Curr Topics Behav Neurosci (2022) 53: 235–254
https://doi.org/10.1007/7854_2021_251
Published Online: 6 September 2021

susceptibility to behaviors that attract attention of the JJ system. Because the ability to intervene prior to system-entrenchment is crucial to disrupting an adverse developmental pathway, we look toward neuroscience to offer insights into how to do so more effectively. In this chapter, evidence is summarized that informs an understanding of how neurodevelopmental pathways may lead to JJ involvement. Because neurodevelopment is malleable in response to both detrimental and positive experiences, there is potential for well-targeted interventions to normalize brain and cognitive development, especially during sensitive periods of maturation. This discussion is followed by a proposed research agenda to determine how to exploit these critical windows of opportunity to divert youth from persistent antisocial behavior and JJ involvement. Lastly, a review of neuroscience findings regarding the ability of intervention to strengthen brain systems that modulate self-regulation is presented. This research has direct practical significance with potential to be translated into meaningful policy change.

Keywords Adolescence · Juvenile justice · Maltreatment · Neurodevelopment · Plasticity

1 Introduction

Late childhood and early adolescence are considered vulnerable periods during which trajectories bifurcate toward success in school and prosocial pursuits or, conversely, toward behavior problems and increasing involvement in high-risk activities, including delinquency (Burchinal et al. 2008). An extensive body of research has distinguished between youth who exhibit an adolescent-limited course versus a lifetime course of delinquency (Moffitt et al. 1996). In large part, the differences between these groups are related to the prevalence of risk factors (e.g., poverty, family dysfunction, child maltreatment) and the relative lack of protective factors (e.g., healthy parenting, social supports, economic stability) in youth who assume the life-course path (Moffitt 2006). A significant body of recent knowledge has been amassed demonstrating the impacts of these factors on neurodevelopment throughout childhood and adolescence. Because the ability to intervene prior to system-entrenchment is crucial to disrupting an adverse developmental pathway, we are compelled to look toward neuroscience to offer insights into how to do so more effectively.

Biobehavioral research in general and neuroscience more specifically have demonstrated the interactive role of neurobiological and social-contextual conditions that influence these developmental pathways. It is now well known that the brain is "experience-dependent," translating to the ability of contextual (e.g., family dynamics, school climate) and experiential (e.g., relationships, adversity) factors to directly impact brain development and functioning, with implications for adult outcomes (Teicher et al. 2016). Although this process plays out across childhood and adolescence, the pre- and early adolescent period is of particular interest for two reasons: (1) during this timeframe, the brain is exquisitely sensitive to environmental experiences (Larsen and Luna 2018) and, intriguingly, (2) earlier childhood exposures commonly manifest in behavioral proclivities once an individual enters adolescence (Ireland et al. 2002; Ryan et al. 2013).

In these respects, studies suggest that experiences have differential effects on social, psychological, and neural processes contingent upon the developmental stage of the child (Andersen 2016; Johnson et al. 2016). For example, exposure to trauma and other adversities have the most deleterious impact on functions that are concurrently developing; however, earlier childhood experiences also predict onset of academic, social, and mental health problems later, such as in adolescence, when affected brain regions that subserve these functions begin to forge connections (Andersen and Teicher 2009; Teicher et al. 2003). Simultaneously, adolescents are becoming increasingly autonomous outside the home and are more susceptible to the influences of their peers (Dishion and Tipsord 2011). These newfound social challenges facing adolescents coincide with complex changes in brain wiring and connectivity taking place throughout this time that have implications for adaptive decision-making and the ability to self-regulate behavior and emotion (Marek et al. 2015).

In this chapter, we begin by presenting the current body of evidence that informs an understanding of how neurodevelopmental pathways may lead to juvenile justice (JJ) involvement. We recognize that youth may attract the attention of the JJ system and be "criminalized" for any number of reasons, and that their involvement is not always due to risky or problematic behaviors, such as associating with delinquent peers, being in the wrong place at the wrong time, or referrals from schools that are underequipped to handle less serious infractions. So, while risk-taking normatively increases during the adolescent years, only a subpopulation of youth engages in recurrent behaviors within the realm of delinquency, including violence, substance abuse, and serious property crimes. Here, we are focused on adolescents who are engaging in recurrent behaviors that are illegal and are likely to result in arrest in the absence of intervention or diversion.

In keeping with this conceptualization, research is reviewed about the numerous ways that adversity can impair particular aspects of neurodevelopment during childhood that can set the stage for poor decision-making, impulsivity, and sensation-seeking in adolescence. Because neurodevelopment is malleable in response to both detrimental and positive experiences, there is potential for well-targeted interventions to normalize brain and cognitive development, especially during sensitive periods of maturation. This discussion is followed by a proposed research agenda to determine how to exploit these critical windows of opportunity to divert youth away from persistent antisocial behavior and JJ involvement. Finally, we review the body of evidence from neuroscience regarding the ability of intervention to strengthen brain systems that modulate self-regulation, which has direct practical significance and potential to be translated into meaningful policy change.

2 Normative Adolescent Brain Development and Risk-Taking

To provide background for discussion of atypical neurodevelopment that increases the likelihood for JJ involvement, we briefly summarize aspects of typical brain development and influences from the social environment that help to explain greater risk-taking that is typical of adolescence. Neurobiological development during adolescence occurs transitionally rather than as a single snapshot in time (Casey et al. 2008). The prefrontal cortex (PFC), responsible for executive cognitive functions (ECF) (e.g., decision-making, impulse control, working memory), is still under construction. A central function of ECFs is to shield long-term goals from temptations afforded by short-term benefits that often lead to negative consequences (Kharitonova and Munakata 2011). Prefrontal "top-down" cognitive regulation over subcortical regions that modulate emotion is somewhat functionally disconnected throughout adolescence (Somerville and Casey 2010). This developmental process translates into the natural bias by adolescents toward acting on emotional stimuli with relatively less cognitive control. Through both the natural course of development and environmental experience, connections between these regions are strengthened, providing a mechanism for increasing top-down regulation of emotional brain systems in adulthood (Tottenham et al. 2011).

During the adolescent years, brain circuits involved in processing rewards (e.g., ventral striatum) also show rapid maturation (Padmanabhan et al. 2011; Geier et al. 2010; Somerville et al. 2010), having the effect of heightening sensitivity to rewarding experiences. This development may play a unique role in normative risk-taking behaviors that emerge in early to mid-adolescence, but that may be exaggerated in the subgroup that escalates into more serious delinquency – the life-course path. Paralleling this normative increase in reward sensitivity is a greater tendency to sensation/novelty seeking (Steinberg 2010). Further compounding these neurological events are early puberty and erratic hormone levels which may also contribute to adolescent's engagement in risky behaviors (Smith et al. 2013).

Overall, what we know about the adolescent brain is consistent with the observation that, though adolescents may physically appear to be as capable as adults of making sound decisions, key regions of their brains are not fully connected until well into the 20s (Giedd 2008; Steinberg et al. 2009). This imbalance between social demands and emergent neurobiological systems during adolescence may lead to heightened vulnerability to engagement in risky behaviors (e.g., truancy, risky sex, running away, substance use), under normal conditions (Casey and Jones 2010). However, the adverse conditions cited above increase vulnerability to more severe and persistent delinquent behavior, often resulting in JJ involvement. Given the high rates of childhood exposure to psychosocial trauma reported among adolescents with delinquency (Kerig et al. 2010) suggests that examining the interplay among trauma exposure, neurodevelopment impacts, and behavioral outcomes is especially critical. Furthermore, delineating these relationships has direct implications for the design of intervention components that target this sensitive period of development with the potential to redirect youth away from a trajectory toward more serious delinquent behavior or, if already system-involved, avert them from a life-course path.

3 The Brain's Experience-Dependence: For Better or for Worse

Taken together, neuroscience has shed light on the interplay between neurobiological and social contextual factors that help to explain why adolescence is typified by a dramatic increase in risky behavior; we now understand that some degree of impulsivity, risk-taking, and sensation-seeking is normative during adolescence (Steinberg 2010). However, a heightened level of risk-taking may extend from social circumstances and experiences that contribute to non-normative neurodevelopmental immaturity or dysfunction. In particular, the experience of toxic stress and trauma places young people at an extreme disadvantage on multiple fronts (Reiss 2013; Sterrett et al. 2014). Children and adolescents exposed to adversity or "toxic stress" – such as maltreatment, poverty, parental addiction, and racism – are at substantial risk for involvement in activities (behaviors) and systems (e.g., JJ, child welfare, mental health), both of which limit their potential to successfully develop into healthy and productive adults. The range of behavioral and mental health (BMH) problems that are often consequent to the experience of toxic stress, including violence, chronic truancy, substance abuse, and property crimes, draw attention of the JJ system. In fact, juvenile offenders in the USA report a very high prevalence and severity of trauma and maltreatment, including the experience of polyvictimization and complex trauma (Dierkhising et al. 2013; Kerig et al. 2010). In a sample of 898 detained youth, 84% had experienced two or more traumas, with a mean average of 14.6 traumas, indicating significantly higher prevalence of trauma among delinquents than in the general community, suggesting that "exposure to trauma is a fact of life for delinquent youth" (Abram et al. 2004, p. 407). Given this confluence of factors, both prevailing opportunities and individual susceptibilities for misbehavior can culminate in more serious delinquency and official police attention.

3.1 *For Worse: Adversity's Impact on Neurodevelopment*

As neurobiological methods have advanced, studies increasingly demonstrate the negative impact of adversity on neurodevelopment across the lifespan. Though the majority of studies using neuroimaging technologies have not included justice-involved youth, the extant literature provides important clues about the relationships among between exposure to adversity, brain development, and poorly regulated behavior that increase risk for JJ involvement.

Table 1 Evidence-based interventions with the potential to impact outcomes associated with delinquency

Program (target age)	Age	Psychosocial impacts
Family foundations	0–2	Antisocial-aggressive behavior, anxiety, conduct problems, depression, externalizing, internalizing, prosocial with peers
Nurse-family partnership	0–2	Child maltreatment, delinquency and criminal behavior, early cognitive development, internalizing, mental health – other, physical health and Well-being, preschool communication/language development, reciprocal parent-child warmth
Family check-up (toddler version)	0–2	Conduct problems, externalizing, internalizing, reciprocal parent–child warmth
Triple P system	0–11	Child maltreatment, mental health, substance use, externalizing behavior
Incredible years – parent	3–11	Antisocial-aggressive behavior, close relationships with parents, conduct problems, depression, externalizing, internalizing, positive social/prosocial behavior
Parent management training – oregon model	3–18	Antisocial-aggressive behavior, conduct problems, delinquency and criminal behavior, externalizing, internalizing
Parent-child interaction therapy (PCIT)	3–11	Antisocial-aggressive behavior, child maltreatment, conduct problems
New beginnings (for children of divorce)	5–18	Antisocial-aggressive behavior, close relationships with parents, externalizing, internalizing, mental health – other, reciprocal parent–child warmth, sexual risk behaviors
Multisystemic therapy (for "deep end" teenagers)	12–17	Antisocial-aggressive behavior, academic problems, sexual risk behaviors, conduct problems, family relationships, substance abuse

that provide opportunities for parenting and family-based interventions to exert their largest effects (Boparai et al. 2018; Luby et al. 2020). For example, targeting interventions to children exposed to maltreatment or poverty has potential to mitigate neurodevelopmental impacts associated with psychopathology and behaviors that place them at risk for later delinquency and JJ involvement (Fisher 2016; Pardini 2016). Table 1 delineates evidence-based family interventions across childhood and adolescence shown to reduce exposure to adversity and outcomes associated with delinquency, including aggressive behaviors, conduct problems, and other forms of internalizing and externalizing psychopathology.

Unfortunately, very few of these interventions have been subjected to research on their ability to alter underlying biological processes. On the other hand, evidence is accumulating to suggest it is possible for psychosocial interventions to improve neurodevelopmental trajectories and stress physiology in youth. As an illustration, Boparai et al. (2018) conducted a scoping review of the ameliorating effects of preventive intervention programs on biological processes in youth exposed to adversity. Their review of 40 intervention studies demonstrated support for the notion that interventions can improve biological functions negatively affected by adversity, including positive change in cortisol release, immune functions, brain development, and epigenetic modifications. Studies included youth from the following groups: previously institutionalized youth, foster care youth, and youth in community settings. The types of interventions employed range from individual and family-based to school and community-based, such as Attachment and Biobehavioral Catch-Up, the Bucharest Early Intervention Project, Multidimensional Treatment for Foster Care Preschoolers, and Strong African American Families (derived from the Strengthening Family program). Though many of these interventions target the early to mid-childhood period of development, there are opportunities to intervene in adolescence.

Contrary to popular belief, windows of opportunity to intervene do not wane during adolescence. As described above, the brain undergoes another sensitive period of neurodevelopment (Fuhrmann et al. 2015; Larsen and Luna 2018) that corresponds with pubertal and hormonal changes. The functions and pathways associated with decision-making, emotion regulation, and reward sensitivity are being refined and result in greater complexity and depth of executive functioning. Thus, although the brain's plasticity in adolescence translates to increased vulnerability to adverse environmental exposures, there is also boundless potential for intervention to positively alter its course. Youth who are at particularly high risk for *early* offending are especially important to target as early offending is a strong predictor of continued offending into adulthood (Loeber and Farrington 2011). Although the impact of intervention on biological processes in this group is a topic largely unexplored, it is possible that interventions specifically designed to target underlying mechanisms may exert greater effects in this subgroup, as compared to youth with later onset of delinquency, for a number of reasons: (a) their earlier and more malleable phase of neurodevelopment; (b) intervening prior to system-entrenchment; (c) the high incidence of childhood adversities that exert negative impacts on development; and (d) malleable individual-level characteristics (e.g., preexisting cognitive deficits, psychological problems, impulsivity).

The extant research also supports the "pubertal stress recalibration hypothesis," providing further evidence for adolescence as a critical developmental window for intervention. This hypothesis purports that, for youth who are no longer facing adversity, puberty is a developmental time period in which the HPA axis can recover from early childhood stress (DePasquale et al. 2019). A line of empirical research (Flannery et al. 2017; Gunnar et al. 2019; Quevedo et al. 2012) has found that, in children who grew up in institutionalized settings and then were later adopted prior to puberty, their stress response was blunted (i.e., cortisol levels before, during, and after a stress task) compared to non-adopted children living with their biological parents. However, by the time these children reached the end of puberty, significant differences dissipated between the two groups, suggesting that the stress-response has recalibrated in those adopted youth.

Relatedly, intervention studies are beginning to shed light on how caregiving may ameliorate the effects of poverty on neurodevelopment. One example is a secondary data analysis (Brody et al. 2017) of a randomized clinical trial of the Strong African

American Families Program, which included a sample of rural African American youth and their primary caregivers from ages 11 to 18. In this sample, Brody and colleagues observed that poverty during childhood was associated with decreased volume in limbic brain regions including the hippocampus and amygdala in adulthood. However, these effects were attenuated for the youth who participated in the intervention at age 11, suggesting that supportive parenting may protect the brain from the effects of poverty. Similarly, in a sample of formally institutionalized youth, Colich et al. (2020) found that adolescents with high-quality caregiving, compared to those with low-quality caregiving, showed better reward processing and executive functioning, and that these associations were stronger in the adolescent period than earlier developmental periods. Notably, this study included a range of caregivers (i.e., biological, adoptive, kinship, etc.) which is promising for JJ-involved youth who oftentimes do not live with biological caregivers. One implication of this work is that when intervening within the caregiving environment, different types of caregivers and settings need to be considered.

Collectively, the research cited herein demonstrates that opportunities abound to prevent delinquency and system involvement for youth, from early in childhood through adolescence. By targeting risk factors that adversely alter neurodevelopment, and bolstering protective factors that strengthen resilience, interventions guided by findings from neuroscience have potential to enhance healthy youth development and, in turn, improve overall outcomes. Interventions that target the caregiving environment are particularly critical across development, including adolescence. However, though adolescence is a highly malleable period and amenable to intervention effects, it is still imperative that we intervene as early as possible to prevent a cascading developmental trajectory toward adolescent delinquency. The evidence suggests that intervention effects are more impactful in early childhood for adversity-exposed children than during adolescence (Boparai et al. 2018). Such findings highlight the need for systematic delivery of programs and interventions that specifically target the malleable underpinnings of risky behaviors, with particular attention to the neurodevelopmental effects of trauma exposure, poverty, child maltreatment, and other forms of toxic stress (Kim et al. 2015). And though it is critical that we strive to ameliorate the deleterious consequences of adversity, greater investments in policies that reduce the exposures in the first place are sorely needed.

4 Proposed Research Agenda

A future research agenda focusing on outstanding etiological and applied questions in the field promises to enhance our knowledge of the neurodevelopmental trajectories of JJ-involved youth, while also guiding us to more precise prevention and intervention targets based on the neuroplasticity of the adolescent brain. To date, very little neuroscience research has been conducted with JJ-involved youth (Caldwell et al. 2019; Lansing et al. 2016) or youth advancing in that direction, to shed light on the role of neurodevelopment in phenotypic behaviors they often exhibit and that could become novel targets for intervention. The following subsection highlights findings from descriptive studies that have begun to elucidate biological characteristics of populations at risk for or entrenched in JJ systems involvement, as well as suggestions for some additional lines of etiological research. The second subsection recommends an overall program of research that applies the neural and biomarkers identified by future descriptive and longitudinal studies as a blueprint for intervention studies that target those markers to more precisely and potently move the proverbial needle.

4.1 Basic Research: Characterizing Youth at Risk for JJ Involvement

In general, biobehavioral studies of JJ-involved youth are typically descriptive and either focus on individuals who exhibit conduct problems (e.g., aggression, substance abuse, impulsivity), but are not explicitly JJ involved (Umbach et al. 2015), or include adults with psychopathy (Yang and Raine 2018). Both vantage points are instructive. An exception is an MRI study of life-course-persistent delinquent boys from Lansing et al. (2016). Results showed that the delinquent boys, who also endorsed high rates of traumatic and loss-related adversity, showed neuroanatomical differences in the fronto-temporal regions compared to matched controls. Other types of biologically-based studies of JJ-involved youth implicate dysregulated physiological responses to stressors, deficits in executive functioning and other processes, which provide clues into mechanisms that may underlie persistent delinquency (Johnson et al. 2015; Lin et al. 2021).

Intriguingly, a more extensive line of research has focused on children and adolescents with callous-unemotional (CU) traits as they present a significant risk for substance use disorders, serious delinquency and JJ involvement, and adult criminal and psychopathic behavior (Frick et al. 2005). They typically fall into the category of "life-course persistent" delinquency in that the expression of these traits tend to emerge well before puberty and persist into adulthood. Youths with high levels of CU traits are often detected within various other disruptive behavioral disorder diagnostic groupings – such as conduct disorder (CD), oppositional defiant disorder (ODD), and substance use disorder (SUD) – distinguishing themselves by the seriousness and stability of their conduct problems (Pardini et al. 2010). They have also been consistently distinguished by neurobiological, psychophysiological, cognitive, and psychological profiles (Blair et al. 2014). For example, several neuroimaging studies that have compared youth with high vs. low CU traits implicate the aberrations in patterns of activation in the amygdala and its circuitry (Ling et al. 2020; Marsh et al. 2008; Waller et al. 2020). Additional neuroimaging research in this subgroup will be informative with respect to etiological underpinnings that point to biomarkers that may serve as novel targets for intervention. It will be especially important to extend this research to the subgroup of CU youth who develop these characteristics through the experience of trauma (Craig et al. 2021).

Although not heretofore a focus of biologically-based research, another subgroup of particular interest for future research includes cross-over youth or youth who are involved in both the child welfare and JJ systems (Herz and Ryan 2008; Herz et al. 2010). Child welfare involvement potentiates a cascade of risk factors that can lead to delinquency. Children who encounter child welfare systems typically have a history of ACEs and other traumatic events, including caregiver abuse and neglect, unstable and chaotic homes, neighborhood disorder, and negative educational experiences (Garcia et al. 2017). These factors alone are known to precipitate conduct problems that can attract the attention of JJ. For those children who end up in the child welfare system, the negative impacts are often compounded and can propel the developmental trajectory toward delinquency and JJ involvement; the two systems are integrally intertwined for these reasons. Hence, the inclusion of cross-over youth in a line of research to elucidate dual-system impacts on neurodevelopment and identify intersections at which intervention can disrupt the child welfare to JJ pathway would be tremendously informative and would provide evidence-based strategies to improve outcomes for these youth. The objective would be to establish ways in which certain child welfare practices interact with histories of adversity to negatively affect brain development and functioning, leading to poor outcomes such as JJ, for the express purpose of policy reform.

Further exploration at this basic science phase of translation will help to delineate points during development when pathways diverge, with some youth becoming increasingly entrenched in risky behaviors and, subsequently, system-involved. The knowledge gained from these discoveries can subsequently be applied to the construction of interventions that more precisely strengthen neural structures and their connections that are damaged by adverse experiences prior to and resulting from system involvement.

4.2 *The Next Phase of Translation: Applying Etiological Information to Program Development*

In general, there remains a great need for research that examines both how and during what stages of maturation interventions impact neuroplasticity and other functional indicators of developmental success. At present, even the most efficacious preventive interventions do not benefit a substantial number of recipients, achieving only low to modest effect sizes. And while effect sizes may be significant, they are not indicative of the extent or nature of response variability and, thus, have low clinical significance. To broaden and strengthen program effects, we must systematically apply what we know about the etiological underpinnings of risk behaviors to the refinement of existing programs by identifying and targeting malleable individual characteristics and contextual processes that affect behavioral change (Fisher

2016; Hyde et al. 2020). Accordingly, for any given intervention, the investigation of social-contextual and neurodevelopmental factors that moderate and mediate a favorable intervention response (e.g., inhibiting inappropriate behaviors, recognizing and regulating emotions in conflict situations, engaging in positive social behavior, developing healthy relationships) has the potential to provide program developers with information critical to optimizing program design. As these fundamental gaps are filled, interventions can be more precisely tailored to strengthen the neural substrates of adaptive behaviors.

The need for this overall line of research is recommended across all intervention and population types but is perhaps even more pressing for youth at risk of JJ involvement given the serious implications of their behavior for their developmental success, adult outcomes, and public safety. Optimizing intervention effects to prevent eventual JJ involvement has been challenging, at best; thus, elucidating individual differences in response to existing interventions via a transdisciplinary research approach is essential. For example, determining neural markers and other variables (e.g., pubertal onset, early trauma, adolescent social stress, or cannabis/alcohol initiation) that signal more severe outcomes has potential to reveal malleable targets for the next generation of novel interventions (Deas and Brown 2006; Boyce et al. 2021). Subtype analyses (e.g., latent class or latent trajectory analyses) can further inform intervention models that account for neurobiological variation across classes of youth with conduct problems. Determining whether neurobiological mechanisms change commensurate with behavioral improvements in response to intervention will be instructive in designing and more effectively targeting interventions for these very high-risk youth. Interventions guided by this blueprint promise to be vastly more effective than non-specific interventions directed toward a heterogeneous population (Scheepers et al. 2011). And because once in the JJ system, these traits may be exacerbated and neurodevelopmental impacts may be compounded, it is important to conduct such studies prior to system involvement.

Interventions that have been informed by etiological knowledge target components to malleable regulatory processes (Stuss 2011; Tracy and Osipowicz 2011; Venkatakrishnan and Sandrini 2012). Although they have not been directly offered to JJ-involved youth, a brief explication is useful in thinking about approaches aimed at this population. For example, pharmacological or psychosocial therapies designed to stimulate activity of the amygdala and its connections (e.g., akin to deep brain stimulation in depression) (Drevets et al. 2008) and reinforce prefrontal inhibitory controls may normalize cognitive and emotional regulatory deficits often seen in JJ-involved youth. Another intriguing possibility is the potential preventative effect of educating caregivers, educators and policymakers regarding approaches that address differential developmental pathways in these youths. Early enrichment, tactile stimulation, stress reduction, and other environmental enhancements early in life may strengthen prefrontal cognitive controls and enlarge the striatum to reduce novelty-seeking and emotional dysregulation (Glenn and Yang 2012). Warm and responsive parenting may also ameliorate social deficits perhaps through effects on amygdala/prefrontal structure and connectivity (Waller et al. 2016). Current therapeutic inefficiencies arise because intervention methods do not map

The manufacturer's authorised representative in the EU is Springer Nature Customer Service Centre GmbH, Europaplatz 3, 69115 Heidelberg, Germany. If you have any concerns regarding our products, please contact ProductSafety@springernature.com

Printed and bound by CPI Group (UK) Ltd, Croydon, CR0 4YY
30/09/2026
02225496-0006